Universitext

Springer
Berlin
Heidelberg
New York
Barcelona
Budapest
Hong Kong
London
Milan
Paris
Santa Clara
Singapore
Tokyo

Springer
Berlin
Heidelberg
New York
Barcelona
Budapest
Hong Kong
London
Milan
Paris
Santa Clara
Singapore
Tokyo

Francine Diener   Marc Diener
Editors

Nonstandard Analysis in Practice

With 34 Figures

Springer

Francine Diener
Marc Diener

Université de Nice
Laboratoire CNRS de Mathématiques
Parc Valrose
F-06108 Nice Cedex 2, France

CIP-data applied for

Die Deutsche Bibliothek - CIP-Einheitsaufnahme

Nonstandard analysis in practice / Francine Diener ; Marc
Diener ed. - Berlin ; Heidelberg ; New York ; Barcelona ;
Budapest ; Hong Kong ; London ; Milan ; Paris ; Santa Clara ;
Singapore ; Tokyo : Springer, 1995
(Universitext)
ISBN 3-540-60297-6 (Berlin ...)
ISBN 0-387-60297-6 (New York ...)
NE: Diener, Francine [Hrsg.]

The cover picture shows the graph of a solution's real part of the Liouville differential equation
(see Fig. 2.3) produced by A. Fruchard.

Mathematics Subject Classification (1991): 26E35, 34E15, 34C35, 60G07, 58F23, 54J05,
28E05, 53A04, 47S20, 30E99, 60F05

ISBN 3-540-60297-6 Springer-Verlag Berlin Heidelberg New York

© Springer-Verlag Berlin Heidelberg 1995

Printed in Germany

SPIN: 10500256   41/3143–543210 – Printed on acid-free paper

Foreword

In the early seventies, Georges Reeb learnt about Robinson's Nonstandard Analysis (NSA). It became quickly obvious to him that a kind of revolution had happened in mathematics: the old dream of actual infinitesimals had been realized. This seemed to him a great event.

He was a mechanics-minded topologist, in the dynamical tradition of Painlevé, Poincaré, Cartan, and the topological tradition of his master Ehresmann. He got convinced that NSA is exactly the right framework within which to study dynamics with small parameters, including problems of asymptotics and bifurcations.

This was the starting point of the second school he created in Strasbourg, the topological school (which was mainly oriented towards foliation theory and dynamical systems) being the first. This new school grew up with the philosophy of using NSA in everyday mathematics, as a tool to get simple and natural proofs and to detect new mathematical phenomena.

The first works were focused on differential equations, but other topics, including perturbation problems in algebra, quickly became of interest. The axiomatic presentation of NSA by E. Nelson within Internal Set Theory in 1977 gave a second impulse to the Alsatian school ; this was partly because this formal setting was in agreement with Reeb's philosophical conviction that infinitesimals were an unexpected benefit of the impossibility of formalising the intuitive feeling that all natural numbers are of the same kind. Calling "naive" the natural numbers obtainable from zero by the successive addition of one, he asked everywhere he went in Europe the disturbing question "Les entiers naïfs remplissent-ils $\mathbb{N}$?"

One of the most famous achievements of Reeb's school was the discovery in 1977 of the "canard solutions" that appear furtively when some one-parameter families of dynamical systems perform a Hopf bifurcation. The phenomenon was brought to light, first theoretically, then by computer, by a group of young mathematicians working at Oran, Algeria. One of them was Jean- Louis Callot, a very bright and clever scientist who unfortunately passed away in August 1993, just three months before Reeb himself. His work in his last years mainly concerned differential equations in one complex variable, a subject that is present in this book because of his inspiration.

In 1981 volume 881 of the Springer Lecture Notes series appeared, the first book to give an account of the state of the art in Reeb's school. Its title was "NSA, a practical guide with applications", and I had the pleasure of writing it in cooperation with my friend, Michel Goze, and with strong encouragement from Reeb. This was a (pleasant) intellectual adventure which highly stimulated our own work on new applications of NSA.

In 1992, the community which developed around Reeb's students organized a meeting at Luminy. It there became clear that the time was ripe to present to mathematicians our collective knowledge which had greatly increased since the early eighties.

In the work that follows, the readers will first find in the tutorial all that they need to become themselves actors. Then nine typical subjects where NSA is deeply helpful are treated with considerable detail. They cover some, but not all matters in which the "reebian network" is interested. For instance, there is almost nothing about infinitesimal stroboscopy and averaging, boundary value problems in singular perturbations of ODE's, perturbation theory of algebraic structures etc... The aim of the book is not to be a catalogue of results, but rather, as was already true of the Springer Lecture Notes publication above, a practical guide that illustrates various possibilities of using NSA. For those who want to introduce NSA in their teaching practice, the last chapter gives a humorous indication, nurtured by effective experience, of what is typical in this respect.

Clearly the authors have written their contributions to this collective work with the aim of motivating the readers and of helping them share the very particular turn of mind which is typical of nonstandardists among mathematicians: to seek simple and natural proofs avoiding artificial subterfuges, to be interested in applying techniques to problems rather than adapting problems to techniques, to jump over the boundaries within mathematics. Reeb and Callot strongly shared this philosophy, which agrees with Abraham Robinson's original purposes.

I am convinced that the present book will incite the readers to use in their everyday mathematical work some pleasant additional tools which, as Leibniz said about his infinitesimals, "facilitent l'art d'inventer".

Robert Lutz
Université de Haute Alsace
Département de Mathématiques
4, rue des Frères Lumière
F- 68093 MULHOUSE
lutz@univ-mulhouse.fr

Editors' Foreword

The project of writing this book in collaboration arose two years ago from the considerations which follow. Firstly, we are often asked by colleagues who are interested in some result of nonstandard analysis, or simply curious to know more of this new approach to infinitesimals, where they might find a practical introduction to the subject. Some ten years ago, we would have suggested to them to look at volume 881 in the Springer Lecture Notes series, by R. Lutz and M. Goze, did they wish to see a development after the manner in which we work. The fact that this book is no longer in print was one good reason for writing an updated introduction to NSA in the spirit of the French school.

The second consideration was that it should comply with the philosophy that we were taught by our mentor, Georges Reeb : NSA must be easy to learn and its use should be intuitive and natural. But what is considered easy, intuitive, or natural depends on the person. One might prefer a picture while another might feel that such a picture gives too fuzzy an idea and so would prefer a formula. Thus, to fulfil the program, we decided that it would be best to have several points of view.

In the book which emerged from these reflections the reader will find, after a brief "tutorial", examples of applications in various domains, written in somewhat different styles. Despite these differences of content and style there is an underlying unity of approach derived from the fact that the various authors have been working together in the same spirit for more than ten years. Only the first chapter is needed as background for the ones that follow which are, by and large, independent of one another.

As promoters of this program, we would like to thank all our friends who agreed to collaborate to bring it to success. We know that they had to be patient to work through the various versions of their contributions, and especially the last one : the translation into English. Special thanks go to Noel Murphy who read all the papers and did his best to improve our poor use of this language. We hope that this venture will help more people to enjoy the use of infinitesimals, now rediscovered.

Nice, April 27th, 1995
Francine and Marc Diener

Table of Contents

1. Tutorial

F. Diener and M. Diener

Nonstandard Analysis, as is the case with Mathematics itself, can be applied to many domains; to illustrate this by means of examples is the object of this book. The reader may perhaps be surprised, and will surely be pleased to learn that very little need be known of the theoretical details of this modern theory of infinitesimals as introduced by A. Robinson [100] in order to apply it well. The purpose of this first chapter is to communicate the common background necessary to the rest of the book. It gathers the results that will be used in the sequel, the following chapters being all independent of each other, more or less. The ease with which it is possible to explain within twenty pages or so what mathematicians sought for in vain during the 19th century is a consequence of an idea of E. Nelson [91]: to resort to an adjective, the predicate *standard*, which is deliberately left undefined. This idea makes it possible to dissociate completely the logical foundation of the nonstandard method from its practical use.

This chapter is a guided tour, using nonstandard eyes to examine basic mathematical objects such as numbers and functions. Its purpose is confined to illustrating a *praxis*, leaving aside proofs. We shall, of course, try to explain the local logic, it being easier to understand a situation if one understands its mechanism. But we shall not prove all properties, more often that not preferring to give a typical example in place of proof. We shall give only those proofs that can be considered to be archetypal. All of the results mentioned in this chapter are proved in, for example, [100, 91, 50] .

1.1 A new view of old sets

Seen from Mathematics as a whole, nonstandard analysis (NSA) is characterized by the use of a single new word, *standard*, employed as what is called a predicate by logicians. A number (and more generally a set) is either standard or not. The abbreviated form is *st* and we shall write $st(x)$ when we want to assert that x is standard.

Let us immediately insist that using this new word will not, in any sense, change the usual number sets, such as $\mathbb{N}$ or $\mathbb{R}$, nor will it change any other classical set that one might built from them, such as the set of even integers, or the set of continuous real functions. "To be an integer" or "to be a real number" keep their usual meanings. We will simply give a meaning to expressions such as "to be infinitely large" or "to be infinitely small", by using the predicate, *st*.

1.1.1 Standard and infinitesimal real numbers, and the Leibniz rules

Standard versus finite When should we say that a real number x is standard ? The answer depends, of course, on the assumptions on x. For example, 0, 1, -2, and π are standard real numbers, and any number that can be characterized in a unique classical way is necessarily standard. Thus, 0 is standard because it is the unit element of the (classical) set of integers, and π is standard as it is the smallest positive zero of the (classical) function sine. But are not all integers therefore standard ? The response is negative and connected with the property that there are infinitely many integers. Indeed, we shall see below that all infinite sets necessarily have nonstandard elements; the fact that this axiom is not contradictory can be understood by reflecting that, at any stage of the mathematical discourse, there will never be more than a finite number of objects that will have been uniquely defined. So there are much too many elements in an infinite set for all of them to be standard. In any event, there exist nonstandard numbers and to be able to distinguish standard ones from nonstandard ones, will make it possible to introduce orders of magnitude.

Orders of magnitude For the integers, things are particularly simple: any integer smaller then some standard integer is itself standard (and thus any integer larger than some nonstandard integer is also nonstandard). This is precisely why nonstandard integers are called *infinitely large*, or better, *unlimited*, as they are larger than any standard integer.

For real numbers, the situation is somewhat richer. A real number ω is called *infinitely large* or *unlimited* if its absolute value $|\omega|$ is larger than any standard integer, n. So a nonstandard integer ω is also an unlimited real number; $\omega + \frac{1}{2}$ is an example of an unlimited real number that is not an integer. A real number ε is called *infinitely small*, or *infinitesimal* if its absolute value $|\varepsilon|$ is smaller than $\frac{1}{n}$ for any standard n. Of course, 0 is infinitesimal but (fortunately) it is not the only one: $\varepsilon = \frac{1}{\omega}$ is infinitesimal, provided ω is unlimited. However 0 is special in that it is the only standard infinitesimal. This trivial but useful remark leads to the *"Carnot principle"* [31, p 19]: two standard numbers whose difference is infinitesimal are equal. It is this principle that allows one, during the computation of a standard quantity such as an integral or a limit, to make estimates up to an infinitesimal, and to neglect this infinitesimal at the end of the computation.

A real number r is called *limited* if it is not infinitely large, and *appreciable* if it is neither unlimited nor infinitesimal. Two real numbers x and y are *infinitely close* (written $x \simeq y$) if their difference $x - y$ is infinitesimal. So, if

ω is a nonstandard integer, the numbers $1/\omega$ and 2ω are non-appreciable as they are respectively infinitesimal and unlimited. On the other hand $\sqrt{2}+1/\omega$ is appreciable, but nonstandard. So we see, by contrast with the integers, that not every limited real number is standard: if $\varepsilon \neq 0$ is infinitesimal and if s is standard, then $r := s+\varepsilon$ is limited but nonstandard (otherwise the difference $\varepsilon = r - s$ would be standard); however any limited real number r is infinitely close to a unique standard number. This real number is called the *standard part* of r or the *shadow* of r; it is denoted by ${}^{\circ}r$. The existence of this standard part is directly related to the fact that $\mathbb{R}$ is complete.

Observe that the notion of infinitely large number has little to do with the notion of infinite set: a set such as $\{0, 1, 2, \ldots, \omega\}$ is finite, and its cardinal, here $\omega + 1$, is infinitely large. On the other hand, the cardinal of an infinite set is by definition not an integer, but it is standard if the set is standard. It is to stress this difference from the classical notion of infinite set that we prefer the words "unlimited" and "infinitesimal" to infinitely large or infinitely small. In the same spirit we use (see [38]) a more compact notation: *i-large* and *i-small* for "ideally" large and "ideally" small. This terminology agrees perfectly with the original ideas of Leibniz, and prevents the confusion with the modern sense of infinity. Moreover, as these words are not in the dictionary, their use underlines that these are technical words, having a precise mathematical definition. In the same spirit, we shall say that x and y are *i-close* if their distance from one another is i-small.

The Leibniz rules The rules for computing with orders of magnitudes are very simple ([39], [41]). The sum of two infinitesimals is infinitesimal, as is the product of an infinitesimal with any limited number. On the other hand, without more precision as to their values, it is not possible to determine the order of magnitude of an infinitesimal and an unlimited number, whence the question marks "?" in the tables below that summarize the *Leibniz rules* for orders of magnitude.

We express these rules using the four symbols $\oslash$, £, @, and $o\!\!\!/\!\!\!/$. Indeed, it turns out to be useful, for the computations, to have a generic notation for a number whose precise value is not of interest, but only its order of magnitude. The notations, $\oslash$, £, @, and $o\!\!\!/\!\!\!/$, introduced by I. van den Berg, are used, respectively, for an infinitesimal, a limited, an appreciable, and an unlimited number. Two occurrences of one of these symbols are usually not equal, but they have the same order of magnitude. See chapter 7 for a more formal sense of the meaning of this notation which leads to the same results.

$+$	$\oslash$	£	@	$o\!\!\!/\!\!\!/$
$\oslash$	$\oslash$			
£	£	£		
@	@	£	£	
$o\!\!\!/\!\!\!/$	$o\!\!\!/\!\!\!/$	$o\!\!\!/\!\!\!/$	$o\!\!\!/\!\!\!/$	?

$-$	$\oslash$	£	@	$o\!\!\!/\!\!\!/$
$\oslash$	$\oslash$			
£	£	£		
@	@	£	£	
$o\!\!\!/\!\!\!/$	$o\!\!\!/\!\!\!/$	$o\!\!\!/\!\!\!/$	$o\!\!\!/\!\!\!/$	?

×	∅	£	@	⌀		↑ / →	∅	£	@	⌀
∅	∅					∅	?	?	∅	∅
£	∅	£				£	?	?	£	∅
@	∅	£	@			@	⌀	?	@	∅
⌀	?	?	⌀	⌀		⌀	⌀	⌀	⌀	?

The following rules give the behaviour of the standard part operator with respect to the elementary operations and the order on $\mathbb{R}$:

$$^{\circ}(x + y) = {}^{\circ}x + {}^{\circ}y \; , \quad {}^{\circ}(xy) = {}^{\circ}x \, {}^{\circ}y \; , \quad {}^{\circ}(1/x) = 1/{}^{\circ}x \; \text{si} \; {}^{\circ}x \neq 0$$

$$x \leq y \Rightarrow {}^{\circ}x \leq {}^{\circ}y \; , \quad {}^{\circ}x < {}^{\circ}y \Rightarrow x < y.$$

Here is some useful notation: we write

– $x \simeq y$ if $x - y$ is i-small
– $x \lesssim y$ if either $x < y$ or $x \simeq y$, that is if ${}^{\circ}x \leq {}^{\circ}y$
– $x \not\simeq y$ if x is not infinitely close to y
– $x \lneq y$ if $x < y$ and $x \not\simeq y$, that is if ${}^{\circ}x < {}^{\circ}y$.

1.1.2 To be or not to be standard

We did not define the term $st()$ (standard), and in fact, we will not do so ! To proceed so is not common in mathematics but is not new either: this is exactly what is done in set theory. The fundamental notions of set and the belonging relation ($\in$) are not defined, but known only from the axioms that apply to them. Of course, we build an intuition about these notions, mostly by considering *naïvely finite* sets (the set of all pupils in the class room, the set of all letters) or a *potentially infinite* set such as the set of integers. Subsequently, having given a name to a formalization of this set[1] ($\mathbb{N}$), one builds many others ($\mathbb{Z}$, $\mathbb{Q}$, $\mathbb{R}$, $\mathbb{C}$, $[0,1]^{\mathbb{R}}$, ...) using set theoretic operations such as $\cap$, $\cup$, $\times$, etc.

Matters are similar here. In section 1.1.1 we gave an outline, the purpose of which was to help the reader form an idea of what the standard numbers are. In the next section we will give the axioms that govern the use of the new predicate $st()$ (and we will then be able to sketch the proofs of some assertions made in section 1.1.1). One remembers that the sets that can be distinguished (as an aside, the adjective *distinguished* would certainly be better than *standard*...) using a precise (standard) definition are standard. It is the meaning of following metaphor due to A. Deledicq: if we look at a "standard-sized" piece of paper (US, of the everyday variety), a very rough

---

[1] The question of the equality of the *potentially infinite set* of the (naïve) integers and the set $\mathbb{N}$, the existence of which is postulated in the (standard) axiom system ZFC is very much an open question. G. Reeb, considering that the former is *strictly* contained in the second, the famous *Les entiers naïfs ne remplissent pas $\mathbb{N}$* [81] (the naïve integers do not fill up $\mathbb{N}$), proposed a still more elementary introduction to infinitesimals than the one of Nelson that we use here.

measure of its width leading to a value of about 21cm will indicate that its actual width is 9ins exactly ! This is a nice example of the use of standard in the process of approximating that characterizes Analysis.

1.1.3 Internal statements (standard or not) and external statements

Recall that any *mathematical statement*(or "formula", in logic) is built of constants, variables, quantifiers, and connectives, displayed in some coherent manner. To enter here into more detail as to what a correctly built formula is, would serve no useful purpose in our current context. Let us merely recall that, roughly speaking, a (logical) *constant* is an object such as 0, 1, π, $\mathbb{N}$, $\mathbb{R}$ that has been defined once for all and to which some notation has been assigned; this terminology is somewhat misleading as functions like sin, ln, $+$, $\times$ are also logical constants, as are predicates such as $\geq$, $\in$, and now *st*. Objects like f or ε that, formally, are *variables*, but that one treats like constants, as though they were objects fixed once for all, are called *parameters*.

We pointed out that the nonstandard language adds to the classical language a new predicate, $st()$. In fact, this allows one to define many others, such as i-small, i-large, limited, appreciable, all of which are *nonstandard* predicates.

A statement is *internal* if it uses none of these nonstandard predicates; otherwise, it is *external*.

So, the statement "for all real numbers $e > 0$ and $r > 0$, there exists some integer n such that $ne \geq r$" is an internal statement, and the statement "there exist real numbers $e > 0$ and $r > 0$ such that for all *standard* integers n, $ne < r$ " is an external statement; both statements are actually true.

An internal statement is *standard* if all its constants and all its parameters are standard. For example, for $\varepsilon > 0$, i-small, the statements [$|x| < \frac{1}{10} \Rightarrow x^2 < \frac{1}{100}$], [$0 < x < \varepsilon \Rightarrow 0 < x^2 < \varepsilon^2$] and [$0 < x < \varepsilon \Rightarrow x^2$ i-small] are respectively standard, internal but nonstandard, and external; all three of them are true. Observe that only internal statements may be standard or not and that an internal statement without any constant or parameter is necessarily standard.

1.1.4 External sets

Given a statement $F(x)$ about the variable x (the only *free variable* in F) and a set $\mathcal{E}$, the mathematician likes to consider the collection of all objects x of $\mathcal{E}$ that satisfy the statement $F(x)$; he usually denotes this by

$$\mathcal{F} = \{x \in \mathcal{E} \mid F(x)\}.$$

The separation axiom of (classical) set theory responds to this demand of the mathematician, giving the title of set to $\mathcal{F}$ and thus allowing the application to $\mathcal{F}$ of all the axioms of the theory and all its theorems.

In the case where $F(x)$ is an external statement, the separation axiom *a priori* no longer applies but we want, nevertheless, to keep the ability to consider the collection $\mathcal{F} \subseteq \mathcal{E}$ of all x of $\mathcal{E}$ that satisfy $F(x)$. As a counterpart to this desire, we have to accept that some classical theorems no longer apply to $\mathcal{F}$; in that case, we shall call such a set an *external set*, although, strictly, it does not merit the title of *set*.

Let us examine this by way of an example, the *halo* of 0 defined by:

$$\mathrm{hal}\,(0) = \{x \in \mathbb{R} \mid x \text{ i-small}\}.$$

Any standard real number $r > 0$ is larger then any $x \in \mathrm{hal}\,(0)$. Yet $\mathrm{hal}\,(0)$ has no least upper bound. Indeed, let s be such a least upper bound, and let $\varepsilon > 0$ be i-small. If s were i-small so would $s+\varepsilon$ be i-small which is impossible because s is the least upper bound of $\mathrm{hal}\,(0)$. Nor can s be appreciable as $s - \varepsilon$ would be so also which would mean that s was not, after all, the least upper bound.

So, we shall call an *external set* any collection of the type $\mathcal{F} = \{x \in \mathcal{E} \mid F(x)\}$, where $\mathcal{E}$ is an internal set and F an external statement, and the element of which do not built a set (what can usually be shown by observing that at least one standard result is wrong). So, the halo of 0 is external because the standard result that any bounded subset of $\mathbb{R}$ has a least upper bound fails to be true for it.

Of course, the simple fact that F is external does not mean that $\mathcal{F}$ is necessarily so: the set of i-small integers, which is just $\{0\}$, would be external according to such a rule.

When we want to insist on the fact that a set E is not external (that is, it is a set in the usual sense), we shall say that E is an *internal set*, which is, formally, redundant.

In practice, one often shows that a collection is an external set by showing that it is possible to construct from it, using internal operations (taking the complement or the preimage by a function, . . .), the halo of zero or some other known external set. Indeed, any set $\mathcal{F}$ that enables one, by classical (internal) operations, to define some external set must itself be external since otherwise set theory would be contradictory. So, the set of infinitely large real numbers $\mathrm{hal}\,(\infty)$ defined by

$$\mathrm{hal}\,(\infty) = \{\omega \in \mathbb{R} \mid \omega \text{ i-large}\}$$

is external because, if it were internal, $\mathrm{hal}\,(0)$ would be so also as $\mathrm{hal}\,(0) = \{0\} \cup inv(\mathrm{hal}\,(\infty))$, with $inv(\omega) = 1/\omega$. In the same way, the *principal galaxy*

$$\mathbb{G} = \{x \in \mathbb{R} \mid x \text{ limited}\}$$

is external since $\mathrm{hal}\,(\infty) = \mathbb{R} - \mathbb{G}$, as is the set of appreciable real numbers

$$\mathbb{A} = \{x \in \mathbb{R} \mid x \text{ appreciable}\}$$

since $\mathbb{G}$ is the convex hull of $\mathbb{A}$. In an analogous manner the sets $\underline{\mathbb{N}}$ of all standard integers, and the set $\underline{\mathbb{R}}$ of all standard real numbers are external, as $\mathbb{G}$ is also the convex hull of both the one and the other.

The previous examples are central; let us summarize by observing that one has the following disjoint unions: $\mathbb{R} = \mathbb{G} \cup \text{hal}(\infty)$ and $\mathbb{G} = \text{hal}(0) \cup \mathbb{A}$. Observe that all these external sets are *absolute*, in the sense that they are defined using only standard sets (here $\mathbb{R}$). Below are some examples of external sets that are very useful in nonstandard asymptotics, and that are defined using an infinitesimal $\varepsilon > 0$:

$$\varepsilon\text{-gal}(0) = \{x \in \mathbb{R} \mid (x/\varepsilon) \text{ limited}\}$$
$$\varepsilon\text{-hal}(0) = \{x \in \mathbb{R} \mid (x/\varepsilon) \text{ i-small}\}$$
$$\varepsilon\text{-microhal}(0) = \{x \in \mathbb{R} \mid x \leq \varepsilon^n \text{ for all standard } n\}$$
$$\varepsilon\text{-microgal}(0) = \{x \in \mathbb{R} \mid \text{there exists a limited } r > 0 \text{ s. t. } |x| < e^{-\frac{1}{r\varepsilon}}\}$$
$$\varepsilon\text{-megagal}(0) = \{x \in \mathbb{R} \mid \text{there exists a limited } n \text{ s. t. } |x| < 1/\varepsilon^n\}.$$

These external sets are called respectively the ε-galaxy, the ε-halo, the ε-microhalo, the ε-microgalaxy and the ε-megagalaxy of 0. Up to the megagalaxy they are all contained in the halo of 0 and form, in that order, a decreasing sequence under inclusion.

Remark 1.1.1. *Since a function $f : E \to F$, from a set theoretical point of view, is nothing other than its graph $\mathcal{G}(f) \subseteq E \times F$, there exist internal functions, that is functions with a graph that is an internal set, and external functions. For example, the characteristic function of the standard numbers in $\mathbb{R}$, that is the function χ defined by $\chi(x) = 1$ iff $st(x)$, is external. Actually,* we shall consider, in the sequel, only internal functions, except when there is explicit mention of *external functions* such as in chapter 7.

We can now imagine that the reader may be a little anxious what with external sets and a predicate which is not defined. But take heart ! The next section gives the tools to manipulate the new predicate (the axioms), and the last one (on permanence principles) will explain how to take advantage of external sets.

1.2 Using the extended language

In order to take the extension of the language into account, and use the predicate *standard* and all its derivatives correctly, we need some axioms. These axioms are added to the usual ones. From the logical point of view two questions arise: do these axioms allow one to prove results that could not be shown without them, and, are they noncontradictory ?

The answer is simple: as long as we consider internal sets, all classical results remain true, so we still have $1 + 1 = 2$ and the ordered field $\mathbb{R}$ is still archimedian. The new theory is said to be an *extension* of the classical one. Moreover, one knows that there will be no logically new theorem, that is, a theorem that could not have been shown without the extension, and so, this extension is said to be a *conservative*.

One may feel cheered by this situation, but also disappointed: why then learn NSA ? The point is that, even if there are no *logically* new theorems, this does not mean that NSA will not help to reveal theorems which are true. Individuals may, of course, also use nonstandard analysis when they think that it may help to understand, formulate, or show new results. Not its least interest is in providing shorter proofs of known results. Using the nonstandard concepts of i-small or i-large number often makes it possible to state significant external results that it would be boring to replace systematically by equivalent standard statements. The example of the ducks of the Van der Pol equation (particular solutions of the singularly perturbed equation $\varepsilon x'' + (x^2 - 1)x' + x = a$, with ε i-small, that were discovered in 1978, see chapter 10) is a significant example of this. It would have been a pity to hide the simplicity of the question considered by introducing, say, function-spaces, to state equivalent results.

Let us finally clarify the status of the constructions by ultraproducts that are often associated with NSA. In his founding approach, A. Robinson [100] constructed what we would call a *model* of the nonstandard axioms considered here: from this somewhat sophisticated construction one might imagine that it was necessary to consider difficult constructions in order to apply NSA. This is not, at all, the case. And Robinson was fully aware of this and explicitly suggested at the end of his book [100] that one should seek an axiomatic approach. The construction of Robinson (or any ad hoc analog) is then just a way of proving the "relative consistency" of the axiom system: as nobody knows whether classical mathematics is noncontradictory, the best we can do is to make sure that the possible contradiction will not come from what has been added by NSA. But such a nice construction (see for example [120]) is as little connected with the *use* of NSA as is the construction of $\mathbb{R}$ from $\mathbb{Q}$, $\mathbb{Z}$ and $\mathbb{N}$ to the practice of real analysis. Nobody (we hope) who is involved in studying real differential equations wants to consider the variable as an element of a quotient field of a ring of Cauchy sequences of rational numbers.

1.2.1 The axioms

Here are the rules as to how to manipulate the predicate "standard". There are three of them: **transfer, idealisation** and **standardisation.** We shall use the following notation:

$$\forall^{st} x \quad \text{for} \quad [\forall x \ st(x) \Rightarrow \ldots] \quad \text{and} \quad \exists^{st} x \quad \text{for} \quad [\exists x \ st(x) \ \& \ldots]$$

Transfer.

Principle 1.1. *For any standard formula $F(x)$, one has:*

$$\forall x \; F(x) \; \Leftrightarrow \; \forall^{st} x \; F(x)$$

Among the three principles, it is the easiest to understand; the transfer principle ensures that, if $F(x)$ is a *standard* statement, $F(x)$ holds if and only if it holds for all standard x. Using the negation $G(x) := \neg F(x)$ of $F(x)$, it is easy to see that an equivalent version of the transfer principle is that for any standard statement $G(x)$, there exists an x satisfying $G(x)$ if and only if there exists a standard x satisfying $G(x)$. The case where the property $G(x)$ is satisfied by a unique x_0 is interesting: it ensures that this x_0 is necessarily standard. Thus, any object that can be characterised uniquely, such as $\emptyset$, 0, 1, 2, π, e, sin, ln,..., +, $\cdot$, $\leq$,...,$[0,1]^{\mathbb{R}}$, $C^{\infty}[0,1]$,... is standard.

An important consequence of this axiom is that *any standard function has standard values at standard points*. Another useful consequence is the following: for n, E_1, E_2, ..., E_n, all standard, the set $E := E_1 \times E_2 \times \ldots \times E_n$ is standard, and $x = (x_1, x_2, \ldots x_n) \in E$ is standard if and only if all its components x_p are standard. This is due to the fact that any $p \leq n$ is itself standard, which follows, independently, from the idealisation principle that we shall consider now.

Idealisation.

Principle 1.2. *For any internal statement $B(x,y)$, one has*

$$[\forall^{st} Y \; Y \, finite \; \Rightarrow \; \exists x \; \forall y \in Y \; B(x,y)] \; \Leftrightarrow \; [\exists x \; \forall^{st} y \; B(x,y)]$$

This principle is a little more difficult to understand. Let us first indicate one of its important consequences: *there exists a finite set $\mathcal{F}$ that contains all standard objects*. As a consequence, all infinite sets have nonstandard elements. Let us now consider a binary (internal) relation $B(x,y)$ that we shall read as "x dominates y"; such a relation will be called *concurrent* if for all standard finite sets Y, there exists an x that dominates any element $y \in Y$. For example, the relation $B(x,y) \equiv x \geq y$ is concurrent on $\mathbb{N}$. The idealisation principle ensures that, for all *internal* concurrent relation B, there exists an x that dominates all the standard y.

Here are some examples of concurrent internal relations and the consequences of the corresponding idealisations.

- The relation $B(x,y) \equiv (x \in \mathbb{N})$ & $(y \in \mathbb{N})$ $\Rightarrow$ $(x \geq y)$ leads to the existence of infinitely large integers.
- The relation $B(x,y) \equiv x \in E$ & $x \neq y$ leads to the fact that any set E is standard and finite if and only if it has only standard elements. A consequence is that any limited integer (i.e. less than some standard integer) is itself standard.

- The relation $B(\mathcal{F}, y) \equiv (\mathcal{F} \text{ finite })$ & $(y \in \mathcal{F})$ leads to the existence of a finite set that contains any standard set as an element, a result that we already mentioned above.

Let us finally indicate that, in our experience, many applications of this principle follow from one of these three examples, and that a new application of the idealisation principle is only necessary to open new domains of application.

Standardisation.

Principle 1.3. *For any formula $F(x)$, internal or external, one has:*

$$\forall^{st}\mathcal{E} \ \exists^{st}\mathcal{S}_F \ \forall^{st}x \ [x \in \mathcal{S}_F \ \Leftrightarrow \ x \in \mathcal{E} \ \& \ F(x)]$$

Let us observe that the transfer principle can be applied only to standard statements, the idealisation principle only to internal ones, but that the standardisation principle applies to any statement, internal or not.

Let $F(x)$ be such a statement; we have already pointed out that if F is not internal, the collection of all x satisfying $F(x)$ has a good chance of not being a set of classical mathematics. The standardization principle ensures that for all standard sets $\mathcal{E}$, *the existence of a standard subset $\mathcal{S}_F$ the standard elements of which are precisely the elements x of $\mathcal{E}$ such that $F(x)$ holds.* It is easy to see that the set $\mathcal{S}_F$ is unique: indeed, the transfer principle implies that any standard set is characterized by its standard elements. One calls it the *standardization* of the (internal or external) set of all x in $\mathcal{E}$ such that $F(x)$ and it is denoted by

$$\mathcal{S}_F := {}^S\{x \in \mathcal{E} \mid F(x)\}.$$

It has to be underscored that if x is nonstandard, it is quite possible that x belongs to $\mathcal{S}_F$ and nevertheless that x does not satisfy $F(x)$, as well as that x does not belong to $\mathcal{S}_F$ yet $F(x)$ holds. For example, if $F(x)$ is the statement $0 \lessgtr x \lessgtr 1$, for which $\mathcal{S}_F =]0, 1]$, any i-small $\varepsilon > 0$ belongs to $\mathcal{S}_F$ even though $F(x)$ is false, and any number $1 + \varepsilon$, with $\varepsilon > 0$ i-small fails to belong to $\mathcal{S}_F$ although $F(1 + \varepsilon)$ is true.

As an example, let us sketch the proof of the property of $\mathbb{R}$ that we gave above that *any limited real number r is infinitely close to some standard one.* By standardization, we may consider the (unique) standard subset $\mathcal{S}_+$ of $\mathbb{R}$, the standard elements of which are precisely the elements x such that $x \geq r$ (here $F(x) \equiv x \geq r$). As r is limited, $\mathcal{S}_+$ is non-empty (there exists a *standard* $n \geq |r|$ as r is by assumption limited); analogously, $\mathcal{S}_- := \mathbb{R} - \mathcal{S}_+$ is non-empty (take $-n$). Applying transfer in order to consider only standard elements of $\mathcal{S}_\pm$ that are the only ones that are "well known", one sees that $(\mathcal{S}_-, \mathcal{S}_+)$ is a cut of $\mathbb{R}$; this cut characterizes a real number s, standard by transfer. It is then easy to show that s has to be i-close to r.

1.2.2 Application to standard objects

We just analysed the three axioms that govern the use of the predicate st and all its derivatives like i-small or i-large. In principle, it suffices to know these axioms to be able to manipulate without error the new concepts introduced by nonstandard analysis. In practice, it is certainly best for the beginner to try to "understand" the "meaning" of the predicate standard. Considering the axioms, we can now see that the standard elements of any standard set are *distinguished elements* that stake out all of its standard structures. They are *witnesses:* there are few of them (they are contained in a finite set) but they suffice to control all the standard properties of the set, and any subcollection of these witnesses characterises a standard subset. So it is that we try to summarize idealization, transfer, and standardization.

But all this may still seem very abstract, and the real meaning will certainly become clearer in considering examples in parts of mathematics that are already familiar. This is what we shall do next.

We will see that the nonstandard language makes it possible to reformulate, in an often simpler setting, the basic definitions of Analysis. We will observe that these new characterizations hold only for standard objects. For nonstandard objects, they no longer apply and we have to return to the classical definitions; but the properties introduced here are often also interesting for nonstandard objects (they are the so-called S-properties), as we shall see in the following chapter.

Definitions. For x and y in $\mathbb{R}$, we have seen that $x \simeq y$ is a notation for $x - y$ is i-small. For x and y in a standard metric space E, the notation $x \simeq y$ means that the distance from x to y is infinitesimal. If there exists in that space a standard x_0 such that $x \simeq x_0$, the element x is called *near-standard* in E and, as in $\mathbb{R}$, the standard point x_0 is called the *standard part* of x (it is unique) and is also denoted by $^\circ x$. The *halo* of x, denoted by $\mathrm{hal}\,(x)$ is the set, usually external, of all y such that $x \simeq y$.

An element x of the standard metric space E is called *near-standard* in some subspace $A \subset E$ if there exists a standard element $x_0 \in A$ such that x belongs to $\mathrm{hal}\,(x_0)$. More generally, still for $A \subset E$, the *halo* of A, denoted by $\mathrm{hal}\,(A)$, is the subset, usually external, of all $x \in E$ that are at an infinitesimal distance from at least one element of A.

In a standard *topological* space E, the *halo* of $x \in E$, denoted by $\mathrm{hal}_T(x)$, is the intersection of all standard neighbourhoods of x. We observed above that, in any metric space, the fact that $\mathrm{hal}\,(x)$ meets $\mathrm{hal}\,(y)$ implies that $\mathrm{hal}\,(x) = \mathrm{hal}\,(y)$, but this is generally not true in a topological space if one replaces hal by hal_T (even if the topology considered is defined by a metric). One is sure to have $\mathrm{hal}\,(x_0) = \mathrm{hal}_T(x_0)$ only if x_0 is *standard*. This is why it is important to distinguish carefully between $\mathrm{hal}\,(x)$ and $\mathrm{hal}_T(x)$ (see chapter 6). In the sequel of this chapter, we shall restrict ourselves when dealing with metric spaces, to the metric $\mathrm{hal}\,(x)$.

Lexicon. Below is a selection of nonstandard rephrasings of basic concepts of Analysis. At the end of the section we shall give an example of an equivalent proof based on classical definitions.

Sequences A standard sequence (u_n) is a *Cauchy sequence* if and only if $x_p \simeq x_q$ for any i-large p and q. It converges towards (a standard) l if and only if $u_n \simeq l$ for all i-large n. It admits a converging subsequence if and only if there exists an i-large n and a standard l such that $u_n \simeq l$.

Functions A standard function f is *continuous* at the standard point x_0 if and only if $f(x) \simeq f(x_0)$ for all $x \simeq x_0$. It is continuous if and only if it is continuous at any standard point. It is *uniformly continuous* if and only if $f(x) \simeq f(y)$ for all $x \simeq y$ (x and y standard or not), so if and only if it is S-continuous at any point.

If f is defined on a standard interval, containing the standard number x_0, the standard number d is the *derivative* of f at x_0 if and only if for any i-small $\varepsilon \neq 0$, $(f(x_0 + \varepsilon) - f(x_0))/\varepsilon \simeq d$ holds. It is *differentiable* if and only if it is differentiable at any standard point. If so, its derivative f' is the standardization of the (external) relation associating to any standard x_0 the derivative d of f at the point x_0.

Let f be a standard function defined on a (standard) open subset of a (standard) normed space that contains the standard point x_0. The standard continuous linear function L is the *differential* of f at x_0 if and only if

$$\frac{1}{\varepsilon}(f(x_0 + \varepsilon X) - f(x_0)) \simeq L[X]$$

for all i-small $\varepsilon \neq 0$ and all X with limited norm. The function $X \mapsto Y = \frac{1}{\varepsilon}(f(x_0 + \varepsilon X) - f(x_0))$ is called the *image of f under the magnifying glass* $x = x_0 + \varepsilon X$, $y = f(x_0) + \varepsilon Y$, of magnification $\frac{1}{\varepsilon}$ centred at the point $(x_0, f(x_0))$. This allows one to give a more geometrical definition of differentiability: f is *differentiable* at some standard x_0 if and only if the image of f under any magnifying glass of i-large magnification centred at $(x_0, f(x_0))$ is infinitely close to the same standard linear function L. The function f is *strictly differentiable* at the some standard point x_0 if, moreover, $\frac{1}{\varepsilon}(f(x + \varepsilon X) - f(x)) \simeq L[X]$ for all $x \simeq x_0$ (and limited X).

Sequences of functions Let f and $(f_n)_{n \geq 0}$ be a standard function and a standard sequence of functions, all with same domain. The sequence (f_n) converges *pointwise* to f if and only if $f_n(x_0) \simeq f(x_0)$ for all i-large n and all standard x_0. It converges *uniformly* to f if and only if $f_n(x) \simeq f(x)$ for all i-large n and for all x, standard or not.

Integration Let us call any finite sequence, $(x_i)_{i \in \{1,...,n\}}$ such that $a = x_0 < x_1 < \ldots < x_n = b$, a *cutting*, D, of the interval $I = [a, b]$. A cutting will be called *thin* if $x_{i+1} \simeq x_i$ for all i in $\{0, \ldots, n-1\}$. A function $\varphi : I \to \mathbb{R}$ is a *step function* on I for the cutting, D, if φ is constant on each subinterval $[x_i, x_{i+1}[$; if so, the *integral* of the step function φ is, by definition, $\int \varphi := \sum_{i=0}^{N-1} \varphi(x_i)(x_{i+1} - x_i)$. A standard function $f : I \to \mathbb{R}$

is *Riemann integrable* if and only if for all thin cuttings D of I, there exists two step functions φ and ψ for D, such that $\int (\psi - \varphi) \simeq 0$ and $\varphi(x) \leq f(x) \leq \psi(x)$ for all $x \in I$. Then, one defines $\int f := {}^{\circ}(\int \varphi) = {}^{\circ}(\int \psi)$.

The function f is *Lebesgue integrable* if and only if for all standard $e > 0$ there exist two step functions φ_e and ψ_e such that $\int (\psi_e - \varphi_e) < e$ and $\varphi_e(x) < f({}^{\circ}x) < \psi_e(x)$ for all $x \in I$. In that case, $\int f$ is the unique standard real number such that $\int \varphi_e \leq \int f \leq \int \psi_e$ for all standard $e > 0$ and all functions φ_e and ψ_e as above.

Topology Let $\mathcal{E}$ be a standard topological space, $\mathcal{A} \subset \mathcal{E}$ a standard subset, and $x_0 \in \mathcal{E}$ a standard element; x_0 is an *interior point* of $\mathcal{A}$ if and only if $\mathrm{hal}_T(x_0) \subset \mathcal{A}$; it is a limit point of $\mathcal{A}$ if and only if $\mathrm{hal}_T(x_0) \cap \mathcal{A} \neq \emptyset$. The (standard) subset $\mathcal{A}$ is *open* if and only if all of its standard elements are interior points of it, and it is closed if and only if all its standard limit points belong to it. The *interior* and the *closure* of $\mathcal{A}$ are the standardization, respectively, of its interior points and of its limit points.

The topological space $\mathcal{E}$ is *Hausdorff* if and only if any element of $\mathcal{E}$ has at most one standard part. It is *compact* if and only if , moreover, any element of $\mathcal{E}$ is *near-standard in $\mathcal{E}$*, that is, if and only if any element of $\mathcal{E}$ has a standard part in $\mathcal{E}$.

Remark 1.2.1. *At the beginning of this section, we wrote that the nonstandard characterizations of this lexicon apply only to standard objects. Let us take, for example, the last characterization: compactness. Let $\varepsilon > 0$ be i-- small. The standard subsets $]0,1[$ and $[0,+\infty[$ of $\mathbb{R}$ are not compact: the first one has elements such as ε and $1 - \varepsilon$ that are not near-standard in this set, and the second has i-large elements that are not i-close to any standard point. On the other hand, if we try to apply this criterion to subsets that are not standard subsets, we observe that it leads to wrong conclusions. For example $[\varepsilon, 1 - \varepsilon]$ is compact (it is closed and bounded) although it has elements like ε that are not near-standard in this set. Analogously, for a set like $] - \varepsilon, 1 + \varepsilon[$ that is of course not compact (it is not closed), any of its elements has a standard part that belongs to this set.*

It is not the purpose of this book to demonstrate all the characterizations of the above lexicon. Here, an example will suffice to understand how these proofs work.

Let us consider the characterization of an interior point x_0 of a subset $\mathcal{A}$. Classically, if x_0 is an interior point of $\mathcal{A}$, there exists a neighbourhood of x_0 contained in $\mathcal{A}$. If x_0 and $\mathcal{A}$ are, as here, assumed to be standard, by transfer there exists also a *standard* neighbourhood of x_0 that is contained in $\mathcal{A}$. This shows that the intersection of all the standard neighbourhoods of x_0 is contained in $\mathcal{A}$, and thus, that $\mathrm{hal}_T(x_0)$ is contained in $\mathcal{A}$.

Conversely, consider first the case where $\mathcal{E}$ is a metric space. Here, $\mathrm{hal}_T(x_0) = \mathrm{hal}(x_0)$ consists of the points that are at an i-small distance from

x_0. If we assume that hal $(x_0) \subset \mathcal{A}$, it suffices to choose an i-small $r > 0$. The ball $\mathcal{V}$ with centre x_0 and radius r is the neighbourhood that we need, as $x_0 \in \mathcal{V} \subset$ hal $(x_0) \subset \mathcal{A}$. In a general topological space, one has to make sure that there exists a neighbourhood $\mathcal{V}$ of x_0 contained in any standard neighbourhood of x_0. This can be obtained by considering the internal relation on the set of all neighbourhoods of x_0 defined by $B(\mathcal{V}, \mathcal{W}) \equiv \mathcal{V} \subset \mathcal{W}$. Here $\mathcal{V}$ "dominates" $\mathcal{W}$ if $\mathcal{V} \subset \mathcal{W}$ (it is a "better" neighbourhood than $\mathcal{W}$); the relation is concurrent as the intersection of any finite family of neighbourhoods $\mathcal{W}_i$ of x_0 is a neighbourhood of x_0 that dominates all the $\mathcal{W}_i$ of the family; by idealization, there thus exists a neighbourhood $\mathcal{V}$ that dominates all the standard neighbourhoods, and thus $\mathcal{V} \subset$ hal (x_0). The conclusion follows as in the case of a metric space.

1.3 Shadows and S-properties

1.3.1 Shadow of a set

As indicated above, for all standard subsets, $\mathcal{A}$, of a standard topological space $\mathcal{E}$, the closure $\overline{A}$ of $\mathcal{A}$ in $\mathcal{E}$ is the unique standard set whose standard elements are precisely all $x_0 \in \mathcal{E}$ such that the halo of x_0 intersects $\mathcal{A}$. For a general $\mathcal{A}$, possibly external, the same definition leads to the notion of shadow:

Definition 1.3.1. *The* shadow *of $\mathcal{A}$, denoted by $^\circ\mathcal{A}$, is the unique standard set whose standard elements are precisely those whose halo intersects $\mathcal{A}$:*

$$^\circ\mathcal{A} := {}^S \{x_0 \in \mathcal{E} \mid hal(x_0) \cap \mathcal{A} \neq \emptyset\}.$$

If $\mathcal{A}$ is standard, we just recalled that $^\circ\mathcal{A} = \overline{A}$. If $\mathcal{A}$ is internal, one shows that $^\circ\mathcal{A}$ is closed. But this is not necessarily true for an external set.

Examples: For any i-small $\varepsilon > 0$, the shadow of the internal set $\mathcal{A} = \{x \mid \varepsilon < x \leq 1 + \varepsilon\}$ is the closed interval $[0, 1]$; but the shadow of the external set $\mathcal{A} = \{x \mid 0 \lneqq x \lneqq 1\}$ is the semi-open interval $(0, 1]$.

The notion of shadow of a subset is thus obtained by extending to sets other than standard ones, a property corresponding to the closure in the case of a standard set: so the shadow is what is also called the "S-closure". This is a first example of a very fruitful technique in nonstandard analysis. We shall examine here some other examples of S-properties.

Definition 1.3.2. *Let $\mathcal{A}$ still be any subset of a standard topological space $\mathcal{E}$. The* interior shadow *of $\mathcal{A}$, denoted by $^i\mathcal{A}$, is the unique standard subset of $\mathcal{E}$ whose standard elements are precisely those whose halo is contained in $\mathcal{A}$:*

$$^i\mathcal{A} := {}^S \{x_0 \in \mathcal{E} \mid hal(x_0) \subset \mathcal{A}\}.$$

We have seen that if $\mathcal{A}$ is standard subset, $^i\mathcal{A}$ is the interior of $\mathcal{A}$: this is why one can interpret the interior shadow as a notion of S-interior. If $\mathcal{A}$ is internal, $^i\mathcal{A}$ is always an open subset of $\mathcal{E}$.

Examples: The interior shadow of the internal set $\mathcal{A} = \{x \mid \varepsilon < x \le 1 + \varepsilon\}$ is the open interval $]0,1[$; but the interior shadow of the external set $\mathcal{A} = \{x \mid 0 \lneqq x \lesssim 1\}$ is the semi-open interval$(0,1]$.

1.3.2 S-continuity at a point

For any map $f : E \to F$ between standard metric spaces (E, d_E) and (F, d_F), we have seen that if f and $x_0 \in E$ are standard, f is continuous at x_0 if and only if for all $x \simeq x_0$, $f(x) \simeq f(x_0)$ holds. It is interesting to consider the corresponding S-notion.

Definition 1.3.3. *Let E and F be two standard metric spaces. An (internal) map f whose domain and target are respectively in E and F is called S-continuous at x_0 if and only if for all $x \simeq x_0$, $f(x) \simeq f(x_0)$ holds.*

Examples: Let $\varepsilon > 0$ be some i-small number. The real function Arctan (x/ε) is continuous but it is not S-continuous at 0. The function $\varepsilon\text{Ent}(x/\varepsilon)$ (where Ent (y) is the largest integer less than y) is S-continuous at any point, but is discontinuous at any point of $\varepsilon\mathbb{Z}$.

It can be shown that f is S-continuous at x_0 if and only if it satisfies the usual quantified formula

$$\forall x \in \mathcal{D}(f) \ \ d_E(x, x_0) < \eta \Rightarrow d_F(f(x), f(x_0)) < \varepsilon$$

with ε and η supposed standard. In other words, if one does not look at f "too closely", then f seems to be continuous. This is why S-continuity is sometimes called "continuity for a short-sighted person".

1.3.3 Shadow of a function

There exists at least two objects that can be considered to be the shadow of a function f. On one hand, there is the shadow of its graph $\mathcal{G}(f)$; however, the example of the function Arctan (x/ε) above shows that this shadow is not always a graph of a function: at the abscissa $x = 0$, that shadow contains all the points with ordinate $y \in [-\pi/2, +\pi/2]$.

On the other hand, one may consider the unique standard set whose standard elements are the $(x_0, {}^\circ f(x_0))$ for all standard x_0 such that $f(x_0)$ is near-standard. This set is always the graph of some standard function called the *standardization* of f and denoted by ^Sf; for example, for $f(x) =$ Arctan (x/ε), $^Sf(x) = -\pi/2$ for $x < 0$, $^Sf(0) = 0$ and $^Sf(x) = \pi/2$ for

$x > 0$. However $^S f$ may be very irregular; for example, if $f(x) = \cos \omega x$ with ω i-large , then for some choice of ω, $^S f$ can be non-measurable. [109, 50].

The nice situation corresponds to the case where both notions coincide. This is the case precisely when f is S-continuous at "all reasonable points": it is the notion that we shall consider now. It corresponds to the case where the shadow of the graph of f is the graph of a standard continuous function f_0, called the shadow of the function f (theorem of the continuous shadow 1.3.1). The example of the function $f(x) = 1/(x^2 + \varepsilon)$ (with i-small $\varepsilon > 0$) shows that the shadow (here $f_0(x) = 1/x^2$) of a function may have a smaller domain than the function itself. The reverse situation may also occur as in the case of $f(x) = \text{Arc} \sin((1 + \varepsilon)x)$ that is defined on $]-\pi/2(1 + \varepsilon), +\pi/2(1 + \varepsilon)[$ while its shadow $f_0(x) = \text{Arc} \sin(x)$ is defined on $]-\pi/2, +\pi/2[$.

The continuous shadow theorem. For any function f let us denote its domain by $\mathcal{D}(f)$, its target by $\mathcal{B}(f)$, and its graph by $\mathcal{G}(f)$. We shall always assume that $\mathcal{D}(f)$ and $\mathcal{B}(f)$ are subspaces of standard metric spaces. Let us recall that for any subset X of a standard metric space E, a point $x \in E$ is called near-standard in X if there exists a standard point $x_0 \in X$ such that $x \simeq x_0$. By "x is near-standard", we shall mean that x is near-standard in $X = E$.

Now we can consider the notion of a function that is S-*continuous*, not only at one point, but *on all of its domain*.

Definition 1.3.4. *Let E and F be standard metric spaces, and f an internal function, such that $\mathcal{D}(f) \subset E$ and $\mathcal{B}(f) \subset F$.*

The function f is called S-continuous in $E \times F$ if and only if it is S-continuous at each point $x \in \mathcal{D}(f)$ such that $(x, f(x))$ is near-standard in $E \times F$.

So, the function $\text{Arctan}(x/\varepsilon)$, $x \in \mathbb{R}$, is not S-continuous (on $\mathbb{R}$), but the function $1/(x^2 + \varepsilon)$, $x \in \mathbb{R}$, is S-continuous (on $\mathbb{R}$). Finally, the restriction of $\text{Arctan}(x/\varepsilon)$ to any subset of $\mathbb{R}$ that does not contain 0 is S-continuous in $\mathbb{R} \setminus \{0\} \times \mathbb{R}$.

Remark 1.3.1. *The choice of terminology is consistent; indeed one can check that if f is a standard function, f is continuous if and only if it is S-continuous in $\mathcal{D}(f) \times \mathcal{B}(f)$. Let us observe that it would not be reasonable to ask for S-continuity of f at every point of $\mathcal{D}(f)$: the function $f(x) = x^2$ would not be S-continuous !*

The shadow in $E \times F$ of the graph of the S-continuous function f is the graph of a (standard) continous function f_0, called the *shadow* of f, and denoted by $^o f$. More precisely, one has:

Theorem 1.3.1 (continuous shadow). *Let E and F be standard metric spaces, and f be a function defined on $\mathcal{D}(f) \subset E$ and with values in F.*

The function f is S-continuous in $E \times F$ if and only if there exists a standard continuous function f_0, with domain $\mathcal{D}(f_0)$ in E, which has the two following properties:

1. *for any $(x, f(x))$ near-standard in $E \times F$, $^{\circ}x$ is in $\mathcal{D}(f_0)$*
2. *for any standard $x_0 \in \mathcal{D}(f_0)$ and any $x \in \mathcal{D}(f)$, if $x \simeq x_0$ then $f(x) \simeq f_0(x_0)$.*

1.3.4 S-differentiability

For a function f defined on an open subset $\mathcal{D}(f)$ of a standard normed space E and with values in a standard normed space F, we have seen that if f is standard, it is strictly differentiable at the standard point x_0 if there exists a standard linear continuous function $L : E \longrightarrow F$ such that for any $X \in E$ limited, any i-small $\varepsilon \neq 0$, and any $x \simeq x_0$, $(f(x + \varepsilon X) - f(x))/\varepsilon \simeq L[X]$ holds. Lets us examine the corresponding S-property.

Definition 1.3.5. *Let E and F be two standard normed spaces. A function with domain and target in E and F respectively is called S-differentiable at the standard point x_0 with differential L, if f is defined at any point $x \simeq x_0$, if $L : E \longrightarrow F$ is a standard linear continuous function, and for all limited X, all i-small $\varepsilon \neq 0$, and all $x \simeq x_0$, one has*

$$\frac{1}{\varepsilon}(f(x + \varepsilon X) - f(x)) \simeq L[X].$$

Definition 1.3.6. *Let f be a function defined on a standard open subset $\mathcal{D}(f)$ of a standard normed space E with values in a standard normed space F. The function f is called S-differentiable if it is S-differentiable at all standard x_0 of its domain for which its value $f(x_0)$ is near-standard in F.*

The next theorem shows that, in an analogous way to S-continuous functions which are those that have a continuous shadow, S-differentiable functions are those that have a differentiable shadow. Jean Louis Callot suggested that one create a new etymology for the "S-" (which actually comes from *standard*): S-continuous functions are those that *seem* continuous, and S-differentiable functions are those that *seem* differentiable.

Theorem 1.3.2 (differentiable shadow). *Let E and F be two standard normed spaces, and let f be an internal function defined and S-differentiable on a standard open subset $\mathcal{D}(f) \subset E$, with values in F.*

The shadow $^{\circ}f$ of f exists, is defined on a standard open subset $\mathcal{D}(^{\circ}f)$ and is C^1. Moreover, for any standard $x_0 \in \mathcal{D}(^{\circ}f)$, the differential $D(^{\circ}f)(x_0)$ is the S-differential of f at the point x_0.

1.3.5 Notion of S-theorem

It is easy to deduce from the continuous shadow theorem the famous Ascoli theorem, that ensures that any uniformly equicontinuous family of functions from a compact space to a complete one has a compact closure for the uniform norm. Practice has shown that, in questions involving convergence of subsequences of uniformly equicontinuous sequences, a nonstandard approach to a problem means that introducing such a sequence leads to,instead of a subsequence, the study of a unique internal function and of its continuous shadow. It is in that sense that the continuous shadow theorem is the S-theorem of Ascoli.

The user of NSA often manipulates such S-theorems. For example, the theorem that we have shown in section 1.2, *any limited real number has a standard part*, can be considered as the Bolzano-Weierstrass S-theorem (any bounded real sequence has a convergent subsequence). One may play the game of finding the classical theorems for which the following statements are the S-theorems: "any standard continuous function is the shadow of some polynomial" and "any holomorphic function, that is limited at all limited points of $\mathbb{C}$, is S-continuous and has a holomorphic shadow".

1.4 Permanence principles

Recall that in the practice a collection can be shown to be an external set by proving that at least one theorem of the classical set theory is wrong for it. So, for example, the set $\mathbb{G}$ of limited real numbers is bounded (by any i-large real number) but has no least upper bound: it is external. These external sets introduce themselves in a natural way when one practices NSA, but one might think, based on the fact that *a priori* no classical theorem can be applied to them, that they would be ticklish to manipulate. The permanence principles are results, specific to external sets, that have precisely the effect of turning their "faults" into qualities.

1.4.1 The Cauchy principle

Here is a very common situation in nonstandard analysis. Let $(a_n)_{n \geq 0}$ be a standard sequence of positive real numbers, and let $\varepsilon > 0$ be any infinitesimal. As long as n is limited, the sequence $b_n = a_n \varepsilon^n$ is decreasing (since $b_n/b_{n-1} = (a_n/a_{n-1})\varepsilon \simeq 0$). The (internal) set D of all indices up to which the sequence $(b_n)_{n \geq 0}$ is decreasing thus contains the (external) set $\underline{\mathbb{N}}$ of all standard integers. Can it be equal to it ? Of course not, because D is internal (the property "to be decreasing up to n") and $\underline{\mathbb{N}}$ is external. So, we are sure that there exists an i-large ω such that $(a_n \varepsilon^n)$ is decreasing for all $n \leq \omega$. And this without any further assumption on the sequence (a_n).

Such a way of reasoning that enables one to extend the range of a property simply by taking advantage of the difference internal versus external, without any other argument, is one of the strengths of nonstandard techniques. It has been given the name of *Cauchy principle*.

Principle 1.4 (Cauchy Principle). *No external set is internal.*

This principle is of course a tautology, and one refers to it to express the notion that the range of a piece of reasoning *overspills* from the domain for which it has been established. Of course, to apply it, one has to check that one of the two sets is really external; an amazing example of inverting the difficulties, in which external sets suddenly become desirable ! To show that a set is really external can often be achieved, as we did for hal (∞), $\mathbb{G}$, and $\mathbb{A}$, by contradiction, expressing a known external set as the result of an internal construction based on the set under consideration: if it were internal, so would the known external set be internal.

1.4.2 Fehrele principle

Do an se ne g'freujt in d' dynamisch Geometrie[2]
Waass het de Fehrele g'sajt
A Halo isch ke' Galaxie,
Des hett de Fehrele g'sajt (bis)

To show that the uniform limit f of a sequence $(f_n)_{n\geq 0}$ of continuous functions is itself continuous, one would like to do the following: first assume, by transfer, that the sequence $(f_n)_{n\geq 0}$ and thus its limit f are standard. To show that the standard function f is continuous one wants to get $f(x) \simeq f(y)$ from $x \simeq y$, using something like

$$f(x) \overset{(a)}{\simeq} f_n(x) \overset{(b)}{\simeq} f_n(y) \overset{(c)}{\simeq} f(y). \tag{1.1}$$

Unfortunately, (a) and (c) hold only for i-large n (from uniform convergence), when (b) *a priori* only holds for standard n (as the nonstandard criterion for continuity only holds for standard functions). Here, the nonstandardist is caught in a dilemma after the manner of Corneille, between what appears attractive and what is demanded by reality. But here too enters a famous lemma of Robinson that has several versions, one of these being

Lemma 1.4.1 (Robinson). *If $(u_n)_{n\geq 0}$ is a sequence such that $u_n \simeq 0$ for all standard n, there exists an i-large N such that $u_n \simeq 0$ for all $n \leq N$.*

---

[2] This is a stanza that was added to a famous alsatian popular song, as suggested by Georges Reeb shortly after the result was stated. Fehrele is the hero of this song.

So, by letting $u_n := f_n(x) - f_n(y)$, this enables one to choose an i-large n such that (1.1) holds, and hence, to conclude as we wished. Let us analyse this Robinson's lemma; there are two kinds of indices involved: the set G of all standard indices, and the set H of all indices for which $u_n \simeq 0$. The hypothesis states that $G \subset H$, and the conclusion gives that G overspills into H. Unfortunately, Cauchy's principle does not apply here (at least not in an immediate way), as, generally, G and H are both external. Nevertheless, by construction, these sets cannot be equal as one, H, gathers all elements for which an internal function $(n \mapsto u_n)$ is i-small and the other, G, gathers all elements such that the internal function $n \mapsto 1/n$ is *not* i-small. And this is impossible, as we shall see. Let us observe, on the other hand, that it is possible that for all standard n, the values of a sequence (v_n) is not i-small and that it is i-small as soon as n becomes i-large: this is for example the case for the sequence $n \mapsto 1/n$. It is the distinction between these two kinds of sets that is systematized by introducing the notions of *halo* and *galaxy*.

Definitions 1.4.1. *Let $H \subset E$; H is a* prehalo *if there exists an internal family $(H_n)_{n \in J}$, J standard, of (internal) subsets of E such that*

$$H = \bigcap_{st(n)} H_n.$$

Let $G \subset E$; G is a pregalaxy *if there exists an internalfamily $(G_n)_{n \in J}$, J standard, of (internal) subsets of E such that*

$$G = \bigcup_{st(n)} G_n.$$

A halo *is a prehalo that is external. A* galaxy *is a pregalaxy that is external.*

Examples:

- hal (a), hal (∞) are halos: Take $J = \mathbb{N}$, $H_n^a = [a - \frac{1}{n}, a + \frac{1}{n}]$ for the first one and $H_n^\infty =]-\infty, -n[\cup]n, +\infty[$ for the second one;
- $\mathbb{G}$, $\mathbb{A}$ are galaxies: take $J = \mathbb{N}$, $G_n^G = [-n, n]$ for the first one and $G_n^A =]-n, -\frac{1}{n}[\cup]\frac{1}{n}, n[$ for the second one.

The next proposition can often be used in applications:

Proposition 1.4.1. *Let $f : E \to \mathbb{R}$ be an internal function.*

If $H = \{x \mid f(x) \text{ is i-small}\}$ is an external set, then it is a halo.

If $G = \{x \mid f(x) \text{ is limited}\}$ is an external set, then it is a galaxy.

Assume E is a connected metric space. If f (resp. g) is continuous or S-continuous, takes both appreciable and i-small (resp. i-large) values, then H (resp. G) is external.

Principle 1.5 (Fehrele's principle). *No halo is a galaxy.*

For the example of Robinson's lemma above, $G = \mathbb{G}$ is a galaxy, and H is *a priori* just a prehalo: u_n could be i-small for all n, and in this case $H = \mathbb{N}$ would of course be an internal set. If H is internal, Robinson's lemma is a consequence of Cauchy's principle. If H is an external set, the lemma follows from Fehrele's principle.

Of course, the complement of a halo in any internal set is a galaxy and conversely; the intersection or the union of a halo and a prehalo is a prehalo and the intersection or the union of a galaxy and a pregalaxy is a pregalaxy. Let us observe that there exist external sets that are neither halos nor galaxies, such as $\{x \in \mathbb{R} \mid 0 \lesssim x \lesssim 1\}$. Nevertheless, if (R_-, R_+) is a cut of $\mathbb{R}$, i.e. $\mathbb{R} = \mathbb{R}_- \cup \mathbb{R}_+$ with $r_- < r_+$ for all $r_- \in R_-$ and all $r_+ \in R_+$, then one of the two is a prehalo and the other a pregalaxy. ([15]). If the cut is internal, it characterises a real number; if it is external, it characterises an "external number" (see chapter 7).

One will find in the other chapters of this book many applications of Fehrele's principle. Here is a proof of this principle, purely formal but very simple; it consists of establishing a *separation principle*, namely, if $G \subseteq H$ with G a galaxy and H a halo, there exists an internal set I such that $G \subset I \subset H$. The conclusion follows then from Cauchy's principle.

Proof. For any set E, let $\underline{E}$ be the subset of all standard elements of E; for E standard, $\underline{E}$ is external as soon as E is infinite. Let X and Y be standard, and let $(G_x)_{x \in X}$ and $(H_y)_{y \in Y}$ be two internal families such that $G = \bigcup_{x \in \underline{X}} G_x$ and $H = \bigcap_{y \in \underline{Y}} H_y$. Let R be the internal set of all $(x, y) \in X \times Y$ such that $G_x \subset H_y$. The hypothesis $G \subset H$ implies that $\underline{X} \times \underline{Y} \subset R$ which, in turn, implies that the relation $B(P, c)$ on $\mathcal{P}(X \times Y) \times (X \times Y)$ defined by the following condition,

$$P \text{ is a cartesian product, } P \subset R, \text{ and } c \in P$$

is concurrent. From the idealisation principle, there thus exists $P = X' \times Y' \subset R$ such that $\underline{X} \times \underline{Y} \subset X' \times Y'$. It then suffices to consider $I = \bigcup_{x \in X'} G_x$ (or $I = \bigcap_{y \in Y'} H_y$). $\qquad\square$

To conclude , let us indicate that Robinson's lemma[3] has also a direct proof that is perfectly elementary, merely by considering the internal set $I := \{n \in \mathbb{N} \mid nu_n \leq 1\}$. From the Leibniz rules, it is easy to see that $G \subset I \subset H$. But the relation $nu_n < 1$ has little to do with the problem considered and $nu_n \leq 1000$ or $nu_n \leq 0.001$ would also do. It was by thinking this over that the definitions of halo and galaxy were introduced and that Fehrele's principle was stated. [52, 19, 13].

---

[3] Keith Stroyan remembers having heard A. Robinson say that if one of his mathematical results were to be carved on his tombstone, it should be this one. Actually, his tombstone simply mentions that he was the founder of Nonstandard Analysis [37].

2. Complex analysis

A. Fruchard

2.1 Introduction

Complex differentiation seems, at first glance, to be simply a formal extension of real derivation but, in fact, this extension is not at all trivial. The single hypothesis of differentiability of a function of one complex variable has unexpected consequences and yields numerous properties of the function ; analytic properties (it is of class C^∞) and algebraic rigidity properties (the values on a set with an accumulation point uniquely characterize the function). Many results follow. The aesthetic nature of these results contributes to the beauty of the theory but their multitude adds to its difficulty, as though you had to fly a Concorde ! Briefly, complex analysis gives complexes. Otherwise excellent courses on complex analysis do not clearly answer the first question one might be tempted to ask: what does an analytic function look like. And, finally, too much information prevents a good understanding of complex analysis.

I[1] propose a new perspective as regards complex analytic functions: the infinitesimal point of view. Of course it is not my purpose to replace entirely a classical course on complex analysis by this infinitesimal approach. For one thing, one loses results of an algebraic nature. As an example, the knowledge of a function up to an infinitesimal in a domain gives no information about the function outside of this domain. In other words there is no "almost analytic continuation".

On the other hand the infinitesimal approach did allow me to distinguish a few results among the vast number available, which were particularly useful in my own work. These results provide a macroscopic description of an analytic function. Together they form the following Robinson-Callot theorem.

Theorem 2.1.1. *Let f be a complex-valued function analytic on a domain which contains the halo of a limited point x_0, and is such that $f(x_0)$ is limited.*

---

[1] Initially this chapter was to have been the work of two authors, one of whom, Jean-Louis Callot, died in August 1993. We had decided earlier that each author would write up the work of the other. Thus, the chapter is devoted to three different topics which are essentially due to Jean-Louis Callot. The part covering my own work on discrete dynamical systems is not included for the reason mentioned. I dedicate this chapter to him.

1. If f takes only limited values in the halo of x_0, then there exists a standard neighbourhood V of x_0 on which:
   a) f is S-continuous: $x_1 \simeq x_2 \Rightarrow f(x_1) \simeq f(x_2)$,
   b) the shadow of f is analytic,
   c) the derivatives of standard order of the shadow of f coincide with the shadows of the derivatives of f.
2. If f is not S-continuous in x_0, then the image of the halo of x_0 by f contains all limited complex numbers except possibly a part of the halo of one point.
3. If f is S-continuous in x_0 and if the shadow of f is not constant, then the image of the halo of x_0 by f is exactly the halo of $f(x_0)$.

From a formal point of view this statement has no simple classical version but, on the other hand, each item corresponds, in essence, to a classical result. These are, in order, the theory of normal families, Picard's theorems and the open mapping theorem. Notice that these results are usually well separated in a classical course. Combined with a few tools such as Cauchy's formula and the order of zeros, this theorem could permit an easy restructuring of an elementary course on complex analysis, but it can also replace such a course. In particular we can do complex analysis without power series.

The statement of the theorem furnishes a *description* of analytic functions. It describes in terms which are simultaneously clear, intuitive, semantical, and strictly mathematical what we see if we look at the graph of an analytic function: at a given point the function is either perfectly regular or appallingly irregular. The statement also teaches us that it is judicious to look at places where the function takes limited values in order to see something more fully, but to examine places where the function is not S-continuous in order that what we see may be *interesting*. This region of the complex plane where f is limited and not S-continuous (or simply not S-continuous if one considers f as a function with values in the Riemann sphere), this *zone of non-S-continuity* of f is, indeed, the significant region of f. We have baptized this zone the *zonosč* of f – denoted $Zn(f)$ – in memory of the well-known Syldavian mathematician who discovered this notion[2]. The zonosč is often a thin set, a curve for instance, and the graph of f looks like a cliff at such places.

The chapter begins with a tutorial which includes a proof of the Robinson-Callot theorem. This proof is based on some classical results of complex analysis which are not proved here: essentially the Cauchy formula, the fact that the holomorphic covering set of the complex plane except for two points is the disk, and Morera and Rouché theorems. Some simple non standard applications are given, followed by exercises together with their solutions which are nothing less than classical results. Please do not consider, dear reader, my decision to present these beautiful results as exercises as a provocation. These

---

[2] Zonosč V. *Sur les ZOnes de NOn S-Continuité des fonctions analytiques complexes*, Comptes Rendus de la Chambre d'Agriculture de Poldévie, 1904 (197)

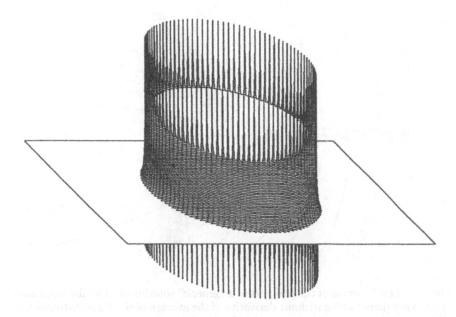

Fig. 2.1. "The birthday cake" ; level curves of the real part of $\frac{1}{1-x^{100}}$ ($-2 < \Re(x) <$ 2, $-1.33 < \Im(x) < 1.33$, $-1 < \Re f(x) < 2$).

results afford a nice opportunity for training in the manipulation of non standard analytic functions; nevertheless it is not my aim to rebuild elementary complex analysis in its entirety from the Robinson-Callot theorem. Complex analysis did not wait for non standard analysis to be invented ! There follows a description of a periodic function with an infinitesimal period.

The second part deals with complex iteration. The main idea is that the shadow of the zonosč of an iterate of i-large order of a standard complex polynomial is the Julia set of this polynomial. (Sorry for the five occurrences of the word "of"). This non standard approach to complex iteration leads to two applications: first a construction of a conformal mapping from the exterior of a connected Julia set to the exterior of the closed unit disk ; secondly the proof of the inferior semi-continuity – according to the Hausdorff metric – of the function which maps the coefficients of a polynomial of second order to the corresponding Julia set.

Finally, the third part solves the problem of the connection between the asymptotic developments of Airy's function at $+\infty$ and $-\infty$. This connection is made by looking at Airy's equation in the complex plane and under the macroscope, and by using a new technique in the field of complex slow-fast vector fields. While the result is classical (it is attributed to Stokes) the method is strongly post-modern, that is to say *simple*. In a few words: Airy's equation has three particular solutions which correspond to three repelling rivers, the asymptotic behaviour of which is known on three sectors, each of

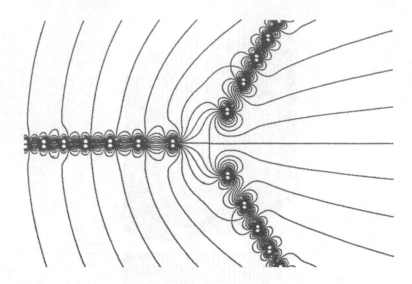

Fig. 2.2. Levels curves of the real part of a "generic" solution of Liouville's equation in overview (here the logarithmic derivative of the average of the three distinguished solutions of Airy's equation, see section 2.4.1).

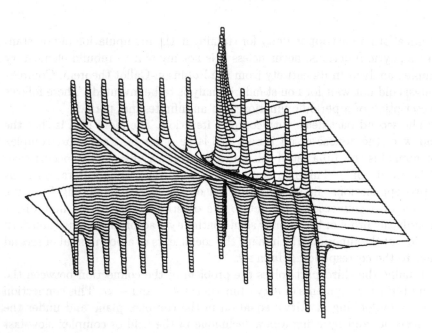

Fig. 2.3. The same graph viewed from above and behind.

them with angle 2π, centred on the directions $0, \dfrac{2\pi}{3}$ and $-\dfrac{2\pi}{3}$ respectively. Using the fact that three solutions of a second order differential equation must satisfy a non-trivial linear relation, one deduces the asymptotic behaviour of each one on the missing ray. Here the shadow of the zonosč of a "generic"[3] solution of the equation seen under a macroscope is the union of the Stokes lines of the equation.

Briefly stated, the zonosč of a periodic function with an infinitesimal period is the union of two lines bounding a horizontal strip in which the function is "flat" and outside of which it is totally irregular ; the zonosč is the Julia set of a polynomial if the function is an i-large iterate of this polynomial ; it is the union of the Stokes lines of a differential equation if the function is the image by a macroscope of a "generic" solution.

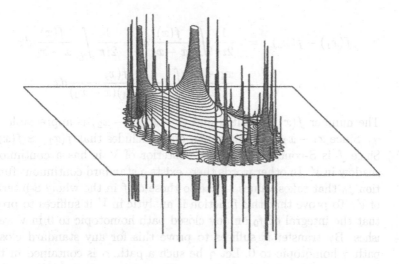

Fig. 2.4. Level curves of the real part of $\frac{1}{P_{50}(x)}$, where P_n is the n-th partial sum of a series $\sum_{n=0} a_n x^n$, where the coefficients a_n have been chosen at random in the unit square $(-2 < \Re(x) < 2,\ -1.33 < \Im(x) < 1.33,\ -4 < \Re f(x) < 4)$.

Moreover it is possible to show [61] that the zonosč of the Taylor polynomial of i-large degree of a standard function is generally the circle of convergence of the series, for which see figure 2.4. In conclusion the zonosč of an analytic function is indeed the interesting thing to examine ; it is its soul, so to speak.

---

[3] The term "generic" remains to be welldefined ; for example the average of the three solutions would be a suitable choice.

2.2 Tutorial

2.2.1 Proof of the Robinson-Callot theorem

1. If f is limited in the halo of x_0, by Fehrele's principle f is limited and analytic in a standard neighbourhood V of x_0. It suffices to prove the results for the S-interior[4] of V since this S-interior contains a standard neighbourhood of x_0.

   a) Let $x_1 \simeq x_2$ be two points in the S-interior and ρ appreciable such that the closed path $\gamma_\rho(t) = x_1 + \rho e^{it}$, $t \in [0, 2\pi]$ is in V. One has

   $$Ind(\gamma_\rho, x_1) = Ind(\gamma_\rho, x_2) = 1$$

   so that, by Cauchy's formula

   $$
   \begin{aligned}
   f(x_1) - f(x_2) &= \frac{1}{2i\pi} \int_{\gamma_\rho} \frac{f(x)}{x - x_1} dx - \frac{1}{2i\pi} \int_{\gamma_\rho} \frac{f(x)}{x - x_2} dx \\
   &= \frac{x_1 - x_2}{2i\pi} \int_{\gamma_\rho} \frac{f(x)}{(x - x_1)(x - x_2)} dx.
   \end{aligned}
   $$

   The number $f(x)$ is limited and $(x - x_1)(x - x_2)$ is appreciable on γ_ρ. Since $x_1 - x_2$ is infinitesimal, one concludes that $f(x_1) \simeq f(x_2)$.

   b) Since f is S-continuous in the S-interior of V it has a continuous shadow in V. In other words there exists a standard continuous function f_0 that takes values i-close to those of f in the whole S-interior of V. To prove that this function is analytic in V it suffices to prove that the integral of f_0 on any closed path homotopic to 0 in V vanishes. By transfer it suffices to prove this for any standard closed path γ homotopic to 0. Let γ be such a path. γ is contained in the S-interior of V, thus $f_0 \simeq f$ on γ. Since f is analytic in V one has $\int_\gamma f = 0$, thus $\int_\gamma f_0 \simeq 0$. Since f_0 and γ are standard one deduces that $\int_\gamma f_0 = 0$.

   c) Let x_1 be in the S-interior of V and ρ appreciable such that the closed path $\gamma_\rho(t) = x_1 + \rho e^{it}$, $t \in [0, 2\pi]$ is in the S-interior of V. Cauchy's formula for the derivative leads to

   $$f_0'(x_1) = \frac{1}{2i\pi} \int_{\gamma_\rho} \frac{f_0(x)}{(x - x_1)^2} dx \simeq \frac{1}{2i\pi} \int_{\gamma_\rho} \frac{f(x)}{(x - x_1)^2} dx = f'(x_1).$$

   The equality of the derivative of f_0 and the shadow of the derivative of f of any higher standard order can be shown with the analogous Cauchy's formula, or follows simply from external induction.

---

[4] The point x is in the S-interior of V if there exists a disk with an appreciable radius, centred in x, which is included in V.

2. If f is not S-continuous in x_0, the proof of the following result is elementary.

   The image of the halo of x_0 by f meets the halo of any limited point.

   Indeed if the halo of one limited point w_0 contains no value of f in the halo of x_0, then the function $g(x) = 1/(f(x) - w_0)$ is analytic and limited, thus S-continuous, in the halo of x_0. Consequently g takes either only appreciable values or only infinitesimal values in the halo of x_0.

   If g takes only infinitesimal values then f takes only i-large values in the halo of x_0 since $f(x) = w_0 + 1/g(x)$. In particular $f(x_0)$ is not limited. If g takes only limited values then f is limited on the halo of x_0. In both cases hypotheses are not satisfied.

   Contrary to the preceding, the proof that the image by f of the halo of x_0 indeed contains any limited point – except possibly a part of the halo of one point – is not elementary. This proof needs the following result: the holomorphic covering set of the complex plane except for two points is the disk. I use this result here without proof.

   First let us suppose that the function f avoids, in the halo of x_0, two limited values $a \not\simeq b$. By permanence there exists an internal neighbourhood V of x_0, containing the halo of x_0 where f avoids a and b. Put $T : \mathbb{C} \to \mathbb{C}, z \to \frac{z-a}{b-a}$ and $g = T \circ f : V \to \mathbb{C} \setminus \{0,1\}$. If the image by g of the halo of x_0 is contained in the halo of 0 or in the halo of 1, then g is S-continuous at x_0, and $f = T^{-1} \circ g$ is S-continuous too. Otherwise, by continuity there exists $x_1 \simeq x_0$ such that $g(x_1)$ is limited and is neither i-close to 0 nor to 1. Let Π be a standard holomorphic covering map of $\mathbb{C} \setminus \{0,1\}$ by the unit disk D, that is to say, Π is an analytic function from D to $\mathbb{C} \setminus \{0,1\}$ such that for any z in the base $\mathbb{C} \setminus \{0,1\}$ there is a neighbourhood W of z such that $\Pi^{-1}(W)$ is a disjoint union of sets all homeomorphic to W. Thus Π is surjective, locally injective and there exists ξ_0 standard with $|\xi_0| < 1$ and $\Pi(\xi_0) = {}^0(g(x_1))$. Thus there exists $\xi_1 \simeq \xi_0$ such that $\Pi(\xi_1) = g(x_1)$. One has $|\xi_1| \lneqq 1$. Let G be the lift of g determined by $G(x_1) = \xi_1$. The function G is analytic in a domain containing the halo of x_1 ; G is limited since its values are in the unit disk, thus G is S-continuous in x_1. The point ξ_1 is nearly standard in the unit disk. Thus Π is S-continuous in ξ_1. Since $f = T^{-1} \circ \Pi \circ G$, f is S-continuous in x_1, thus also in x_0.

   Suppose now that f is not S-continuous in x_0. From the preceding we have the following description: either f takes, in the halo of x_0, all limited values without exception, or f avoids some limited value a. Therefore the only values that f can avoid are in the halo of a, which proves item 2. Note that, since $f(\mathrm{hal}\,(x_0))$ is a halo, it contains also i-large values and values i-close to a. Moreover the image of the halo of x_0 by f contains a ring $R = \{z \in \mathbb{C} : \alpha < |z - a| < 1/\alpha\}$ for some α infinitesimal.

3. One has to prove that $f(\mathrm{hal}\,(x_0)) = \mathrm{hal}\,(f(x_0))$. Since f is S-continuous in x_0, one has already $f(\mathrm{hal}\,(x_0)) \subset \mathrm{hal}\,(f(x_0))$; thus it suffices to prove

that any point $w \simeq f(x_0)$ is the image of a point $x \simeq x_0$. Let f_0 be the shadow of f and $w_0 = f_0(^0x_0)$ $(w_0 = {}^0(f(x_0)) = {}^0w)$. Since the zeros of the non-constant standard function $f_0(x) - w_0$ are isolated, there exists a standard number $r > 0$ such that $f_0(x) - w_0$ does not vanish if $0 < |x - x_0| < r$. If ρ is appreciably between 0 and r, then $f_0 - w_0$ cannot take an infinitesimal value at any point of the circle $C(x_0, \rho) = \{x \in \mathbb{C} : |x - x_0| = \rho\}$ since in that case $f_0 - w_0$ vanishes at the shadow of this point, and this shadow belongs to the set $\{x \in \mathbb{C} : 0 < |x - {}^0x_0| < r\}$. Therefore $f_0 - w_0$ takes only appreciable values on $C(x_0, \rho)$. Since $f - w$ and $f_0 - w_0$ take values which are i-close one to another on this circle, one deduces from the Rouché theorem that both functions have the same number of zeros in the disk $D(x_0, \rho) = \{x \in \mathbb{C} : |x - x_0| < \rho\}$, taking multiplicity into account. The function $f - w$ has therefore a zero in any disk $D(x_0, \rho)$ with ρ appreciably between 0 and r. By permanence this function must have a zero at some point $x \simeq x_0$.

2.2.2 Applications

Proposition 2.2.1. *The shadow of a limited injective analytic function is injective or constant.*

Proof. Let f_0 be the shadow of f given by item 1 of theorem 2.1.1 and suppose that f_0 is neither injective nor constant. Then by transfer there exist two different standard points x_1 and x_2 which have the same image by f_0. Since f_0 is not constant, by part 3 of theorem 2.1.1 one has $f(\mathrm{hal}\,(x_1)) = \mathrm{hal}\,(f(x_1))$. But $f(x_1) \simeq f(x_2)$, therefore $f(\mathrm{hal}\,(x_1)) = \mathrm{hal}\,(f(x_2))$; this means that there exists $x \simeq x_1$ such that $f(x) = f(x_2)$. One has necessarily $x \neq x_2$ since $x \simeq x_1 \not\simeq x_2$. Thus f is not injective.

The next proposition shows that one sees the zonošč of a function if one looks at the zonošč of its real part. This fact demonstrates the appropriateness of illustrating only the real parts.

Proposition 2.2.2. *If the real part of an analytic function is limited on the halo of a point at which the function itself is limited then the function is S-continuous there.*

Proof. Let f be such a function and x be such a point. If f is not S-continuous at x then by part 2 of theorem 2.1.1 f takes all limited values – except possibly a part of the halo of one point – on the halo of x. Therefore $\Re(f)$ takes all limited real values (without exception) in the halo of x. By Fehrele's principle $\Re(f)$ must take i-large values in the halo of x, which contradicts the hypothesis.

2.2.3 Exercises with answers

Exercise 1: (Liouville theorem) A bounded entire function is constant.

Exercise 2: (Maximum modulus principle) Let f be analytic in a domain D. If the modulus of f assumes its maximum at an interior point of D then f is constant.

The answers are at the end of this chapter.

2.2.4 Periodic functions

Let ε be an infinitesimal positive real number and f be an ε-periodic complex function. Suppose f takes limited values in the halo of a given limited point. By permanence f is limited in a standard neighbourhood of x_0. By periodicity f is limited in a standard horizontal strip S containing x_0 in its interior. By part 1 of theorem 2.1.1:

1. f is S-continuous in the S-interior of S,
2. its shadow ${}^0f$ is analytic in S,
3. the derivatives of ${}^0f$ of any standard order coincide with the shadows of the corresponding derivatives of f.

On the other hand, since for any n in $\mathbb{Z}$ $f(x_0 + n\varepsilon) = f(x_0)$ and since ε is infinitesimal, the function ${}^0f$ is constant and equal to ${}^0f(x_0)$ on the line $\{x \in \mathbb{C} : \Im(x) = \Im({}^0x_0)\}$. By the principle of isolated zeros, ${}^0f$ is constant in S. Thus all derivatives of f of standard order take only infinitesimal values in the S-interior of S.

Consider the maximal (possibly external) strip $S_{\pounds}$ on which f is limited and the maximal strip $S_{\varnothing}$ where f takes values i-close to $f(x_0)$. More precisely, put:

$$
\begin{aligned}
G_f \;=\; & \{t \le \Im x_0 : \forall x \in \mathbb{C} \;\; (t \le \Im x \le \Im x_0) \Rightarrow f(x) \; limited\} \\
& \bigcup \{t \ge \Im x_0 : \forall x \in \mathbb{C} \;\; (\Im x_0 \le Im\, x \le t) \Rightarrow f(x) \; limited\} \\
H_f \;=\; & \{t \le \Im x_0 : \forall x \in \mathbb{C} \;\; (t \le \Im x \le Im\, x_0) \Rightarrow f(x) \simeq f(x_0)\} \\
& \bigcup \{t \ge \Im x_0 : \forall x \in \mathbb{C} \;\; (\Im x_0 \le \Im x \le t) \Rightarrow f(x) \simeq f(x_0)\} \\
S_{\pounds} \;=\; & \{x \in \mathbb{C} : \Im x \in G_f\} \\
S_{\varnothing} \;=\; & \{x \in \mathbb{C} : \Im x \in H_f\}.
\end{aligned}
$$

Notice that S-int$(S_{\pounds}) = \{x \in \mathbb{C} : \Im x \in$ S-int$(G_f)\}$.

The strips $S_{\pounds}$ and $S_{\varnothing}$ have non-infinitesimal "height" i.e. they contain an internal strip with non-infinitesimal height), $S_{\pounds}$ is a galaxy (internal or strictly external) and $S_{\varnothing}$ is a halo. Moreover $S_{\pounds}$ contains $S_{\varnothing}$ and $S_{\varnothing}$ contains the S-interior set of $S_{\pounds}$ as we have seen above.

Fig. 2.5. Level curves of the real part of $\sin(x)$ in overview.

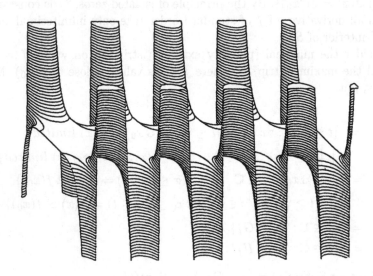

Fig. 2.6. Level curves of the real part of $\sin(x)$ viewed from above and behind.

Suppose now that f is entire. There are two possibilities. If $S_{\mathcal{L}}$ and $S_\varnothing$ coincide then they are internal by Fehrele's principle, and thus equal to $\mathbb{C}$ by the continuity of the function F. Therefore f is bounded in $\mathbb{C}$, thus constant by Liouville theorem. If $S_{\mathcal{L}}$ and $S_\varnothing$ do not coincide then both are strictly external by the continuity of f. One has the following result.

Proposition 2.2.3. *With the preceding hypotheses and notation, if x belongs to $S_{\mathcal{L}}$ and not to $S_\varnothing$ then $hal(x)$ contains points of both $S_\varnothing$ and $S_{\mathcal{L}}^C$.*

Proof. We will need the following lemma.

Lemma 2.2.1. *If G is a galaxy (internal or strictly external) then its S-interior is also a galaxy. Moreover, one has*

$$g \in S - int(G) \Leftrightarrow hal(g) \subset G. \qquad (2.1)$$

Proof. Put $G = \bigcup_{st(t) \in T} A_t$ with $(A_t)_{t \in T}$ internal. One has

$$x \in S - int(G) \Leftrightarrow \exists^{st} r > 0 D(x, r) \subset G$$

where $D(x, r)$ is the disk of centre x and radius r, thus

$$x \in S - int(G) \Leftrightarrow \exists^{st}(r, t) \in (\mathbb{R}^+ \backslash \{0\}) \times T : D(x, r) \subset A_t$$

which proves that $S - int(G)$ is a galaxy.

As to (2.1), since the implication left to right is obvious let us show the converse. Let g be such that $hal(g) \subset G$ and put

$$R(g) = \{r > 0 : D(g, r) \subset G\}.$$

One has $R(g) = \bigcup_{stt \in T} \{r > 0 : D(g, r) \subset A_t\}$ which shows that $R(g)$ is a galaxy. Since $R(g)$ contains all infinitesimal numbers, it must contain also appreciable numbers by Fehrele's principle; thus g is in the S-interior of G, which proves lemma 2.2.1.

To prove proposition 2.2.3, it suffices to prove the corresponding result on G_f and H_f as above. One has $S - int(G_f) \subset H_f \subset G_f$, and from lemma 2.2.1 above and Fehrele's principle these inclusions are strict. Let t be in $G_f \backslash H_f$. Then $t \notin S - int(G_f)$ thus $hal(t) \cap G_f^C \neq \varnothing$. Moreover $hal(t) \cap H_f \neq \varnothing$ since, otherwise, the halo of any point of II_f would be included in G, which would mean that $H \subset S - int(G_f)$. This proves proposition 2.2.3.

Note that the function f is not S-continuous at such a point and that part 2 of theorem 2.1.1 applies.

Hence a non-constant ε-periodic analytic function is "flat" (all its derivatives of standard order are infinitesimal) in the S-interior of a strip, and is extremely irregular at the boundary of the strip. In [62] it is shown that such a function is exponentially flat with respect to ε in this S-interior.

2.3 Complex iteration

Let c be a standard complex number and $f(x) = x^2 + c$. We denote by f^n the n-th iterate of f. The *fulfilled Julia set* of f is

$$K = \{x \in G : (f^n(x))_{n \in \mathbb{N}} \text{ is bounded}\}.$$

The *Julia set* J is the boundary of K. These classical sets have had their letters patent attesting to their nobility for almost a century now and have earned a considerable reputation especially with the proliferation of computers. This reputation is perfectly justified: they provide an opportunity to create beautiful drawings. Since Julia sets are limit objects, how can a computer calculate them ? How can a computer calculate an infinite number of iterations in a finite time ? In fact the computer did not calculate indefinitely; I simply programmed it to calculate a level line of the modulus of an iterate of f of sufficiently great order. Here is a excellent opportunity to use a non standard tool: below I have simulated this iterate by an i-large integer ω.

Fig. 2.7. The Julia set corresponding to $c = -0.77 + 0.116i$.

In fact non standard analysis is not only useful for showing the link between computer drawings and theoretical Julia sets, it is also a well-adapted approach for understanding these sets. First I will give a non standard characterisation of standard Julia sets, secondly I will use this characterisation to build a conformal mapping of the exterior of those which are connected, and

Fig. 2.8. The Julia set corresponding to $c = 0.26 + 0.002i$.

thirdly I will show the lower semi-continuity of Julia sets. Of course these results are not new, but the proofs presented here are interesting for their simplicity.

Let $c \in \mathbb{C}$ be standard and ω an i-large integer. Let

$$
\begin{aligned}
R &= 1 + \sqrt{1 + |c|} \\
K_\omega &= \{x \in \mathbb{C} = |f^\omega(x) \le R\} \text{ and} \\
J_\omega &= \{x \in \mathbb{C} = |f^\omega(x)| = R\}.
\end{aligned}
$$

The number R is chosen in order to be sure that an orbit which leaves the disk of radius R tends to infinity. Indeed

$$|x| \ge R \Rightarrow |f(x)| \ge |x|^2 - |c| \ge 2|x|.$$

One has therefore $K \subset K_\omega$ and $J_\omega \subset K^c$. The set J_ω is an ordinary Jordan curve with possibly many components. The next statement claims that it is closely connected to the Julia set J.

Theorem 2.3.1. *The halos of J_ω and J coincide.*

Proof. It suffices to prove that the halo of any point in J_ω meets J and the halo of any point in J meets J_ω. Let x be in J_ω. For any standard integer n, $f^n(x) \simeq f^n(x)$ but $|f^n(x)| \le R$, so that $|f^n(^0x)| \le R + 1$. Since $f, {}^0x$ and R are standard, this remains true for any n by transfer. Hence ^0x belongs to K. On the other hand, x does not belong to K since it is in J_ω which is included in K^c. Thus the halo of x meets the boundary of K, which is J.

Conversely let x be in J. In any standard neighbourhood of x there exists a standard y which belongs to K^c. Since f and y are standard and since $f^n(y)$ tends to infinity as n tends to infinity, one has $f^\omega(y) \simeq \infty$, thus y is in K_ω^c. Therefore any standard neighbourhood of x contains a point of K_ω^c. Since K_ω^c is an internal set, one deduces by permanence that the halo of x contains an element of K_ω^c. Conversely x is in K, thus in K_ω. Hence the halo of x meets the boundary of K_ω, which is J_ω.

Assume again that c is standard and suppose now that 0 belongs to K. By the stability of K under f, the set K contains all the images and inverse images of 0. Since K is in the interior set of K_ω, all the critical points of f^ω are in the interior of K_ω. Indeed one has $(f^\omega)'(x) = \Pi_{k=0}^{\omega-1} f'(f_{(x)}^k)$ and 0 is the only critical point of f, hence:

$$(f^\omega)'(x) = 0 \Rightarrow (\exists k \in \{0..\omega-1\} : f^k(x) = 0) \Rightarrow x \in \overset{\circ}{K}_\omega .$$

Lemma 2.3.1. *If 0 is in K then the algebraic curve J_ω is a simple Jordan curve (ie homeomorphic to $\mathbb{S}^1$).*

Proof. This follows from a general result on polynomials. A curve $\{z \in \mathbb{C} : |P(z)| = R\}$ where P is a polynomial of degree n is called a lemniscate [73]. Such a curve consists of at most n separate components, each of which containing in its interior a zero of P by the principle of the maximum. These components are exterior to each other for the same reason. If P' does not vanish on the lemniscate then each component is homeomorphic to $\mathbb{S}^1$.

The Riemann-Macdonald theorem [73] asserts that if f is holomorphic inside and on a simple curve γ on which $|f|$ is constant and f' does not vanish, and if f has k zeros inside γ, then f' has $k - 1$ zeros there.

Using this result we deduce that the number of components is equal to $n - m$ where m is the number of critical points of P which are in the lemniscatic region $\{z \in \mathbb{C} : |P(z)| < R\}$. If 0 is in K then all critical points are in the lemniscatic region $Int(K_\omega)$, hence the lemniscate J_ω consists of one component.

Proposition 2.3.1. *If 0 belongs to K then f^ω has a 2^ω-th root φ_ω which is a conformal mapping from K_ω^c to the complement of the closed disk of centre 0 and radius $R^{(2^{-\omega})}$, $\overline{D}(0, R^{(2^{-\omega})})^c$.*

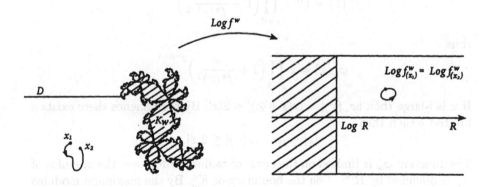

Fig. 2.9. A connected Julia set and its conformal mapping to a strip. The points x_1 and x_2, which occur in the text, are assumed to be distinct in the figure. Since I show that, in fact, they coincide, the figure gives a slightly false impression.

Proof. Let δ be the left horizontal ray starting from one of the points of J_ω the furthest to the left of J_ω, and let x_0 be in $K_\omega^c \backslash \delta$. Let us define

$$\log f^\omega(x) = \int_{x_0}^x \frac{(f^\omega)'(\xi)}{f^\omega(\xi)} d\xi$$

which maps $K_\omega^c \backslash \delta$ on a strip with height a multiple of 2π since the function f^ω is single valued. Therefore this height is constant by continuity of $\log f^\omega$. Since it tends to $2^\omega 2\pi$ at infinity, it is equal to $2^\omega 2\pi$.

The function $\log f^\omega$ is locally injective from a simply connected open set to a simply connected open set. It is therefore an injective function. Indeed, if $\log f^\omega(x_1) = \log f^\omega(x_2)$ then the image by $\log f^\omega$ of a path joining x_1, to x_2 in $K_\omega^c \backslash \delta$ is a contractile path. After contraction, one obtains a path on which $\log f^\omega$ is constant. Hence this path is constant and we have $x_1 = x_2$.

For x in $K_\omega^c \backslash \delta$ define $\varphi_\omega(x) = \exp(2^{-\omega} \log f^\omega(x))$. Since the exponential function is injective on a strip with height 2π, φ_ω is injective. The function φ_ω can be continued on the cut δ to an analytic function from K_ω^c to $\overline{D}(0, R^{(2^{-\omega})})^c$. By the way φ_ω can also be continued to a continuous function from $\overline{K_\omega^c}$ to $D(0, R^{(2^{-\omega})})^c$.

Theorem 2.3.2. *If 0 is in K then the shadow of $\varphi_\omega(x) = (f^\omega(x))^{2^{-\omega}}$ is a conformal mapping from K^c to $\overline{D}(0,1)^c$.*

Proof. We first have to show that φ_ω takes limited values at limited points in K_ω^c. Using the formula $f(y) = y^2(1 + \frac{c}{y^2})$ for $y = f^k(x)$, one obtains

$$f^\omega(x) = x^{(2^\omega)} \prod_{k=0}^{\omega-1} \left(1 + \frac{c}{f^k(x))^2}\right)^{2^{\omega-k-1}}$$

thus

$$\varphi_\omega(x) = x \prod_{k=0}^{\omega-1} \left(1 + \frac{c}{f^k(x)^2}\right)^{2^{-k-1}}.$$

If x is i-large then $|\varphi_\omega(x)| = |x(1 + \oslash)| \leq 2|x|$. By permanence there exists a limited ρ such that

$$|x| \leq \rho \Rightarrow |\varphi(x)| \leq 2|x|.$$

The function φ_ω is limited on the circle of radius ρ. Moreover the modulus of φ_ω is bounded by $R^{(2^{-\omega})}$ on the boundary of K_ω^c. By the maximum modulus principle φ_ω is therefore limited on $K_\omega^c \cap D(0, \rho)$. Since for x limited not in $D(0, \rho)$ one has $|\varphi(x)| \leq 2|x|$, the function φ_ω is limited at any limited point of K_ω^c. By item 1 of theorem 2.1.1 φ_ω is S-continuous and its shadow φ is an analytic function from the shadow of K_ω^c which is K^c to the shadow of $\overline{D}(0, R^{(2^{-\omega})})^c$ which is $\overline{D}(0, 1)^c$. To conclude, by proposition 2.2.1, an injective limited analytic function has a shadow which is injective or constant. Thus φ is injective.

So, despite appearances, Julia sets are simple from a conformal viewpoint, much simpler than the square for instance. Let us examine how changes J when c varies. In general the relationship is continuous but is can arise that two values of the parameter c close to each other give Julia sets which are

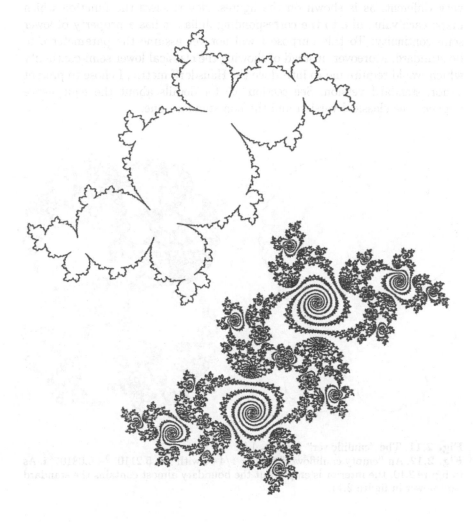

Fig. 2.10. Two Julia sets corresponding to two values of c which are close to each other. The first, which can be considered as a standard value $c_0 = -\frac{1}{8} - 0.6495i$, leads to "Douady's rabbit, with a thin boundary and a non empty interior. The second value is $c = c_0 + \varepsilon$ with $\varepsilon = 3.3 \ 10^{-2} - 4.2 \ 10^{-3}$. It yields a Julia set with a much more complicated boundary and an empty interior. A superimposition of these figures shows that the first Julia set is almost contained in the second one; conversely the first fulfilled Julia set almost contains the second one. It is suggested to the reader to photocopy this page (as well as the following two "standard" Julia sets) on a transparent sheet in order to see the result of this superimposition.

very different, as is shown on the figures. Nevertheless the function which maps each value of c to the corresponding Julia set has a property of lower semi-continuity. To this purpose I will not yet assume the parameter c to be standard. Moreover, instead of showing the classical lower semi-continuity which would require me to introduce the Hausdorff metric, I chose to present a non standard version. See section 6.5 for details about the equivalence between the classical version and the non standard one.

Fig. 2.11. The "cauliflower" for $c_0 = 1/4$.
Fig. 2.12. An "empty cauliflower" for $c = 1/4 + \varepsilon$ with $\varepsilon = 5.2110^{-3} + 6.0810^{-4}i$. As in figure 2.10, the interior is empty but the boundary almost contains the standard cauliflower in figure 2.11.

Theorem 2.3.3. *Let c be a limited complex number and c_0 its shadow. Suppose $c_0 \neq 1/4$. Then the halo of $J(c)$ contains the halo of $J(c_0)$.*

Proof. Let us denote by f and J the polynomial and the Julia set corresponding to c (i.e. $f(x) = x^2 + c$) and f_0, J_0 those corresponding to c_0. Since $c_0 \neq 1/4$, f_0 has one repelling fixed point x_0, the two inverse images of which are distinct. By structural stability (or simply by continuous dependance of the roots of a polynomial with respect to the coefficients) f has one repelling fixed point x which is i-close to x_0.

Lemma 2.3.2. *Let f be a complex polynomial. Then the Julia set of f contains all the repelling fixed points of f.*

Proof. By transfer we can suppose f standard and show the statement for a standard repelling fixed point of f. One has $f(x) = x$ and $|f'(x)| \gtrless 1$; therefore, if ω is i-large then $(f^{\omega})'(x) = \prod_{k=0}^{\omega-1} f'(f^k(x)) = (f'(x))^{\omega}$ is i-large. The function f^{ω} is not S-continuous at x, and thus it takes i-large values in the halo of x. Therefore $\mathrm{hal}\,(x) \cap K^c$ is non-void ; but x is in K and x is standard, hence x belongs to the boundary of K which is the Julia set J.

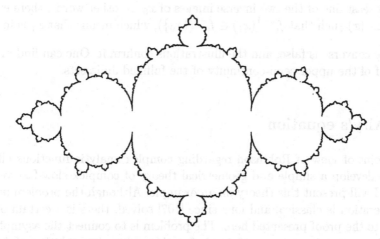

Fig. 2.13. "St. Mark's Basilica, Venice, Italy, reflected in the Grand Canal" for $c_0 = -3/4$.

Fig. 2.14. A "destroyed Basilica" in the fashion of the 70's for $c = -3/4 + \varepsilon$ with $\varepsilon = -5.88 \; 10^{-3} - 6.17 \; 10^{-2}i$.

Let us return to the proof of theorem 2.3.3. We deduce that x belongs to J. Put $A_n^0 = f_0^{-n}(\{x_0\})$ and $A_n = f^{-n}(\{x\})$. By continuity of the roots of polynomials one has for any standard n, hal (A_n^0) = hal (A_n). By Fehrele's principle there exists ω i-large such that hal (A_ω^0) = hal (A_ω). Since x is in J, by the stability of Julia sets under inverse images one has $A_\omega \subset J$. Hence, to show that the halo of J contains the halo of J_0 it suffices to prove that J_0 is included in the halo of A_ω^0. So let z be in J_0. The $\omega - 1$-th iterate $f_0^{\omega-1}$ is limited and not S-continuous at z. By item 2 of the Robinson-Callot theorem 2.1.1, $f_0^{\omega-1}$ takes, in the halo of z, all standard values except at most one, thus at least one of the two inverse images of x_0. In other words, there exists x_1 in hal (x) such that $f_0^{\omega-1}(x_1) \in f_0^{-1}(\{x_0\})$, which means that x_1 is in A_ω^0.

The converse is false, and the illustrations confirm it. One can find in [26] a proof of the upper semi-continuity of the fulfilled Julia sets.

2.4 Airy's equation

The point of view of Robinson regarding complex analytic functions allows one to develop a simple and geometrical theory of complex slow-fast vector fields. I will present this theory on an example. Although the problem under consideration is classical and long since [107] solved, there is a certain originality to the proof presented here. The problem is to connect the asymptotic behaviours at $+\infty$ and $-\infty$ of Airy's function, which is a solution of Airy's equation. The asymptotic study of a standard equation begins by contemplating its standard solutions at i-large values of the variable. The use of the complex plane permits one to join the i-large positive numbers and the i-large negative numbers without passing through the limited numbers ; in others words, it permits one to link the asymptotic behaviour at $+\infty$ and $-\infty$ inside the asymptotic domain.

Airy's equation

$$Y'' = XY \tag{2.2}$$

is the simplest equation the solutions of which have both oscillatory and exponential behaviour. Locally, the solutions of (2.2) behave like the solutions of

$$Y'' = aY, \ a \in \mathbb{R} \tag{2.3}$$

For $a < 0$, the solutions of (2.3) are

$$Y = A\sin(\sqrt{-a}X + \varphi)$$

and have all the same oscillatory behaviour. On the other hand, if a is positive, the solutions are

$$Y = A\exp(\sqrt{a}X) + B\exp(-\sqrt{a}X)$$

and have exponential growth, except for a one-dimensional subspace of solutions which are an exponential decreasing, for $A = 0$.

One will not be surprised to find an analogous situation concerning Airy's equation: all the solutions have an oscillatory behaviour when the variable x tends to $-\infty$, while a one dimensional subspace of solutions tend exponentially to zero, the other solutions increase exponentially to infinity. This subspace of particular solutions contains Airy's function. A rigorous study yields the following result:

1. given a real solution Y of (2.2), there exist two real constants A and φ in $\mathbb{R}$ such that

$$Y(X) = A(-X)^{-1/4} \sin(\tfrac{2}{3}(-X)^{3/2} + \varphi + O(X^{-3/2}))(1 + O(X^{-3/2}))$$

as X tends to $-\infty$,

2. there exists a one dimensional subspace V of solutions of (2.2) such that
   a) if Y belongs to V then there exists B in $\mathbb{R}$ such that

$$Y(X) = BX^{-1/4} \exp(-\tfrac{2}{3}X^{3/2} + O(X^{-3/2}))$$

   as X tends to $+\infty$ ($B = \frac{1}{2\sqrt{\pi}}$ for Airy's function),
   b) if Y does not belong to V then there exists B in $\mathbb{R}$ such that

$$Y(X) = BX^{-1/4} \exp(\tfrac{2}{3}X^{3/2} + O(X^{-3/2}))$$

   as X tends to $+\infty$.

Now we can formulate our problem precisely. The Airy function has the following asymptotic behaviours

$$Ai(X) = \frac{1}{2\sqrt{\pi}} X^{-1/4} \exp(-\tfrac{2}{3}X^{3/2} + O(X^{-3/2}))$$

as X tends to $+\infty$, and

$$Ai(X) = A(-X)^{-1/4} \sin(\tfrac{2}{3}(-X)^{3/2} + \varphi + O(X^{-3/2}))(1 + O(X^{-3/2}))$$

as X tends to $-\infty$. *What are the values of A and φ ?*

From now on, Airy's equation is considered in the complex plane and observed from a great distance.

Let ε be a positive infinitesimal number and put $x = \varepsilon X$ and $y = \varepsilon Y$. Equation (2.2) yields

$$\varepsilon^3 y'' = xy \qquad (2.4)$$

The use of the associated Riccati equation decreases the order by one. Putting $u = \varepsilon^{3/2} \frac{y'}{y}$ and $\alpha = \varepsilon^{3/2}$ one obtains

$$\varepsilon^{3/2} u' = x - u^2. \qquad (2.5)$$

This is a slow-fast complex differential equation of the form

$$\alpha u' = f(x, u). \tag{2.6}$$

with α infinitesimal.

Let me present a short exposition of the general theory of these equations. Equation (2.6) shows that generally u' is i-large, and that u' can be limited only if $f(x, u) = \alpha\pounds$. In good situations (and Airy's equation is such a case) it is possible to calculate a certain number of "wise curves" with equations $u = \varphi_k(x)$ such that $f(x, u) = \alpha\pounds$ if and only if $u = \varphi_k(x) + \alpha\pounds$ for a certain k.

Let $u = \varphi(x)$ be the equation of some wise curve (ie $f(x, \varphi(x)) = \alpha\pounds$) and let (x_0, u_0) be a limited point of $\mathbb{C}^2$ such that $u_0 = \varphi(x_0) + \alpha\pounds$. Let $\gamma : [0, 1] \to \mathbb{C}$ be a C^1 path with limited derivative and such that $\gamma(0) = x_0$. Consider $\overline{u}$ the solution of (2.6) with initial condition $\overline{u}(x_0) = u_0$ and let us look at the restriction of $\overline{u}$ on the path γ. Let δ be infinitesimal such that $\frac{\alpha}{\delta}$ is infinitesimal and put $\overline{u} = \varphi(x) + \delta v$ and $w(t) = v(\gamma(t))$. One has $w(0) \simeq 0$. Developing $f(x, \varphi(x) + \delta v)$ up to first order in δ with Taylor's formula and using (2.6), one obtains

$$\alpha(\varphi'(x) + \delta v') = f(x, \varphi(x)) + \delta v(\partial_u f(x, \varphi(x)) + \oslash)$$

for v limited, where $\partial_u f$ denotes the derivative of f with respect to the second variable. Since we have $\alpha\varphi'(x) - f(x, \varphi(x)) = \alpha\pounds = \delta\oslash)$, we obtain, for v limited:

$$\alpha v' = \partial_u f(x, \varphi(x))v + \oslash.$$

Therefore, if w is limited then

$$\alpha\dot{w}(t) = (\partial_u f(\gamma(t), \varphi(\gamma(t)))w(t) + \oslash)\dot{\gamma}(t)$$

where the dot denotes differentiation with respect to t. For w appreciable, one deduces $\alpha\frac{\dot{w}}{w} = (\partial_u f(\gamma, \varphi(\gamma))\dot{\gamma} + \oslash$, thus

$$\frac{d}{dt}\log|w| = \frac{1}{\alpha}\Re(\partial_u f(x, \varphi(x))\dot{\gamma} + \oslash).$$

Suppose that the path γ is such that the function $Re(\partial_u f(x, \varphi(x))\gamma')$ is appreciably negative on $[0, 1]$; then at a place where w is limited the function $|w|$ is decreasing or infinitesimal. Since $w(0)$ is infinitesimal, we deduce that $w(t)$ is infinitesimal on $[0, 1]$.

Let us introduce the *landscape* function defined by

$$R(x) = \Re(\int_0^x \partial_u f(\xi, \varphi(\xi))d\xi).$$

One has $\frac{d}{dt}R(\gamma(t)) = \Re(\partial_u f(\gamma(t), \varphi(\gamma(t)))\dot{\gamma}(t))$, therefore along any path which descends the landscape, the solution $\overline{u}$ remains i-close to the wise curve $u = \varphi(x)$.

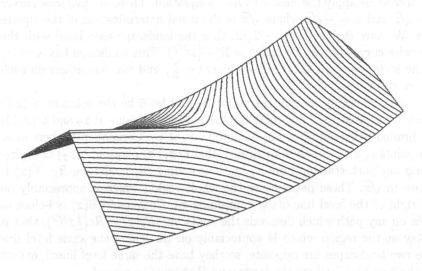

Fig. 2.15. Level curves of the landscape function R_1.

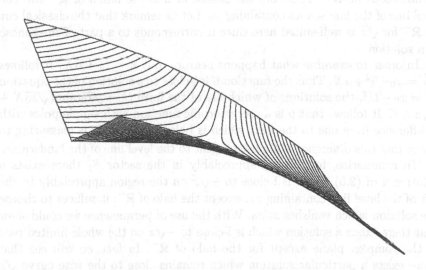

Fig. 2.16. Level curves of the second landscape function R_2. As previously, it is suggested to the reader to photocopy this pager onto a transparent sheet in order to superimpose it with figure 2.15. This will show prominently the complete two-leaf surface given by the two-valued function $\frac{4}{3}x^{3/2}$.

Now let us apply the theory to Airy's equation. There are two wise curves $u = \sqrt{x}$ and $u = -\sqrt{x}$ where $\sqrt{x}$ is the usual determination of the square root. We have $\partial_u f(x, \sqrt{x}) = -2\sqrt{x}$, thus the landscape associated with the first wise curve is given by $R_1(x) = \Re(-\frac{4}{3}x^{3/2})$. This landscape has a valley in the sector $S_1 = \{x \in \mathbb{C} : -\frac{\pi}{3} < Arg(x) < \frac{\pi}{3}\}$ and two mountains on each side of this valley.

Let x_0 be appreciably in the sector S_1 and let $\overline{u}$ be the solution of (2.5) which vanishes at x_0. The function $\overline{u}$ takes a limited value at x_0 and $\overline{u}'(x_0)$ is not limited ; therefore $\overline{u}$ is not S-continuous at x_0. Consequently there exist two points x_1 and x_2 i-close to x_0 such that $\overline{u}(x_1) \simeq \sqrt{x_1}$ and $\overline{u}(x_2) \simeq -\sqrt{x_2}$. Along any path starting from x_1 and descending the landscape R_1 , $\overline{u}(x)$ is i-close to $\sqrt{x}$. These paths cover the whole region which is appreciably on the right of the level line of R_1 containing x_0. Conversely, $\overline{u}(x)$ is i-close to $-\sqrt{x}$ on any path which descends the landscape $R_2(x) = Re(\frac{4}{3}x^{3/2})$, that is to say on the region which is appreciably on the left of the same level line (the two landscapes are opposite, so they have the same level lines), except for the halo of $\mathbb{R}^-$ where the landscape R_2 begins to ascend.

Although the wise curve $u = \sqrt{x}$ is multivalued, the solution $\overline{u}$ is single-valued. Indeed any solution of (2.4) is entire since (2.4) is a linear differential equation without singularity, thus $\overline{u}$ is the logarithmic derivative of an entire function and therefore is meromorphic. Thus the solution $\overline{u}$ is inevitably non S-continuous on $\mathbb{R}^-$. Therefore the zonosč of $\overline{u}$ is the union of $\mathbb{R}^-$ and the level line of the landscapes containing x_0. Let us remark that the classical cut on $\mathbb{R}^-$ for $\sqrt{x}$ is well-suited here since it corresponds to a part of the zonosč of a solution.

In order to examine what happens near x_0, put $x = x_0 + \alpha X$. It follows $\frac{d\overline{u}}{dX} = x_0 - \overline{u}^2 + \alpha X$. Thus the function $\overline{u}$ is i-close to a solution of the equation $U' = x_0 - U^2$, the solutions of which are of the form $U = \sqrt{x_0}\tanh(\sqrt{x_0}X + a), a \in \mathbb{C}$. It follows that $\overline{u}$ is a meromorphic function with simple poles with a difference from one to the next which is i-close to $\frac{i\alpha\pi}{2\sqrt{x_0}}$. It is reassuring to verify that this difference is almost tangent to the level line of the landscapes.

To summarize, for any x_0 appreciably in the sector S_1 there exists a solution $\overline{u}$ of (2.5) which is i-close to $-\sqrt{x}$ on the region appreciably to the left of the level line containing x_0, except the halo of $\mathbb{R}^-$: it suffices to choose the solution which vanishes at x_0. With the use of permanence we could show that there exists a solution which is i-close to $-\sqrt{x}$ on the whole limited part of the complex plane except for the halo of $\mathbb{R}^-$. In fact we will see that there exists a particular solution which remains close to the wise curve $\sqrt{x}$ on $\mathbb{C}\backslash\mathrm{hal}\,(\mathbb{R}^-)$, even for x i-large.

2.4.1 The distinguished solutions

A standard complex differential equation which leads to a slow-fast vector field under a macroscope, has for each wise curve a landscape admitting a saddle point at 0. This saddle point consists of a series of mountains and valleys and we have the following result:

Each mountain corresponds to a unique solution, called a distinguished solution, which remains close to the corresponding wise curve up to infinity.

I will prove this result only in the particular case of Airy's equation with the caution that the proof uses the properties of the equation in an essential way; for instance the existence of a second wise curve will be used.

Theorem 2.4.1. *There exists a solution u_0 of (2.5) such that for any x in $\mathbb{C} \setminus hal(\mathbb{R}^-)$, $u_0(x) = -\sqrt{x}(1 + \oslash)$.*

Proof. Let $\overline{u}$ be the solution of (2.5) with initial condition $\overline{u}(1) = -1$. Remark that, according to the landscape R_2, if x is appreciable with $|Arg(x)| \lessgtr \pi$ and $|x| < \frac{1}{2}$, then $\overline{u}(x) \simeq -\sqrt{x}$. The change of variables

$$\begin{cases} x = \varepsilon X \\ u = \sqrt{\varepsilon} U \end{cases} \tag{2.7}$$

yields the solution $\overline{U}$ of the Liouville equation

$$U' = X - U^2 \tag{2.8}$$

with initial condition $\overline{U}(\frac{1}{\varepsilon}) = -\frac{1}{\sqrt{\varepsilon}})$.

The idea of the proof is to consider the shadow U_0 of this solution $\overline{U}$ and the corresponding solution u_0 of (2.5) under the macroscope (2.7). For that purpose we have to show that $\overline{U}$ takes a limited value at a limited point X which is the aim of the following lemma. It will make the proof easier if we take some sector Σ and some internal tube which avoids the second wise curve $\sqrt{X}$.

Let δ be a standard number in $]0, \frac{\pi}{2}[$ and consider the sector Σ given by $\Sigma = \{X \in \mathbb{C} : |Arg(X)| < \pi - \delta\}$.

Lemma 2.4.1. *There exists $a > 0$ limited such that for any X in Σ, $a < X < \frac{1}{2\varepsilon} \Rightarrow |\overline{U}(X) + \sqrt{X}| < |\sqrt{X}|$.*

Proof. The remark above shows that the statement is true for any $a < \frac{1}{2\varepsilon}$ such that εa is appreciable. Let $H = \{r \in [0, \frac{1}{2\varepsilon}[: \forall X \in \Sigma, (r < |X| < \frac{1}{2\varepsilon}) \Rightarrow (|\overline{U}(X) + \sqrt{X}| < |\sqrt{X}|)\}$. The set H is an interval of the form $[a, \frac{1}{2\varepsilon}[$, and it remains to prove that a is limited. Indeed if a is i-large then there exists X_1 with $|X_1| = a$ and $|\overline{U}(X_1) + \sqrt{X_1}| = |\sqrt{X_1}|$. Put $\alpha = 1/a$. The macroscope

$$\begin{cases} x = \alpha X \\ u = \sqrt{\alpha} U \end{cases} \tag{2.9}$$

transforms the solution $\overline{U}$ into a solution $\hat{u}$ of the equation $\alpha^{3/2} \frac{du}{dx} = x - u^2$. Since $|\hat{u}(x_1) + \sqrt{x_1}| = 1$ (with $x_1 = \alpha X_1$) and $|\sqrt{x_1}| = 1$, we deduce that $x_1 - \hat{u}(x_1)^2$ is not infinitesimal. Thus $\hat{u}$ is limited and non-S-continuous at x_1.

Using the presence of a second wise curve for the equation and the Robinson-Callot theorem, there exists $x_2 \simeq x_1$ such that $\hat{u}(x_2) \simeq \sqrt{x_2}$. Now consider the landscape $R_1(x) = \Re(-\frac{4}{3}x^{3/2})$ associated with the second wise curve $u = \sqrt{x}$. There exists a path γ starting from x_2 which descends this landscape and which leaves the unit disk. Along this path one has $\hat{u}(\gamma(t)) \simeq \sqrt{\gamma(t)}$. Therefore there exists x_3 with $\hat{u}(x_3) \simeq \sqrt{x_3}$ and $1 < |x_3| < \frac{\alpha}{2\varepsilon}$. The inverse of the change of variables (2.9) gives a point X_3 with $a < |X_3| < \frac{1}{2\varepsilon}$ such that $|\overline{U}(X_3) + \sqrt{X_3}| > |\sqrt{X_3}|$. This contradicts the assumption that a is in H.

Notice that we can suppose a to be standard in the statement of lemma 2.4.1. Since the solution $\overline{U}$ takes a limited value at some standard point X_0, we can consider the solution U_0 of (2.8) with initial condition $U_0(X_0) = (\overline{U}(X_0))$. This solution is standard and moreover, by the short shadow lemma, U_0 takes a value i-close to that of $\overline{U}$ at any limited point where $\overline{U}$ is limited. Consequently for any X standard in Σ with $|X| > a$ one has $|U_0(X) + \sqrt{X}| < \frac{3}{2}|\sqrt{X}|$, where the number $\frac{3}{2}$ is chosen to have an internal tube larger than the first one but not containing the second wise curve $u = \sqrt{x}$. By transfer (U_0 and a are standard) this holds for any X such that $|Arg(X)| < \pi - \delta$ and $|X| > a$. Now consider u_0 the solution of (2.5) given by U_0 after the change of variables (2.7). For any appreciable x such that $|Arg(x)| < \pi - \delta$ one has $|u_0(x) + \sqrt{x}| < \frac{3}{2}|\sqrt{x}|$; thus u_0 takes a limited value at these points, which implies that u_0 is S-continuous on $G = \{x \in \mathbb{C} : |x| \gtrapprox 0$ and $|Arg(X)| < \pi - 2\delta\}$, so that the derivative u_0' is limited on G, and, hence, $u_0(x) \simeq -\sqrt{x}$ on G. Since u_0 does not depend on δ, the same holds on $\mathbb{C} \setminus \mathrm{hal}(\mathbb{R}^-)$.

Using the repelling property of the wise curve $-\sqrt{x}$ on $\mathbb{R}^+$, one can also show that u_0 is unique with this property. The proof is exactly the same as in the real case.

By the Robinson-Callot theorem one has $u_0'(x) \simeq \frac{1}{2\sqrt{x}}$ for any appreciable x in $\mathbb{C} \setminus \mathrm{hal}(\mathbb{R}^-)$; but $\varepsilon^{3/2}u_0'(x) = x - u_0(x)^2 \simeq 2\sqrt{x}(\sqrt{x} - u_0(x))$ for such x, thus $u_0(x) = -\sqrt{x} - \frac{\varepsilon^{3/2}}{4x}(1+\oslash)$. A more precise study yields the development, for x in $\mathbb{C} \setminus \mathrm{hal}(\mathbb{R}^-)$:

$$u_0(x) = -\sqrt{x} - \frac{\varepsilon^{3/2}}{4x} + \varepsilon^3 \pounds,$$

which gives for the standard function U_0:

$$U_0(x) = -\sqrt{X} - \frac{1}{4X} + O(X^{-5/2}), \quad X \to \infty, \ |Arg(X)| < \pi - \delta.$$

Since U_0 is the logarithmic derivative of a solution Y_0 of the linear equation, one has for this solution the following asymptotic formula. For any $\delta > 0$:

$$Y_0(X) = KX^{-1/4} \exp\left(-\frac{2}{3}X^{3/2} + O(X^{-3/2})\right), X \to \infty, |Arg(X)| < \pi - \delta.$$

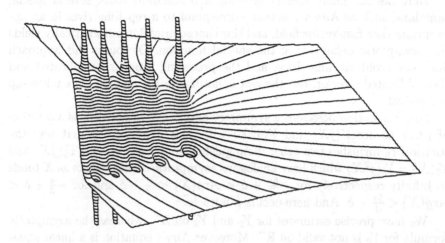

Fig. 2.17. Graph of the real part of the logarithmic derivative of Airy's function $\Re\left(\dfrac{Ai'(x)}{Ai(x)}\right)$, seen from above.

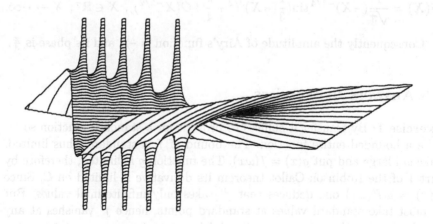

Fig. 2.18. The same graph seen from below.

Choosing the constant K equal to $\frac{1}{2\sqrt{\pi}}$, one finds again Airy's function $Ai(x)$.

Here lies the main interest of using non standard tools: several special functions, such as Airy's function, correspond to a repelling river in an appropriate slow-fast vector field, and this characterisation automatically yields the asymptotic behaviour of the special function. Of course this approach does not avoid all difficulties, and the preceding proof is complicated and thus of limited interest, but the rest is natural in my opinion. The follow-up is classical.

Airy's equation possesses a symmetry of order 3. Indeed if Y is a solution of (2.2), so are $Y(jX)$ and $Y(j^2X)$ with $j = \exp(\frac{2i\pi}{3})$. In that way the solution Y_0 furnishes two other distinguished solutions $Y_1(X) = Y_0(jX)$ and $Y_2(X) = Y_0(j^2X)$ which have analogous asymptotic developments as X tends to infinity respectively for $-\frac{5\pi}{3} + \delta < Arg(X) < \frac{\pi}{3} - \delta$ and for $-\frac{\pi}{3} + \delta < Arg(X) < \frac{5\pi}{3} - \delta$. And here occurs a miracle.

We have precise estimates for Y_1 and Y_2 on $\mathbb{R}^-$, whereas the asymptotic formula for Y_0 is not valid on $\mathbb{R}^-$. Moreover Airy's equation is a linear equation of second order, hence its solution space is determined by two independent solutions. The asymptotic expression of Y_1 and Y_2 will give information about Y_0 on $\mathbb{R}^-$. Indeed the three functions Y_0, Y_1 and Y_2 are linearly dependent: there exist two complex numbers λ and μ such that $Y_0 + \lambda Y_1 + \mu Y_2 = 0$. At the point 0 the three functions coincide, hence $1 + \lambda + \mu = 0$, and the derivative in 0 gives $1 + j\lambda + j^2\mu = 0$; thus $\lambda = j$ et $\mu = j^2$. Finally one obtains

$$Ai(X) = \frac{1}{\sqrt{\pi}}(-X)^{-1/4}\sin(\frac{2}{3}(-X)^{3/2} + \frac{\pi}{4} + O(X^{-3/2})), \ X \in \mathbb{R}^-, \ X \to -\infty.$$

Consequently the amplitude of Airy's function is $\frac{1}{\sqrt{\pi}}$ and its phase is $\frac{\pi}{4}$.

2.5 Answers to exercises

Exercise 1: By transfer it suffices to prove it for a standard function so let f be a bounded entire function. The bound of f is standard, thus limited. Take ω i-large and put $g(x) = f(\omega x)$. The function g is limited, therefore by part 1 of the Robinson-Callot theorem its derivative is limited on $\mathbb{C}$. Since $g'(x) = \omega f'(\omega x)$ one deduces that f' takes only infinitesimal values. But f' must take standard values at standard points, hence f' vanishes at any standard point, thus at any point, and f is constant.

Exercise 2: By transfer we can suppose that f and D are standard, as well as the hypothetic point x_0 in the interior of D where f reaches its maximum. The image by f of the halo of x_0 does not even contain a half of the halo of $f(x_0)$ since it contains only values with modulus less than $|f(x_0)|$, hence f is constant by part 3 of Robinson-Callot's theorem 2.1.1.

3. The Vibrating String

Pierre Delfini and Claude Lobry

3.1 Introduction

The mathematical interpretation of physical evolution problems often leads
to a study of partial differential equations that very soon needs the use of
distribution theory. Thus, some phenomena must be interpreted by the use
of this theory, and not by the use of classical function theory. Our purpose
is to consider a physical example, like a vibrating string, and to construct
a mathematical theory that describes its dynamical evolution without any
notion of functional spaces. This example is very simple, but it is interesting
because it can be generalised. The generally used theory represents the string
position at the point x, at the time t, by a continuous function $u(x,t)$, with
x and t real. This function is a solution of the equations (PDE):

$$(PDE) \qquad \begin{cases} u''_{tt}(x,t) = K^2 u''_{xx}(x,t) \\ u(0,t) = u(1,t) = 0 \\ u(x,0) = f(x) \\ u'_t(x,0) = g(x) \end{cases}$$

Like every theory this one is based on the modelling of reality. Let us see
what is written about this subject in the classical physics course: the "Berkeley" [35].

• *If a system contains a very large number of moving parts, and if these
parts are distributed within a limited region of space, the average distance
between neighboring moving parts becomes very small. As an approximation,
one may wish to think of the number of parts as becoming infinite and the
distance between neighboring parts as going to zero. One then says that the
system behaves as if it were "continuous". Implicit in this point of view is
the assumption that the motion of near neighbors is nearly the same. This
assumption allows us to describe the vector displacement of all the moving
parts in a small neighborhood of a point x, y, z whith a single vector quantity
$\psi(x, y, z, t)$. Then the "displacement" $\psi(x, y, z, t)$ is a continuous function of
position, x, y, z and of time t.(p. 48).*
• *According to the discussion above, a truly continuous system has an
infinite number of independent moving parts, although they occupy a finite*

*space. There are therefore an infinite number of degrees of fredom, and hence
an infinite number of modes. This is not literally true for a real material
system. One liter of air does not contains an infinite number of moving parts,
but only $2,7.10^{22}$ molecules, each of which has three degrees of fredom (for
motion along x, y and z directions). Thus a bottle containing 1 liter of air
does not have an infinite number of possible vibrational modes of the air, but
only 8.10^{22} at most. Anyone who has practiced blowing a bottle or a flute
knows that it is not easy to excite more than the first few modes... In practice
we are often concerned only with the first few (or few dozen or few thousand)
modes. As we shall see, it turns out that the lowest modes behave as if the
system were continuous.(p. 49)*

*   *In Sec. 2.2 we considered a continuous string, which is a system with
infinitely many degrees of freedom. No real mechanical system has an infinite number of degrees of freedom, and we are interested in real systems. In
this section we will find the exact solution for the modes of a uniform beaded
string having N beads and with fixed ends. In the limit that we take the number of beads N to be infinite (and maintain the finite length L), we shall find
the standing waves that we studied in Sec. 2.2. Our purpose is not merely
that, however. Rather, we shall find that, in going to the limit of a continuous string, we discarded some extremely interesting behavior of the system.
Remember that in order to use the smooth function $\psi(z,t)$ to describe the
displacement when N is huge but not infinite, we have to prohibit ourselves
from considering the highest modes, i.e., the modes $m = N, N-1, N-2$,
etc... (p. 72).*

It appears clearly from these three extracts that, for a physicist, a tight-
ened string is not similar to a continuous function (in particular it has a finite
number of "modes"). Mathematicians call the following differential system an
"approximation" of (PDE) , denoted (DENC), (C for classic):

(DENC)

$$\begin{cases} u''_{tt}(nh,t) = K^2 \dfrac{u((n+1)h,t) - 2u(nh,t) + u((n-1)h,t)}{h^2} & \\ u(0,t) = u(1,t) = 0 & n = 1, ..., N-1 \\ u(nh,0) = f(nh) & h.N = 1 \\ u'_t(nh,t) = g(nh) & \end{cases}.$$

They mean that the behaviour of the solution of (DENC) is near (in a
sense to be defined) to that of (PDE) when the step h is small enough. That
does not mean that the behaviour of (PDE) is truer that the one of (DENC),
for a given h. For example, a set of 1000 small masses connected by springs
of negligible weight, representing a beaded string, is better approximated by
(DENC) with $N = 1000$ than by (PDE).

On the contrary, the success of (PDE) is due to the fact that it is a
good approximation of reality when there is a very large number of degrees
of freedom. One of the advantages of (PDE) in comparison with (DENC) is

universality. As the number of degrees of freedom need not be known, there is no problem concerning the choice of N when N is an unknown large number. Another advantage of (PDE) is a greater simplicity of calculation. The price to pay for these advantages, universality and simplicity, is the difficulty met in defining what we mean by a solution of (PDE) and in showing that this solution exists. Thus, every theory of (PDE) has to be able to talk about propagation, and therefore it has to take into account some nondifferentiable or noncontinuous initial conditions; only Distribution Theory allows one, as we have said before, to do this properly for a partial differential equation.

An alternative to this theory is the following: we consider the ordinary differential system (DEN),

(DEN)

$$\begin{cases} U_n''(t) = K^2 \dfrac{U_{n+1}(t) - 2U_n(t) + U_{n-1}(t)}{h^2} & n = 1, ..., N-1 \\ U_0(t) = U_N(t) = 0 & h = \frac{1}{N} \\ U_n(0) = F_n \\ U_n'(0) = G_n \end{cases}$$

where N is an i-large integer in the sense of Nonstandard Analysis. The two variable function $u(x, t)$ has been replaced by a $\mathbb{R}^{N+1}$ vector:

$$U(t) = {}^T(U_0(t), U_1(t), ..., U_N(t))$$

where the two $\mathbb{R}^{N+1}$ vectors F and G describe the initial deformation and speed.

As this system is an ordinary second order differential system with constant coefficients, even though nonstandard, there is no problem in proving the existence and uniqueness of a solution which follows from an elementary classical theorem. We will try to show that (DEN) has as many qualities as (PDE), while inheriting some qualities of the finite dimensional case, lost by the passage to the limit in the continuous model.

Our theory is therefore an "elementary theory" of vibration of a system with a large number of degrees of freedom, in the sense used by Nelson in "Radically Elementary Probability Theory" [91]. E. Nelson considers that all probability space is a finite set (i-large) and he shows how the Wiener process can be seen as a random walk with i-small step.

3.2 Fourier analysis of (DEN)

3.2.1 Diagonalisation of A

The resolution of the system (DEN) is elementary. The boundary conditions being equal to zero we consider the $\mathbb{R}^{N-1}$ vector:

$$U(t) = {}^T(U_1(t), ..., U_{N-1}(t)).$$

Likewise for the vectors F and G associated with the initial conditions. The system leads then to a matrix representation:

$$\begin{cases} U''(t) = AU(t) \\ U(0) = F \\ U'(0) = G \end{cases}$$

where

$$A = \frac{1}{h^2} \begin{pmatrix} -2 & 1 & . & . \\ 1 & -2 & 1 & . \\ . & . & . & . \\ . & . & 1 & -2 \end{pmatrix}$$

is the finite difference second order matrix (the constant K is supposed equal to 1). Resolution of (DEN) becomes a simple study of the eigenvalues and eigenvectors of the matrix A. This matrix is diagonalizable and its $N-1$ eigenvalues are the reals λ_k given by:

$$\lambda_k = -\frac{4}{h^2} \sin^2(\frac{k\pi h}{2}) \text{ for } k = 1, ..., N-1. \tag{3.1}$$

In $\mathbb{R}^{N-1}$, with the classical scalar product

$$< U, V > = \frac{1}{N-1} \sum_{n=1}^{N-1} U_n V_n. \tag{3.2}$$

We consider the eigenvectors V^k, associated to the λ_k, and given by their components:

$$V_n^k = \sin(k\pi n h) \text{ for } n = 1, ..., N-1. \tag{3.3}$$

They constitute an orthogonal basis in which a vector U is written $U = \sum_{k=1}^{N-1} u_k V^k$, with

$$u_k = \frac{< U, V^k >}{\| V^k \|^2} \text{ and } \| V^k \|^2 = \frac{N}{2(N-1)}. \tag{3.4}$$

3.2.2 Interpretation of N i-large

Everything said in the previous paragraph about the matrix A is internal and therefore is also true with an i-large integer N.

Let us try to compare the discrete operator A with N i-large and the Laplacian continuous operator. First, we draw on the same graph (figure 3.1), the eigenvalues λ_k of A and the eigenvalues $\mu_k = -k^2\pi^2$, $k \in \mathbb{N}$, of the operator $\frac{\partial^2}{\partial x^2}$.

These two curves are similar along a nonnegligible part of the graph. More precisely, it can be proved that:

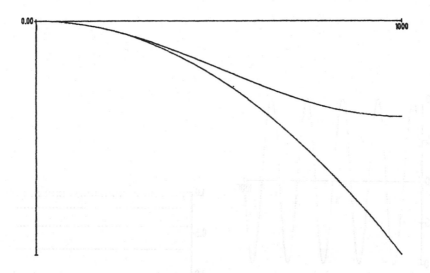

Fig. 3.1. Eigenvalues λ_k of matrix A and eigenvalues μ_k of the operator $\frac{\partial^2}{\partial x^2}$.

$$\forall\, k \in \{1,...,N-1\},\ k\ limited,\ \lambda_k \simeq \mu_k.$$

Thus, for such a k, $k\pi h/2 \simeq 0$, so that

$$\sin\left(\frac{k\pi h}{2}\right) = \frac{k\pi h}{2}(1+\varepsilon)\ with\ \varepsilon \simeq 0$$

and

$$\lambda_k = -\frac{4}{h^2}\frac{k^2\pi^2 h^2}{4}(1+\varepsilon)^2 = -k^2\pi^2(1+\varepsilon)^2 \simeq -k^2\pi^2 = \mu_k.$$

In the same way as we did for the eigenvectors of matrix A, and apart from the fact that they constitute an orthogonal $\mathbb{R}^{N-1}$ basis (a strictly algebraic property), it is possible to compare them with "functions". We associate to the vector U of $\mathbb{R}^{N-1}$ with components U_n (we have added the two components $U_0 = U_{N+1} = 0$) the subset of $[0,1] \times \mathbb{R}$, called a dotted-line:

$$U = \{(nh, U_n); n = 0, 1, ..., N\}.$$

Definition 3.2.1. *The dotted-line or vector U of $\mathbb{R}^{N+1}$ is said to be S-continuous if:*

      1) For all n, U_n is limited   (i.e. $\| U \|_\infty$ is limited)
      2) $(n - m)h \simeq 0$ implies $U_N - U_m \simeq 0$.

Therefore, we can state the following adaptation of the "theorem of the continuous shadow".

Theorem 3.2.1. *If the dotted-line U is S-continuous, its shadow is the graph of a continuous function on $[0,1]$.*

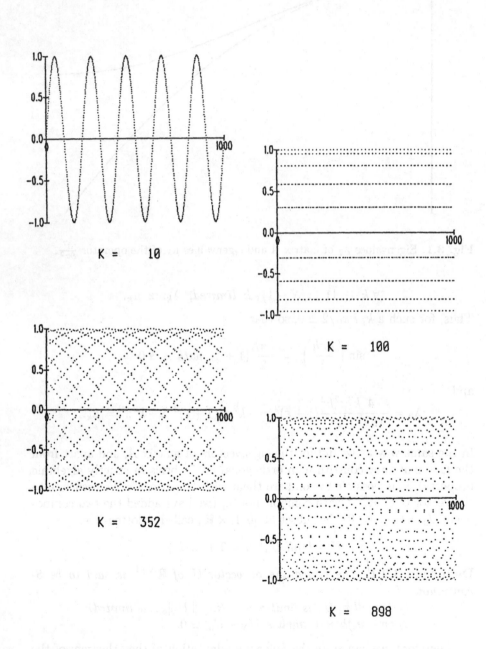

Fig. 3.2. Some dotted-line eigenvectors of A.

If we draw, see figure 3.2, a few of these dotted-line vectors V^k, we may note that for "small" values of k, they look like graphs of the continuous functions

$$x \mapsto sin(k\pi x)$$

which are the eigenfunctions of the continuous Lapacian operator. But, for larger values of k we can't observe even simple functions.

As a matter of fact, it is easy to prove that for limited k, V^k is S-continuous and i-close to the previous functions. In order to obtain more information about this subject, particularly about what happens for i-large k, one may refer to [22]. All these results seem to be very satisfying when compared to the above extracts from the Berkeley course emphasizing the importance of the early modes.

Let us note that only the S-continuity of dotted-lines will be used in the greater part of this work; the notion of shadow will appear only at the end, when we'll try to make a comparison with the classical analysis of (PDE).

3.2.3 Resolution of (DEN)

If the $\mathbb{R}^{N-1}$ vector $U(t)$, the unique solution of the Cauchy problem (DEN), is written as a linear combination of the vectors V^k,

$$U(t) = \sum_{k=1}^{N-1} u_k(t) V^k$$

then, for all k, $u_k(t)$ is the solution of the equation:

$$u_k''(t) - \lambda_k u_k(t) = 0.$$

Setting

$$\omega_k = \sqrt{-\lambda_k}$$

we obtain

$$U(t) = \sum_{k=1}^{N-1} \{u_k \cos(\omega_k t) + \frac{v_k}{\omega_k} \sin(\omega_k t)\} V^k \qquad (3.5)$$

where the components u_k and v_k are given by:

$$u_k = \frac{< F, V^k >}{\| V^k \|^2} = 2h \sum_{n=1}^{N-1} F_n \sin(k\pi nh)$$

and

$$v_k = \frac{< G, V^k >}{\| V^k \|^2} = 2h \sum_{n=1}^{N-1} G_n \sin(k\pi nh).$$

This solution which we call the "formal solution" of (DEN) makes no physical sense. The physical sense of a solution is only defined with respect to a real and concrete system. For example, in the case of a "continuous" string a solution $U(t)$ which is not S-continuous at a time t will be regarded as nonphysical because it describes a breaking of the string. On the other hand if the system represents a "plumbed string" or "masses coupled by springs" or an "electric circuit" such a phenomenon can be physically observed. We have chosen to study the S-continuity of the solution $U(t)$. What type of condition implies S-continuity? Is it possible to obtain S-continuous solutions without S-continuous initial conditions? Is it possible to observe non-S-continuous solutions with S-continuous initial conditions? Many questions remain...

3.3 An interesting example

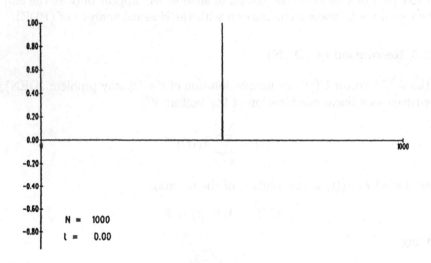

Fig. 3.3. An example of discontinuous initial condition.

In order to answer these questions, we are particularly interested in a "discontinuous" initial condition. We consider a dotted-line F which may be compared with a "Dirac" distribution (see figure 3.3 for the initial string position). The initial speed described by the dotted-line G is set equal to zero.

What can be said about the S-continuity of $U(t)$, for $t > 0$? A computer simulation, with $N = 10^3$, using the solution $U(t)$ given by formula (3.5), leads for various different values of t to the dotted-lines of figure 3.4.

The solution seems to become stable near zero. This fact will be confirmed by the following mathematical study. For technical reasons the string is described on the interval [-1,1] instead of [0,1], with the step $h = \frac{1}{N}$, and $U(t)$

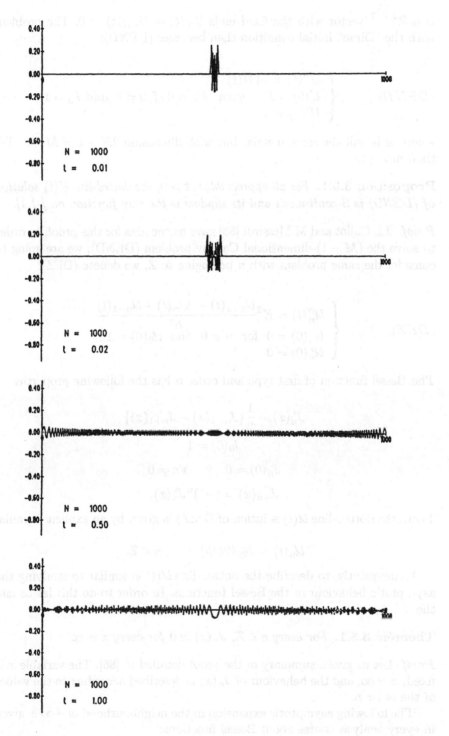

Fig. 3.4. Evolution of the dotted-line solution $U(t)$.

is a $\mathbb{R}^{2N-1}$ vector with the fixed ends $U_N(t) = U_{-N}(t) = 0$. The problem with the "Dirac" initial condition then becomes (DEND):

$$(DEND) \quad \begin{cases} U''(t) = A\,U(t) \\ U(0) = F \quad \text{with} \quad F_n = 0 \text{ if } n \neq 0 \text{ and } F_0 = 1 \\ U'(0) = 0 \end{cases}$$

where A is still the same matrix, but with dimension $2N - 1 = M - 1$. We then have the

Proposition 3.3.1. *For all appreciable t, $t > 0$, the dotted-line $U(t)$ solution of (DEND) is S-continuous and its shadow is the zero function on $[-1,1]$.*

Proof. J.L.Callot and M.Messirdi [86] gave us the idea for this proof. In order to solve the $(M-1)$-dimensional Cauchy problem (DEND), we are going to consider the same problem with n belonging to $\mathbb{Z}$, we denote (DEZ):

$$(DEZ) \quad \begin{cases} \mathcal{U}_n''(t) = K^2 \dfrac{\mathcal{U}_{n+1}(t) - 2\mathcal{U}_n(t) + \mathcal{U}_{n-1}(t)}{h^2} \\ \mathcal{U}_n(0) = 0 \text{ for } n \neq 0 \text{ and } \mathcal{U}_0(0) = 1 \\ \mathcal{U}_n'(0) = 0 \end{cases}$$

The Bessel function of first type and order n has the following properties

$$J_n'(x) = \frac{1}{2}\{J_{n-1}(x) - J_{n+1}(x)\}$$

$$J_0(0) = 1$$

$$J_n(0) = 0 \qquad \forall n \neq 0$$

$$J_{-n}(x) = (-1)^n J_n(x).$$

Then, the dotted-line $\mathcal{U}(t)$ solution of (DEZ) is given by an explicit formula:

$$\mathcal{U}_n(t) = J_{2n}(2t/h) \qquad n \in \mathbb{Z}.$$

Consequently, to describe the dotted-line $\mathcal{U}(t)$ is similar to studying the asymptotic behaviour of the Bessel functions. In order to do this let us cite the

Theorem 3.3.1. *For every $n \in \mathbb{Z}$, $J_n(x) \simeq 0$ for every $x \simeq \infty$*

Proof. Let us give a summary of the proof detailed in [86]. The variable x is fixed, $x \simeq \infty$, and the behaviour of $J_n(x)$ is described according to the values of the order n.

The following asymptotic expansion in the neighbourhood of $+\infty$ is given in every analysis course about Bessel functions:

$$J_n(x) = \sqrt{\frac{2}{\pi x}} \cos(x - n\frac{\pi}{2} - \frac{\pi}{4}) + \mathcal{O}\left(\frac{1}{7}x^{\frac{3}{2}}\right).$$

Therefore, for the standard integer n, J_n is a standard function, so if x is i-large

$$J_n(x) \simeq 0.$$

If n is no longer standard, which happens as soon as n is an i-large integer, this argument ceases to be available. In a book such as the "Theory of Bessel functions" by G.N.Watson [115], more accurate asymptotic expansions in the neighbourhood of infinity are given. Let us notice that in Watson's book, a whole chapter is assigned to the asymptotic behaviour of $J_n(x)$ when n and x are large. The author distinguishes the following cases: n small with regard to x translated by $x > n$ and $\frac{x}{n} = 1 + @$ (@ for appreciable), n large with regard to x translated by $x < n$ and $\frac{x}{n} = 1 - @$, and n and x close translated by $\frac{x}{n} \simeq 1$. So, using the infinitesimals the asymptotic expansions given in all these cases lead to the expected results.

This theorem implies that $\mathcal{U}_n(t)$ is i-small for all $n \in \mathbb{Z}$, as soon as $\frac{2t}{h}$ is i-large; therefore, the dotted-line $\mathcal{U}(t)$ is S-continuous and its shadow is the zero function on $\mathbb{R}$ as soon as t is i-small enough, for example, whenever $t \geq \sqrt{h}$.

In order to prove now that the solution $U(t)$ of (DEND) is S-continuous too and that its shadow is the zero function on [-1,1], it will be sufficient to prove that the difference between $\mathcal{U}(t)$ and $U(t)$ is i-small. Let us consider the dotted-line $Y(t)$:

$$Y_n(t) = U_{n-N}(t) - \mathcal{U}_{n-N}(t) \qquad for\ n = 1,...,M-1.$$

It is a solution of:
$$\begin{cases} Y_n''(t) = AY(t) - \Psi(t) \\ Y(0) = Y'(0) = 0 \end{cases}$$

where $\Psi(t) \in \mathbb{R}^{2N-1}$ with $\Psi(t) = {}^T(\mathcal{U}_{-N}(t), 0, ..., 0, \mathcal{U}_N(t))$.

Remark 3.3.1. *This system looks like a small perturbation of the system*

$$\begin{cases} Y_n''(t) = AY(t) \\ Y(0) = Y'(0) = 0 \end{cases}$$

whose solution is zero (from the previous theorem, $\Psi_n(t) \simeq 0$, for all $n = 1,...,M-1$, as soon as $t \geq \sqrt{h}$). We may be tempted to conclude that the solution is close to the zero function. This would be right if our system were standard but it is not standard and we have to build the proof directly.

Again, we solve this system with a diagonalization of A, and some inequalities which lead to the stated result.

Eigenvalues and eigenvectors are given by formulas (3.1) and (3.3) where the dimension $N-1$ is replaced by $M-1$. If we put:

$$Y(t) = \sum_{k=1}^{M-1} y_k(t)V^k \quad and \quad \Psi(t) = \sum_{k=1}^{M-1} \psi_k(t)V^k,$$

each component, $y_k(t)$, is a solution of

$$\begin{cases} y''(t) + \omega_k^2 y(t) = -\psi_k(t) \\ y(0) = y'(0) = 0. \end{cases}$$

This means that

$$y_k(t) = -\frac{1}{2} \int_0^t e^{i\omega_k(t-s)} \psi_k(s) ds - \frac{i}{2\omega_k} \int_0^t e^{i\omega_k(t-s)} \psi_k(s) ds.$$

We have that

$$| y_k(t) | \le \frac{1}{2}(1 + \frac{1}{\omega_k}) \int_0^t | \psi_k(s) | ds.$$

Since the sequence (ω_k) is strictly increasing,

$$\frac{1}{2}(1 + \frac{1}{\omega_k}) \le \frac{1}{2}(1 + \frac{1}{\omega_1}) = \frac{1}{2}(1 + \frac{1}{\pi}) < 1.$$

we have the inequality:

$$| y_k(t) | \le \int_0^t | \psi_k(s) | ds. \tag{3.6}$$

The k-th component of the vector $\Psi(t)$ in the basis of eigenvectors is given by formulas (3.4).

$$\psi_k(s) = \frac{< \Psi(s), V^k >}{\| V^k \|^2} = \frac{2}{M} \{ \mathcal{U}_{-N}(s) \sin\left(\frac{k\pi}{M}\right) + \mathcal{U}_N(s) \sin\left(\frac{(n-1)k\pi}{M}\right) \}.$$

But $\mathcal{U}_{-N}(s) = \mathcal{U}_N(s)$ and if we transform the second sine we get:

$$\psi_k(s) = \frac{2}{M} \sin(\frac{k\pi}{M}) J_{2N}(\frac{2s}{h})\{1 - (-1)^k\}.$$

Thus,

$$| \psi_k(s) | \le \frac{4}{M} | J_{2N}(\frac{2s}{h}) |$$

what leads us to write (3.6) in the form:

$$| y_k(t) | \le \int_0^{\sqrt{h}} | \psi_k(s) | ds + \int_{\sqrt{h}}^t | \psi_k(s) | ds.$$

The Bessel functions being bounded by 1 the first integral is i-small. We have seen before that $J_{2N}(\frac{2s}{h})$ is i-small as soon as $\frac{2s}{h}$ is i-small, hence as soon as s is greater than $\sqrt{h}$, so we conclude that the second integral is i-small too. We deduce the existence of an i-small real α such that:

$$\forall\, k = 1, ..., M-1 \qquad \mid y_k(t) \mid\, \le \frac{4}{M}\alpha$$

from which:

$$\mid Y_n(t) \mid\, =\, \mid \sum_{k=1}^{M-1} y_k(t) V_n^k \mid\, \le \sum_{k=1}^{M-1} \mid y_k(t) \mid$$

$$\le \frac{4}{M}\alpha(M-1) \le 4\alpha \simeq 0.$$

Thus, the components of $Y(t)$ are i-small. We may conclude the argument at this point.

The differential equation describing the string model can be integrated in the opposite direction so, we may, as a consequence, choose as initial condition the value at time $t = 100$ of the solution $U(t)$ given by (5). So, on the whole time interval $[0, 100 - \varepsilon]$ with ε appreciable, the shadow of $U(t)$ is the zero function on [-1,1]. The string seems to be in an equilibrium state and suddenly at time $t = 100$ we observe a "Dirac". This discontinuity can be explained by the breaking of the string. Hence, the S-continuity of F does not imply the S-continuity of $U(t)$.

In [46], we studied the control problem for a vibrating string (fixed at the extremity 0, controlled at 1), and we illustrated with the above example the difference between a "formally controllable" system and a "physically controllable" system: physical controls are limited controls, physical observations do not distinguish i-close states. We don't control a string which is apparently in an equilibrium state and so we cannot avoid it breaking: the string is not physically controllable, nevertheless, it is mathematically, formally controllable.

3.4 Solutions of limited energy

In the penultimate paragraph above, we saw that if the initial dotted-line F is S-continuous it does not imply S-continuity for the solution $U(t)$, $t > 0$. We now search for the types of condition (with physical meaning) which would give an S-continuous solution of (DEN).

3.4.1 A preliminary theorem

The following result gives the S-continuity of a sum of a finite number (i-large) of vectors under some particular conditions (which will be interpreted later on).

Proposition 3.4.1. Let M be an i-large integer, let M dotted-lines X^k and M limited reals a_k, $k = 1, ..., M$ be given, and let X be the sum $X = \sum_{k=1}^{M} a_k X^k$. If

1) the vectors X^k are limited (i.e. for all k, $\| X^k \|_\infty$ is limited)

2) vectors X^k are S-continuous for limited k

3) the sum $\sum_{k=1}^M | a_k |$ is S-Cauchy, i.e.

$$\forall \, \omega, \omega' \; i - large, \;\; \omega \leq \omega' \leq M \;\; \Rightarrow \;\; \sum_{k=\omega}^{\omega'} | a_k | \simeq 0$$

then X is S-continuous.

Proof. The result follows from the Cauchy principle (see chapter 1). Setting

$$C = Max_{k=1,\ldots,M} \, \| X^k \|_\infty,$$

it follows from 1) that C is limited. We can write

$$X = \sum_{k=1}^{l-1} a_k X^k + \sum_{k=l}^M a_k X^k.$$

Let E be the internal set

$$E = \{l \in \{1, \ldots, M\}; \; \| \sum_{k=l}^M a_k X^k \|_\infty \leq 1\}.$$

For all l, we have:

$$\| \sum_{k=l}^M a_k X^k \|_\infty \leq \sum_{k=l}^M | a_k | \, \| X^k \|_\infty \leq \sum_{k=l}^M | a_k | . Max_{k=l,\ldots,M} \, \| X^k \|_\infty \, .$$

Therefore

$$\| \sum_{k=l}^M a_k X^k \|_\infty \leq C \sum_{k=l}^M | a_k |,$$

and hypothesis 3) leads, for all l i-large, to:

$$\sum_{k=l}^M | a_k | \simeq 0.$$

Therefore all the i-large integers are in the set E and, consequently, E contains also a limited integer l_0. We can write

$$\| X \|_\infty \leq \| \sum_{k=1}^{l_0-1} a_k X^k \|_\infty \; + \; \| \sum_{k=l_0}^M a_k X^k \|_\infty \leq C . \sum_{k=1}^{l_0-1} | a_k | \, + 1.$$

The sum on the right of the above inequality is limited (limited sum of limited numbers, multiplied by a limited number C), so $\| X \|_\infty$ is limited. This is the first condition of S-continuity.

Let n, m be integers such that

$$n, m \in \{0, 1, ..., N\}, \quad (n-m)h \simeq 0.$$

We have

$$X_n - X_m = \sum_{k=1}^{l-1} a_k (X_n^k - X_m^k) + \sum_{k=l}^{M} a_k (X_n^k - X_m^k).$$

Let us consider the sequence (α_i) such that

$$\alpha_i = \sum_{k=1}^{i-1} a_k (X_n^k - X_m^k).$$

As the X^k are S-continuous for all limited k (hypothesis 3)), α_i is i-small for all limited i, so the Robinson lemma implies there exists an i-large integer ω_0 satisfying α_{ω_0}. Therefore $X_n - X_m$ is i-close to:

$$\sum_{k=\omega_0}^{M} a_k (X_n^k - X_m^k).$$

This real number is bounded by

$$2C \sum_{k=\omega_0}^{M} |a_k|$$

and hypothesis 3) leads to the second condition of S-continuity.

3.4.2 Limited energy: S-continuity of solution

The string energy at a given time t is the sum of the potential and kinetic energies:

$$E(t) = \frac{1}{2} \sum_{n=0}^{N-1} \left(\frac{U_{n+1}(t) - U_n(t)}{h} \right)^2 h + \frac{1}{2} \sum_{n=0}^{N-1} (U_n'(t))^2 h.$$

If we differentiate this function, we see that it is a constant function. The energy can also be written as the sum of the energies of each mode. Thus, discrete integration by parts of

$$\sum_{i=0}^{n-1} (\Delta X)_i Y_i = X_n Y_n - X_0 Y_0 - \sum_{i=0}^{n-1} (\Delta Y)_i X_{i+1},$$

where X and Y are two $\mathbb{R}^{n+1}$ vectors, and

$$(\Delta X)_i = X_{i+1} - X_i \quad \text{and} \quad (\Delta Y)_i = Y_{i+1} - Y_i,$$

gives

$$E(t) = E(0) = \frac{1}{4} \sum_{k=1}^{N} u_k^2 \omega_k^2 + v_k^2.$$

Theorem 3.4.1. *If the string energy is limited, then for all $t > 0$, $U(t)$ is S-continuous and its shadow is the graph of a continuous function.*

Proof. We simply need to verify the three hypotheses of proposition 3.4. For all $t > 0$, the sum of the general term

$$\mid u_k \cos(\omega_k t) + \frac{v_k}{\omega_k} \sin(\omega_k t) \mid$$

is S-Cauchy. This means that the sequences of general term $\mid u_k \mid$ and $\frac{|v_k|}{\omega_k}$ are S-Cauchy.

For all $\omega \in \{1, ..., N-1\}$ we have:

$$\sum_{k=\omega}^{N-1} \mid u_k \mid = \sum_{k=\omega}^{N-1} \mid u_k \mid \omega_k \frac{1}{\omega_k} \leq \left(\sum_{k=\omega}^{N-1} u_k^2 \omega_k^2 \right)^{\frac{1}{2}} \left(\sum_{k=\omega}^{N-1} \frac{1}{\omega_k^2} \right)^{\frac{1}{2}}$$

$$\leq E^{\frac{1}{2}} \left(\sum_{k=\omega}^{N-1} \frac{1}{\omega_k^2} \right)^{\frac{1}{2}}.$$

As $\omega_k \geq \frac{k\pi}{2}$, $\frac{1}{\omega_k^2} \leq \frac{4}{k^2\pi^2}$ which is the general term of a S-Cauchy sum. So

$$\forall \omega \in \{1, ..., N-1\}, \ \omega \ i-large \ \sum_{k=\omega}^{N-1} \frac{1}{\omega_k^2} \simeq 0.$$

and the sum of the general term $\mid u_k \mid$ is S-Cauchy. The same proof is used for the other sequence.

Since the eigenvectors V^k have been shown to be S-continuous for limited k and bounded, the three needed conditions are proved. The dotted-line solution of (DEN) is S-continuous.

The above theorem means that if the energy is not too large (limited) then an initial condition such as F S-continuous (in other words F is i-close to a continuous function) gives, for any initial speed condition, a solution which is macroscopically continuous.

In the Callot-Messirdi example, the i-large discrete energy,

$$E = E(0) = \frac{1}{2} \sum_{n=0}^{N-1} \left(\frac{U_{n+1}^0 - U_n^0}{h} \right)^2 h = \frac{1}{2} \frac{2}{h} = \frac{1}{h} \simeq \infty$$

may be concentrated at a point and lead to a break. We don't know if, in reality, the probability of such an event is considerable or even not infinitesimal.

3.4.3 Limited energy: propagation and reflexion

The above paragraph ensures the S-continuity of $U(t)$ for all $t > 0$ but it doesn't give any information about the evolution of the initial condition. In this paragraph we show that our mathematical model for the vibrating string allows the observation of classical propagation and reflexion phenomena. Some numerical examples using the solution $U(t)$ (3.5) confirm this. We will prove that the initial dotted-line $U(0) = F$ divides itself into two semi-dotted-lines moving in opposite directions and reflecting towards the extremities 0 and 1.

Let $F = U(0)$ be the dotted-line defined on $[0, 1]$. We extend it to the infinite dotted-line $\mathcal{F}$ defined on $\mathbb{R}$ using unevenness and periodicity:

$$\mathcal{F}_n = F_n \text{ for } n = 0, ..., N$$

$$\mathcal{F}_n = -F_n \text{ for } n = -N, ..., 0$$

$$\mathcal{F}_{n+2N} = \mathcal{F}_n \text{ for } n \in \mathbb{N}.$$

First, in order to simplify the calculation, let us suppose that the initial speed is equal to zero. So, we have

$$F_n = \sum_{k=1}^{N-1} u_k \sin(k\pi nh).$$

The components of the solution $U(t)$ are

$$U_n(t) = \sum_{k=1}^{N-1} u_k \cos(\omega_k t) \sin(k\pi nh),$$

which can be written as

$$U_n(t) = \frac{1}{2} \sum_{k=1}^{N-1} u_k \sin(k\pi nh + \omega_k t) + \frac{1}{2} \sum_{k=1}^{N-1} u_k \sin(k\pi nh - \omega_k t).$$

Since $\omega_k \simeq k\pi$ for limited k, $\omega_k t \simeq k\pi t$ for limited k and t, then

$$k\pi nh \pm \omega_k t \simeq k\pi(nh \pm t).$$

Let m be the "integer part" of $\frac{t}{h}$, $m = [\frac{t}{h}]$. We have, for limited k and t:

$$k\pi nh \pm \omega_k t \simeq k\pi(n \pm m)h,$$

and

$$\sin(k\pi nh \pm \omega_k t) \simeq \sin(k\pi(n \pm m)h).$$

If we write

$$\sin(k\pi nh \pm \omega_k t) = \sin(k\pi(n \pm m)h) + \{\sin(k\pi nh \pm \omega_k t) - \sin(k\pi(n \pm m)h)\}$$

we get:

$$U_n(t) = \frac{1}{2} \left\{ \sum_{k=1}^{N-1} u_k \sin(k\pi(n+m)h) + \sum_{k=1}^{N-1} u_k \sin(k\pi(n-m)h) \right\}$$

$$+ \frac{1}{2} \sum_{k=1}^{N-1} u_k [\sin(k\pi nh + \omega_k t) - \sin(k\pi(n+m)h)]$$

$$+ \frac{1}{2} \sum_{k=1}^{N-1} u_k [\sin(k\pi nh - \omega_k t) - \sin(k\pi(n-m)h)].$$

From the previous remark the general term of the last two sums are i-small for limited k. Using again the Cauchy principle, we deduce the existence of an i-large integer ω_0 so that each one of the two last sums can be written as two sums: the first one

$$\sum_{k=1}^{\omega_0-1} u_k [\sin(k\pi nh \pm \omega_k t) - \sin(k\pi(n \pm m)h)]$$

is i-small, and the second one

$$\sum_{k=\omega_0}^{N-1} u_k [\sin(k\pi nh \pm \omega_k t) - \sin(k\pi(n \pm m)h)]$$

is also i-small because the sine functions are bounded, and because the sum of the general terms $| u_k |$ is S-Cauchy under the hypothesis of limited energy. Consequently,

$$U_n(t) \simeq \frac{1}{2} \left\{ \sum_{k=1}^{N-1} u_k \sin(k\pi(n+m)h) + \sum_{k=1}^{N-1} u_k \sin(k\pi(n-m)h) \right\}.$$

On account of periodicity and the unevenness of the sine function and the dotted-line $\mathcal{F}$ we have shown that for all $n \in \mathbb{N}$:

$$U_n(t) \simeq \frac{1}{2}\{\mathcal{F}_{n+m} + \mathcal{F}_{n-m}\}$$

i.e.

$$U_n(t) \simeq \frac{1}{2}\{\mathcal{F}_{n+[\frac{t}{h}]} + \mathcal{F}_{n-[\frac{t}{h}]}\}.$$

If we consider now the general case $U(0) = F$ and $U'(0) = G$, the same type of proof leads to the following result:

Theorem 3.4.2. *If the energy is limited, then for all $t > 0$, limited:*

$$U_n(t) \simeq \frac{1}{2}\{\mathcal{F}_{n+[\frac{t}{h}]} + \mathcal{F}_{n-[\frac{t}{h}]}\} + \frac{1}{2} \sum_{i=n-[\frac{t}{h}]}^{n+[\frac{t}{h}]} \mathcal{G}_i h.$$

Therefore, the evolution of the dotted-line $U(t)$ is the one we would have expected (reflexion and propagation).

3.4.4 A particular case: comparison with classical model

If we suppose, still speaking of the case of limited energy, that the two initial dotted-lines F and G are S-continuous (let us note that F is S-continuous whenever the energy is limited), then from the theorem of the continuous shadow, their shadows are the graphs of continuous functions on $[0, 1]$, respectively designated by f and g. We extend these to the functions $\tilde{f}$ and $\tilde{g}$ defined on $\mathbb{R}$ as we did before for F and G. From the previous formula, $U(t)$ can be written

$$U_n(t) \simeq \frac{1}{2}\{\tilde{f}((n + [t/h])h) + \tilde{f}((n - [t/h])h)\} + \frac{1}{2} \sum_{i=n-[\frac{t}{h}]}^{n+[\frac{t}{h}]} \tilde{g}(ih)h,$$

which we transform into

$$U_n(t) \simeq \frac{1}{2}\{\tilde{f}(nh + t) + \tilde{f}(nh - t)\} + \frac{1}{2} \int_{nh-t}^{nh+t} \tilde{g}(s)ds.$$

We know, see theorem 3.4.1, that for all $t > 0$ the dotted-line $U(t)$ is S-continuous, so it has a shadow which is the graph of a continuous function on $[0, 1]$,

$$u_t : x \longrightarrow u_t(x).$$

Therefore,

$$u_t(x) \simeq \frac{1}{2}\{\tilde{f}(x + t) + \tilde{f}(x - t)\} + \frac{1}{2} \int_{x-t}^{x+t} \tilde{g}(s)ds.$$

so that we get equality for standard t. We cannot have equality for all t because u_t is the shadow of $U(t)$ for each t. This little defect may be removed if we define the function u as the shadow of the two variable function:

$$(t, x) \longmapsto U_n(t) \quad \text{where} \quad nh \leq x < (n + 1)h$$

We now get the classical formula:

$$u(t, x) = \frac{1}{2}\{\tilde{f}(x + t) + \tilde{f}(x - t)\} + \frac{1}{2} \int_{x-t}^{x+t} \tilde{g}(s)ds.$$

This also proves that the shadow of $U(t)$ is the solution of the classical string equation.

3.5 Conclusion

We have presented an elementary Fourier theory of propagation in a vibrating string based on the explicit calculation of the spectrum of the matrix A. We may well wonder whether the same results may be available for more general operators.

It had been shown in [6] that for the inverse matrix of a "compact" matrix, which means a matrix transforming limited vectors (in the sense of the maximum norm) into S-continuous vectors, the eigenvectors of limited rank (absolute values of eigenvalues arranged in an increasing order) are S-continuous. From this we get a discrete version of the Sturm-Liouville theory [5].

Finally, let us conclude by a remark with regard to the "elementary theory" we evoked in the introduction.

In "Radically elementary probability theory" [91], Nelson insists upon the following point: the nonstandard analysis of [91] uses a weak version of I.S.T., only the idealisation axiom which allows the introduction of infinitesimals being necessary. However, he uses all the power of I.S.T. in the last chapter, in order to prove that any classical result on continuous time processes has an elementary analogue. In the same way, in this text, we used only the consequences of idealisation except in the last paragraph, when we needed the theorem of continuous shadow which follows from the standardisation axiom.

J.L. Callot [27] and R. Lutz [84] talk about the power of weak infinitesimal theory. The existence of a radically elementary theory of partial differential equations in which any classical theorem would have an elementary analogue remains to be proved. What we have seen above concerning the string equation and some more general situations, shows that it is not stupid to try one's luck.

4. Random walks and stochastic differential equations

Eric Benoît

4.1 Introduction

The Wiener random walk is often studied in books on probability. It is a discrete process and classically, we cannot ask any question about the continuity or derivability of the trajectories or about the density of the process. The two results below are obtained by using discrete methods, but we are interested in the case of an infinitely large number of steps. With this hypothesis on the size of the variables, we prove some results which are similar to theorems on classical brownian motion. We want to take advantage of the simplicity of discrete concepts together with the simplicity of the analytical calculus. The link between the two classical methods is provided by nonstandard analysis.

In this paper, the Wiener random walk with infinitely small steps is first studied; thereafter, the study is generalized to that of any real diffusion process. We will compute the density of the solution of a stochastic differential equation. This is a density on a space of trajectories where the canonical measure is the Wiener measure. The result is known as Girsanov's theorem.

Throughout, processes are assumed to be defined on $\mathbb{R}$.

4.2 The Wiener walk with infinitesimal steps

This is the simplest random walk, which models the game of *heads or tails*; the basic random variables z_t are independent. They obey the same law given by

$$P(z_t = +1) \; = \; 1/2 \qquad P(z_t = -1) \; = \; 1/2$$

The index t takes its values in the finite set $\{0, dt, 2dt, \ldots, (N-1)dt\}$. The *Wiener walk* w_t is a family of random variables, w_t, defined by

$$\left\{ \begin{array}{rcl} w_{t+dt} & = & w_t + z_t\sqrt{dt} \\ w_0 & = & 0 \end{array} \right.$$

By definition, we denote by dw_t the random variable $w_{t+dt} - w_t$. With this notation, we have

$$dw_t = z_t\sqrt{dt} \qquad \text{and} \qquad w_t = \sum_{i=0}^{\frac{t-dt}{dt}} dw_{i\,dt}$$

Fig. 4.1. One trajectory of the Wiener walk

To stress the particular values of the variables z_t, we will write in this paragraph ± 1 instead of z_t. Remark though that this could lead to confusion since

$$(\pm 1).(\pm 1) = +1 \quad \text{or} \quad (\pm 1).(\pm 1) = \pm 1$$

so that we must be careful with the use of this notation.

The study below is made in the discrete finite context (as for example in [60]), but we assume that dt is infinitesimal and that Ndt is appreciable; we will make approximations, and we will find continuous results for *brownian motion* which is the classical limit of the Wiener walk.

4.2.1 The law of w_t for a fixed t

Let n be a fixed natural number such that $t = n\,dt$ is appreciable. The question now posed is to study the law of the random variable w_t given by

$$w_t = \sum_{i=0}^{n-1} dw_{i\,dt} = \sum_{i=0}^{n-1} \pm\sqrt{dt}$$

This number is a multiple of $\sqrt{dt}$ and the probability $P(w_{n\,dt} = k\sqrt{dt})$ is given by the binomial law. This is the first part of the proposition below.

Proposition 4.2.1. *Let* $t = n\,dt$, $x = k\sqrt{dt}$ *where* k *and* n *are natural numbers with the same parity and* $|k| \leq n$. *The probability that a trajectory of the Wiener walk goes through the point* (t, x) *is given exactly by*

$$P(w_{n\,dt} = k\sqrt{dt}) = \frac{2^{-n}\, n!}{\left(\frac{n+k}{2}\right)!\,\left(\frac{n-k}{2}\right)!}$$

If n *is unlimited and* k/n *infinitesimal, we have the approximation*

$$P(w_{n\,dt} = k\sqrt{dt}) = \sqrt{\frac{2}{\pi}}\,\frac{1}{\sqrt{n}}\,\exp\left(-\frac{k^2}{2n}(1+\oslash)\right),$$

where $\oslash$ *is an infinitesimal real number. Consequently, if* t *is appreciable and* x *limited, the approximation is valid, and we have :*

$$P(w_t \in [x - \sqrt{dt}, x + \sqrt{dt}[) = 2\sqrt{dt}\,\frac{(1+\oslash)}{\sqrt{2\pi t}}\,\exp(-\frac{x^2}{2t}).$$

Definition 4.2.1. *The density of a real random variable* X *is, if it exists, the* <u>*standard*</u> *function* ρ *obtained by the following method of averaging :*

$$\rho(x) \simeq \frac{P(X \in [x - \varepsilon, x + \varepsilon[)}{2\varepsilon}$$

for all t *and* x *standard and all sufficiently large infinitesimal* ε. *(The reader interested to the details of the theory of averaging on monads may consult [69, 97].)*

Thus, the density of the random variable w_t is obviously given by

$$\rho(t, x) = \frac{1}{\sqrt{2\pi t}}\,\exp(-\frac{x^2}{2t})$$

Proof. The first part of the proposition (the exact formula) follows from a straightforward computation using the binomial law. The second part (the approximation formula) will be proved in two steps: Stirling's formula and its application to the approximation of the binomial coefficients. The reader will find a proof of Stirling's formula in van den Berg's contribution to this book (see chapter 8). The formula can be written as

$$\ln(n!) = (n + \frac{1}{2})\ln n - n + \frac{1}{2}\ln(2\pi) + \oslash.$$

We need to compute the binomial coefficients in:

$$\ln P(w_{n\,dt} = k\sqrt{dt}) = -n\ln 2 + \ln(n!) - \ln\left(\frac{n+k}{2}\right)! - \ln\left(\frac{n-k}{2}\right)!$$

If n, $n + k$ and $n - k$ are unlimited, it is possible to use Stirling's formula, and we can simplify the formula above to obtain that $\ln P(w_{n\,dt} = k\sqrt{dt})$ is equal to :

$$\ln \left(\frac{2}{\sqrt{2\pi n}} \right) - \frac{n+k+1}{2} \ln \left(1 + \frac{k}{n} \right) - \frac{n-k+1}{2} \ln \left(1 - \frac{k}{n} \right) + \oslash$$

However, if k/n is infinitesimal, a Taylor expansion of order two of the logarithm gives the desired formula .

Remark 4.2.1. It is possible to make the approximation of proposition 4.2.1 more precise, if we use a Taylor expansion of higher order. These improved approximations are useful for studying finer properties of the brownian motion, the case of large deviations, for example.

4.2.2 Law of w

The results of the section above do not give any idea of the trajectories of the Wiener walk. We have studied the law of an arbitrary w_t, but, as yet, we know nothing about the correlation between w_t and w_s for different t and s. We will now study the random variable w which takes its values in the set of all functions. To find the "density" of w, we will study the following question.

Let $\gamma(t)$ be a path, for $t \in [0,T]$. Let ε be a real positive infinitesimal number. We suppose that ε is not too small (we will see later that it is a good idea to suppose that $dt = \oslash \varepsilon^4$). What is the value of the probability

$$P \left(\sup_{t \in [0,T]} |w_t - \gamma(t)| < \varepsilon \right) \quad ?$$

The question can also be phrased as follows: how many trajectories of the Wiener process are included in the ε-neighbourhood of γ ?

For a general γ, we will solve the problem in the section 4.7. Here we study only the case $\gamma = 0$.

Proposition 4.2.2. *Let $T = N\,dt$ and $\varepsilon = k\sqrt{dt}$. The number of trajectories of the Wiener walk included in the ε-neighbourhood of zero is given exactly by*

$$2^N \sum_{p=1}^{2k-1} \frac{1}{k} \cot \frac{\pi p}{4k} \sin \frac{\pi p}{2} \left(\cos \frac{\pi p}{2k} \right)^N$$

When T is appreciable, and ε infinitesimal such that $dt = \oslash \varepsilon^4$, this number can be approximated and we have

$$P(\sup_{t \in [0,T]} |w_t| < \varepsilon) = \frac{4}{\pi} (1 + \oslash) \exp \left(\frac{-\pi^2}{8} \frac{T}{\varepsilon^2} \right)$$

Proof. We can prove the first "exact" formula by using three very different methods : The first (see [60, chapter 3]) uses the reflexion principle. It is a purely combinatorial proof. The second relies on Fourier expansions. We explain the third one which uses computations on matrices.

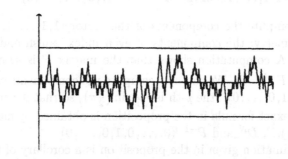

Fig. 4.2. One of the trajectories in the ε-neighbourhood of 0.

Let $X(n,j)$ be the number of trajectories of the Wiener walk defined for t in $[0, ndt]$, included in the ε-neighbourhood of zero, and going from $(0,0)$ to $(ndt, j\sqrt{dt})$. The penultimate point of such a trajectory is either the point $\left((n-1)dt, (j-1)\sqrt{dt}\right)$ or the point $\left((n-1)dt, (j+1)\sqrt{dt}\right)$, (one of them, at least, must lie in the ε-neighbourhood of 0). Accordingly, we have

$$X(n,j) = \begin{cases} X(n-1,j+1) & \text{if } j = -k+1 \\ X(n-1,j-1) + X(n-1,j+1) & \text{if } -k+1 < j < k-1 \\ X(n-1,j-1) & \text{if } j = k-1 \end{cases}$$

Put $V_n = (X(n,j))_{j=-k+1\ldots k-1}$. The relation above can be written as follows using matrices :

$$V_n = M.V_{n-1} \quad \text{with} \quad M = \begin{bmatrix} 0 & 1 & 0 & \ldots & 0 & 0 \\ 1 & 0 & 1 & \ldots & 0 & 0 \\ 0 & 1 & \ldots & \ldots & \cdot\cdot & \cdot\cdot \\ \cdot\cdot & \cdot\cdot & \ldots & \ldots & 1 & 0 \\ 0 & 0 & \ldots & 1 & 0 & 1 \\ 0 & 0 & \ldots & 0 & 1 & 0 \end{bmatrix}$$

After a long but straightforward computation, we can diagonalize the matrix M to give $M = P.D.P^{-1}$; The eigenvalues of M and the matrix P are given by :

$$\lambda_p = 2\cos\left(\frac{\pi p}{2k}\right) \qquad P = \left(\frac{1}{\sqrt{k}}\sin\left(\frac{\pi}{2k}ij\right)\right)_{1 \leq i,j \leq 2k-1}$$

The vector V_0 is evidently given by

$$V_0 = {}^t(0,\ldots,0,1,0,\ldots,0)$$

The number of paths of the Wiener walk on $[0, T]$, , included in the ε-neighbourhood of zero is the sum of the $X(N, j)$ for all j. This is

$$(1, 1, \ldots, 1, 1).P.D^N.P^{-1}.{}^t(0, \ldots, 0, 1, 0, \ldots, 0)$$

We can now compute the components of the vector $(1, 1, \ldots, 1, 1).P$, using a little trigonometry: the components of even index vanish and the others are $\frac{1}{\sqrt{k}} \cot \frac{\pi p}{4k}$. A computation shows that the matrix P is symmetric and orthogonal. So $P^{-1} = P$. We can now compute the components of the vector $P^{-1}.{}^t(0, \ldots, 0, 1, 0, \ldots, 0)$: the p-th is $\frac{1}{\sqrt{k}} \sin \left(\frac{p}{2}\pi\right)$, so that it vanishes when p is even. The exact formula in the proposition is obtained by multiplication of $(1, 1, \ldots, 1, 1).P$, D^N and $P^{-1}.{}^t(0, \ldots, 0, 1, 0, \ldots, 0)$.

The approximation given in the proposition is a corollary of the approximation lemma 4.2.2 below. One ought to remark that $\frac{1}{k} \cot \frac{\pi p}{4k}$ is always limited and equivalent to $\frac{4}{\pi p}$ since p is limited, and we can apply the lemma to the sum $\sum_{p=1}^{k}$, and to the "backward" sum $\sum_{p=2k-1}^{k}$.

Lemma 4.2.1. *If α_p is limited and $\beta_{p+1}/\beta_p \simeq 0$ for all $p \leq k$, we have*

$$\sum_{p=1}^{k} \alpha_p \beta_p = (\alpha_1 + \oslash)\beta_1.$$

Proof. Let η be the maximum of $|\beta_{p+1}/\beta_p|$. It is an infinitesimal positive number. We have that

$$\sum_{p=1}^{k} \alpha_p \beta_p = \left(\alpha_1 + \sum_{p=2}^{k} \alpha_p \frac{\beta_p}{\beta_1}\right) \beta_1$$

and the inequality

$$\left|\sum_{p=2}^{k} \alpha_p \frac{\beta_p}{\beta_1}\right| \leq \max|\alpha_p| \sum_{p=2}^{k} \eta^{p-1} < \max|\alpha_p| \frac{\eta}{1 - \eta} = \oslash.$$

Lemma 4.2.2. *If α_p is limited for all p, if $1/k$, k^2/N and N/k^4 are all infinitesimal, we have*

$$\sum_{p=1}^{k} \alpha_p \left(\cos \frac{\pi p}{2k}\right)^N = (\alpha_1 + \oslash) \left(\cos \frac{\pi}{2k}\right)^N = (\alpha_1 + \oslash) \exp\left(-\frac{\pi^2 N}{8k^2}\right)$$

Proof. First, the reader should check the formulas below:

- if $x \in [0, \frac{\pi}{2}]$, $0 \leq \cos x \leq 1 - x^2/4$
- if $u \in [0, 1]$, $0 \leq (1 - u)^N \leq \exp(-Nu)$
- if $x \simeq 0$, $\cos x = 1 - \frac{x^2}{2} + \pounds x^4$
- if $u \simeq 0$ and $Nu^2 \simeq 0$, $(1 - u)^N = (1 + \oslash) \exp(-Nu)$

With these formulas, we can

– evaluate the first term of the sum:

$$\left(\cos\frac{\pi}{2k}\right)^N = \left(1 - \frac{\pi^2}{8k^2} + \frac{\pounds}{k^4}\right)^N = (1+\oslash)\exp\left(-\frac{\pi^2 N}{8k^2}\right)$$

– show that the remainder of the sum is negligible:

$$\left|\sum_{p=2}^{k}\alpha_p\left(\cos\frac{\pi p}{2k}\right)^N\right| \leq \max|\alpha_p|\sum_{p=2}^{k}\left(\cos\frac{\pi p}{2k}\right)^N$$

$$\leq \max|\alpha_p|\sum_{p=2}^{k}\left(1 - \frac{1}{2}\frac{\pi^2 p^2}{8k^2}\right)^N$$

$$\leq \max|\alpha_p|\sum_{p=2}^{k}\exp\left(-\frac{1}{2}\frac{\pi^2 N}{8k^2}p^2\right)$$

We can apply lemma 4.2.1 to this last series because we have

$$\frac{\exp\left(-\frac{\pi^2 N}{16k^2}(p+1)^2\right)}{\exp\left(-\frac{\pi^2 N}{16k^2}p^2\right)} = \exp\left(-\frac{\pi^2 N}{16k^2}(2p+1)\right) \simeq 0$$

and the majoration becomes

$$\left|\sum_{p=2}^{k}\alpha_p\left(\cos\frac{\pi p}{2k}\right)^N\right| \leq \max|\alpha_p|\,(1+\oslash)\exp\left(-\frac{\pi^2 N}{4k^2}\right) = \oslash\exp\left(-\frac{\pi^2 N}{8k^2}\right)$$

4.3 Equivalent processes

4.3.1 The notion

A process can be "close" to another one in the sense that all the interesting properties are the same. The interesting properties are, for us, properties that we can see without distinction between infinitely near numbers. To render this precise, we will define in this section the notion of *equivalent processes*. We will first give the properties that equivalence should satisfy, then give the formal definition of equivalence and, finally, prove these properties.

We must keep in memory that a stochastic process is a random variable whose values belong to the set B of bounded functions from $[0, T]$ to $\mathbb{R}$. Limiting the range of our processes to the set of bounded functions (bounded but sometimes non-limited) is not restrictive because these processes are defined by discrete formulas, so that the trajectories cannot become infinite (although they may sometimes reach unlimited numbers). The set B will be fitted with the norm $\|f\| = \sup_{t\in[0,T]}|f(t)|$. According to this definition of the processes, we consider the trajectories of the Wiener walk as continuous paths, piecewise linear as in figure 4.1.

Proposition 4.3.1. *Let x_t and y_t be two processes defined on $[0,T]$. In the following two cases, the processes are equivalent :*

1. *Equivalence due to the values of the processes :*
   the two processes are defined on the same probability space Ω, and ,
   almost surely[1], $||x - y|| \simeq 0$.
2. *Equivalence due to neighbourhood probability laws :*

$$\sum_{\gamma \in B} |P(x = \gamma) - P(y = \gamma)| \simeq 0$$

Definition 4.3.1. *A functional f from B to $\mathbb{R}$ is called* uniformly S-continuous *if and only if*

$$\forall \gamma_1 \; \forall \gamma_2 \; ||\gamma_1 - \gamma_2|| \simeq 0 \Rightarrow f(\gamma_1) \simeq f(\gamma_2)$$

Definition 4.3.2. *The processes x_t and y_t are called* equivalent, *and we write $x_t \sim y_t$ if and only if, for all functionals f from B to $\mathbb{R}$ which are uniformly S-continuous and limited, the expectations $E(f(x))$ and $E(f(y))$ are infinitely close.*

Remark 4.3.1. This definition is given by "duality", with test-functionals, as is the classical definition of equality of probability laws.

Proof. **(of proposition 4.3.1)** Let f be a uniformly S-continuous, limited functional from B to $\mathbb{R}$.

1. Assume that point 1 of proposition 4.3.1 is satisfied by the processes x and y. The random variables $f(x)$ and $f(y)$ are almost surely infinitely close. Moreover, they are always limited. Lebesgue's dominated convergence theorem[2] shows that their expectations are infinitely close.
2. Assume now that point 2 of the proposition 4.3.1 is satisfied by the processes x and y. Then :

$$
\begin{aligned}
|E(f(y)) - E(f(x))| &= \left| \sum_{\gamma \in B} (P(y = \gamma) - P(x = \gamma)) f(\gamma) \right| \\
&\leq \sum_{\gamma \in B} |P(y = \gamma) - P(x = \gamma)| \; ||f|| \simeq 0
\end{aligned}
$$

---

[1] For a precise definition of external almost sure properties, see [11] ou [8]; here it is sufficient to know that an external event is almost sure if and only if it contains internal events of probability α, with $1 - {}^\circ\alpha$ arbitrarily small.

[2] or better its nonstandard version : see [92] or [9].

4.3.2 Macroscopic properties

Definition 4.3.3. *A property $A(\gamma)$, often external, of the variable γ in $\mathcal{B}$ is called* macroscopic *if and only if*

$$\forall \gamma_1 , \; \forall \gamma_2 , \; \gamma_1 \simeq \gamma_2 \; \Rightarrow \; (A(\gamma_1) \Leftrightarrow A(\gamma_2))$$

Examples:

- "γ is S-continuous", "γ is limited", "$\gamma \simeq 0$" are macroscopic properties.
- "γ is S-continuous and its shadow is derivable" is also a macroscopic property although the norm on $\mathcal{B}$ is not dependent on the derivative.

Proposition 4.3.2. *If x_t and y_t are two equivalent processes, and if $A(x)$ is an almost surely true, macroscopic property, then $A(y)$ is almost surely true.*

Proof. It is not conceptually difficult, but it requires a certain facility because of the duality in the definition of the equivalence of processes. The reader will find the entire proof in [9].

4.3.3 The brownian process

Any process equivalent to the Wiener walk is called a *brownian process*. The Wiener walk will be called *the canonical* brownian process with step dt. If a property is macroscopic, we can say that it is almost surely satisfied by *the* brownian process, even if we don't make precise the process we are studying. The theorem below (due to Nelson: see [92]) builds many brownian processes. We won't give here the proof of this theorem, since it is somewhat long and difficult, although it requires no special functional analytic or measure theoretic knowledge. The main tools belong to combinatorics and discrete probabilities and the arguments are close to those found in [60]. The reader can find the proof in [92] or [9].

Theorem 4.3.1. *Let z_t be a family of random variables, $t \in \{0, dt, 2dt, \ldots, Ndt\}$ such that* [3]

$$E(z_t \mid z_0 \ldots z_{t-dt}) = 0$$
$$E(z_t^2 \mid z_0 \ldots z_{t-dt}) = 1$$

---

[3] If X and Y are two random variables defined on the same space Ω, the conditional expectation $E(X|Y)$ is the random variable defined by

$$E(X|Y)(\omega) = \frac{\sum_{Y(\alpha)=Y(\omega)} p(X = \alpha)\, X(\alpha)}{\sum_{Y(\alpha)=Y(\omega)} p(X = \alpha)}$$

$$z_t \quad \text{is always limited}$$

Then the processes

$$\begin{cases} x_{t+dt} &= x_t + z_t\sqrt{dt} \\ x_0 &= 0 \end{cases} \quad \text{and} \quad \begin{cases} w_{t+dt} &= w_t \pm \sqrt{dt} \\ w_0 &= 0 \end{cases}$$

with values in the set of bounded functions on $[0, Ndt]$ are equivalent: the process x_t is brownian.

Examples:

– If the z_t are independent random variables such that $z_t = 10$ with probability $1/101$ and $z_t = -1/10$ with probability $100/101$, the hypotheses of the theorem are satisfied. The reader can see in figure 4.3 one trajectory of the process. Since this process is equivalent to the Wiener walk, we cannot see any difference between this figure and figure 4.1 representing the Wiener walk, unless we use a microscope to enlarge infinitesimal details.

Fig. 4.3. A trajectory of a non-canonical brownian process.

– We give a generalization of the previous example : let φ be an appreciable positive function. Let z_t be the family of random variables given by :

$$\begin{cases} P(z_t = \varphi(x_t) \mid z_0 \ldots z_{t-dt}) &= \frac{1}{\varphi^2(x_t)+1} \\[2mm] P(z_t = \frac{-1}{\varphi(x_t)} \mid z_0 \ldots z_{t-dt}) &= \frac{\varphi^2(x_t)}{\varphi^2(x_t)+1} \end{cases}$$

$$\text{with} \quad x_t = \sum_{\tau=0}^{t-dt} z_\tau \sqrt{dt}$$

This again gives a brownian process.

- If the z_t are independent identically distributed random variables with expectation 0 and standard deviation 1, the hypotheses of the theorem are satisfied. Accordingly, $\sum z_t$ converges to a gaussian random variable, so that the central limit theorem appears as a special case of Nelson's theorem.
- If the z_t are independent random variables with a normalized binomial law :

$$P(z_t = \frac{k}{\sqrt{n}}) = \frac{1}{2^n} \begin{pmatrix} n \\ \frac{n+k}{2} \end{pmatrix} \quad , \quad k \in \{-n, -n+2, \ldots, n-2, n\} \ n \text{ limited,}$$

the process defined in the previous example is now the same as the Wiener walk with step $\frac{dt}{n}$. Consequently, the theorem 4.3.1 builds random walks of different steps, all equivalent to the brownian process.
- Next we give an example of a non-brownian random walk satisfying the two hypotheses on mean and standard deviation. It shows that the third hypothesis[4] is needed. Let z_t be independent random variables with the same law :

$$z_t = \begin{cases} \frac{1}{\sqrt{dt}} & \text{with probability} \quad \frac{dt}{2} \\ 0 & \text{with probability} \quad 1 - dt \\ \frac{-1}{\sqrt{dt}} & \text{with probability} \quad \frac{dt}{2} \end{cases}$$

The z_t are sometimes unlimited. The process $x_t = \sum z_\tau \sqrt{dt}$ is a Poisson random walk. Its values are always integral, and the macroscopic property

$$\forall t \subset [0, T] \quad , \quad x_t \text{ is unlimited or } {}^{\circ}x_t \in \mathbb{N}$$

is always true. Since this is almost surely false for the Wiener walk, the Poisson random walk is not brownian.

4.4 Diffusions. Stochastic differential equations

4.4.1 Definitions

Definition 4.4.1. *Let z_t be a family of random variables satisfying the hypotheses of the theorem 4.3.1. Let x_0 be a random variable which is independent[5] with respect to the z_t. Let $b(x, t)$ and $s(x, t)$ be two functions of*

---

[4] In fact, this hypothesis is too restrictive: it can be replaced by "z_t is a $\mathcal{L}^2$ random variable" (see [92]).

[5] In fact, the needed hypothesis is only that convenient conditional expectations are vanishing. If the reader is not too familiar with these notions, he could suppose that x_0 is a given non-random real number, for example 0.

class S^0 having a C^∞ shadow. We suppose[6] that there exist some standard constants C_1 and C_2 such that

$$\forall x \in \mathbb{R} \quad \forall t \in [0, T] \qquad |b(x, t)| \leq C_1 x + C_2$$

The induction formula

$$\boxed{x_{t+dt} = x_t + b(x_t, t)\, dt + s(x_t, t)\, ,.\, z_t \cdot \sqrt{dt}}$$

is called a stochastic differential equation. *With the initial condition x_0, it defines an unique random walk: the* solution. *A* diffusion *is the equivalence class of a solution of a stochastic differential equation. Two stochastic differential equations are called* equivalent *if they define the same diffusion.*

Remarks 4.4.1. – *If $b = 0$, $s = 1$, $z_t = \pm 1$ and $x_0 = 0$, the solution is the Wiener walk and the diffusion is the brownian diffusion (called briefly "the brownian").*

– *If $s = 0$, we have an Euler discretization of the ordinary differential equation $\dot{x} = b(x, t)$.*

– *Let us write $dx_t = x_{t+dt} - x_t$. When z_t is the canonical family $z_t = \pm\sqrt{dt} = dw_t$, the stochastic differential equation can be written as*

$$\frac{dx_t}{dt} = b(x_t, t) + s(x_t, t)\frac{dw_t}{dt}$$

In this expression, $\frac{dw_t}{dt}$ is infinitely large and, to avoid computing with unlimited numbers, I will always prefer to write

$$dx_t = b(x_t, t)dt + s(x_t, t)dw_t$$

– *In the following sections, I want to study the macroscopic properties of a diffusion; consequently, when it is convenient, I will replace a stochastic differential equation by another in the same equivalent class. The good choice is the main tool of the proofs.*

4.4.2 Theorems

The two theorems below are very natural. The proof is a little bit technical due to the definition "by duality" of the equivalence of random walks. It can be found in [9].

Theorem 4.4.1. *If z_t is a family of random variables satisfying the hypothesis of the theorem 4.3.1, the stochastic differential equation built from the z_t is equivalent to the same one built from the canonical $\pm\sqrt{dt}$.*

Theorem 4.4.2. *If $\beta(x, t)$ and $\sigma(x, t)$ are two functions having the same shadow as b and s, the stochastic differential equation built from β and σ is equivalent to the same one built from the canonical b and s.*

---

[6] With this strong hypothesis, a trajectory of the diffusion will be almost surely limited on $[0, T]$.

4.4.3 Change of variable

In the following computation, the reader should note the usage of the Taylor formula *of order 2*. It shows that the properties of a random walk are not invariant with respect to the differential structure. The riemannian metric is needed for the study of a diffusion.

Consider the stochastic differential equation

$$dx_t = b(x_t, t)dt \pm s(x_t, t)\sqrt{dt}$$

We will make the change of variable given by $y = \varphi(x, t)$, $x = \psi(y, t)$ where φ and ψ are standard C^∞ functions. A straightforward computation gives :

$$
\begin{aligned}
dy_t &= \varphi(x_t + dx_t, t + dt) - \varphi(x_t, t) \\
&= \varphi(x_t \pm s(x_t, t)\sqrt{dt} + b(x_t, t)dt, t + dt) - \varphi(x_t, t) \\
&= \pm \frac{\partial\varphi}{\partial x}s\sqrt{dt} + \frac{\partial\varphi}{\partial x}bdt + \frac{1}{2}\frac{\partial^2\varphi}{\partial x^2}s^2 dt + \frac{\partial\varphi}{\partial t}dt + \oslash dt
\end{aligned}
$$

where the derivatives of φ are to be evaluated at the point $(x_t, t) = (\psi(y_t), t)$. We may thus write another equivalent stochastic differential equation :

$$dy_t = \left(\frac{\partial\varphi}{\partial x}b + \frac{\partial\varphi}{\partial t} + \frac{1}{2}\frac{\partial^2\varphi}{\partial x^2}s^2\right)dt \pm \frac{\partial\varphi}{\partial x}s\sqrt{dt}$$

Proposition 4.4.1. *If, on a domain $\mathcal{D}$, the function s is not vanishing, a change of variable can transform a stochastic differential equation into a new one with $s = 1$.*

Example 4.4.1. $dx = \pm x\sqrt{dt}$ On the domain $x > 0$, we choose $y = \varphi(x, t)$ such that $\frac{\partial\varphi}{\partial x}x = 1$. It is the change of variable $y = \ln x$ which gives $dy = -\frac{1}{2}dt \pm \sqrt{dt}$. The reader should remark that the same change of variable transforms the ordinary differential equation $\dot{x} = 0$ into $\dot{y} = 0$ and not into $\dot{y} = -\frac{t}{2}$.

4.5 Probability law of a diffusion

Let x be a process defined on $[0, T]$ by the stochastic differential equation :

$$\begin{cases} dx_t &= b(x_t, t)dt \pm \sqrt{dt} \\ x_0 &= 0 \end{cases}$$

The restriction on the case $x_0 = 0$ and $s = 1$ is for practical reasons which we have discussed above.

In this section, I will prove a nonstandard theorem, equivalent to the classical Girsanov's theorem. The purpose is to compare the "density" on $\mathcal{B}$ of the two diffusions x and w. The first thing we have to do is to choose a "good" process $\hat{x}$ equivalent to the given one x.

Proposition 4.5.1. *The process $\hat{x}$ defined below is equivalent to the process x.*

$$\begin{cases} P(d\hat{x}_t = +\sqrt{dt} \,|\, \hat{x}_0 \ldots \hat{x}_{t-dt}) & = & \frac{1}{2}(1 + b(\hat{x}_t, t)\sqrt{dt}) \\ P(d\hat{x}_t = -\sqrt{dt} \,|\, \hat{x}_0 \ldots \hat{x}_{t-dt}) & = & \frac{1}{2}(1 - b(\hat{x}_t, t)\sqrt{dt}) \\ \hat{x}_0 & = & 0 \end{cases}$$

Remark 4.5.1. The choice of $\hat{x}$ is made so that any trajectory of $\hat{x}$ is exactly a trajectory of w. Consequently, it is possible to compare the probability $P(\hat{x} = \lambda)$ and $P(w = \lambda)$ for each of the $2^{T/dt}$ trajectories of the Wiener walk.

Proof. In the definition above of the process $\hat{x}$, the increments $d\hat{x}_t$ are not of mean 0 and standard deviation 1. We can write the process $\hat{x}$ in a different way (where the function b is evaluated at the point $(\hat{x}_t, t)$) :

$$\begin{cases} d\hat{x}_t & = & b\,dt + \sqrt{1 - b^2 dt}\, z_t \sqrt{dt} \\ \hat{x}_0 & = & 0 \end{cases}$$

with
$$\begin{cases} P(z_t = +\sqrt{\frac{1 - b\sqrt{dt}}{1 + b\sqrt{dt}}} \,|\, z_0 \ldots z_{t-dt}) & = & \frac{1}{2}(1 + b\sqrt{dt}) \\ P(z_t = -\sqrt{\frac{1 + b\sqrt{dt}}{1 - b\sqrt{dt}}} \,|\, z_0 \ldots z_{t-dt}) & = & \frac{1}{2}(1 - b\sqrt{dt}) \end{cases}$$

The hypothesis of theorem 4.3.1 are satisfied by the z_t and the function $\sqrt{1 - b^2 dt}$ is infinitely close to 1. Thus theorems 4.4.1 and 4.4.2 may be applied.

Proposition 4.5.2 (Temporary Girsanov's formula). *If λ is a trajectory of the Wiener walk on $[0, T]$, we have :*

$$P(\hat{x} = \lambda) \;=\; P(w = \lambda)\, \exp(M(\lambda))$$

where M is a functional defined on the set of all trajectories of the Wiener walk by :

$$M(\lambda) \;=\; \sum_{t=0}^{T-dt} \ln(1 + b(\lambda_t, t)d\lambda_t)$$

Proof. The proof is a straightforward computation. The proposition 4.5.1 gives :

$$P(d\hat{x}_t = \pm\sqrt{dt}) \;=\; \frac{1}{2}(1 \pm b(\hat{x}_t, t)\sqrt{dt}).$$

If we remember that $d\lambda_t = \pm\sqrt{dt}$, we write :

$$P(d\hat{x}_t = d\lambda_t) \;=\; \frac{1}{2}(1 + b(\hat{x}_t, t)d\lambda_t).$$

The increments of the process have the good properties of independence and
so give :

$$P(\hat{x} = \lambda) = \prod_{t=0}^{T-dt} P(d\hat{x}_t = d\lambda_t)$$

$$= \prod_{t=0}^{T-dt} \frac{1}{2}(1 + b(\lambda_t, t)d\lambda_t)$$

$$= \frac{1}{2^N} \prod_{t=0}^{T-dt} (1 + b(\lambda_t, t)d\lambda_t)$$

and proposition 4.5.2 is now obtained by writing the product $\prod u_n$ as
$\exp \sum \ln u_n$.

4.6 Ito's calculus – Girsanov's theorem

The proposition 4.5.2 above is precise, without approximation, but the expression for $M(\lambda)$ is not usable for two reasons. First, we have to sum N
terms of order $\sqrt{dt}$ and the result may be of order $N\sqrt{dt} = 1/\sqrt{dt}$. Secondly,
it appears to depend on details of λ and not merely on $^\circ\lambda$. The Ito formula
below helps.

Proposition 4.6.1 (Ito's formula). *Let $b(x, t)$ be a standard C^∞ function. Let B be the function defined by the integral*

$$B(x, t) = \int_0^x b(\xi, t)dt$$

Let λ be a trajectory of the Wiener walk. We have

$$\sum_{t=0}^{T-dt} b(\lambda_t, t)d\lambda_t \simeq B(\lambda_T, T) - \int_0^T \left(\frac{\partial B}{\partial t} + \frac{1}{2}\frac{\partial^2 B}{\partial x^2} \right) dt$$

Proof. The proof is essentially an integration by parts. We have to take care
of the terms of order two in the Taylor formulas, and to remember that
$d\lambda^2 = dt$. We have the following computations :

$$B(\lambda_T, T) = \sum_{t=0}^{T-dt} (B(\lambda + d\lambda, t + dt) - B(\lambda, t))$$

$$= \sum_{t=0}^{T-dt} \left(\frac{\partial B}{\partial x}d\lambda + \frac{1}{2}\frac{\partial^2 B}{\partial x^2}d\lambda^2 + \frac{\partial B}{\partial t}dt + \oslash dt \right)$$

$$\simeq \sum_{t=0}^{T-dt} b(\lambda, t)d\lambda + \sum_{t=0}^{T-dt} \left(\frac{\partial B}{\partial t} + \frac{1}{2}\frac{\partial^2 B}{\partial x^2} \right) dt$$

The second occurrence of the symbol $\sum$ is a Riemann sum. It can be replaced by the symbol $\int$.

I remark that the sums $\sum$ are very easy to use conceptually, but the computations are laborious. Conversely, computations with integrals demand more theoretical lemmas but can be done more rapidly.

Remark 4.6.1. In the classical books on stochastic calculus, the Ito calculus and the Stratonovitch calculus are introduced. They correspond to computations of sums such as, respectively,

$$\sum_{t=0}^{T-dt} b(\lambda_t, t)d\lambda_t \quad \text{or} \quad \sum_{t=0}^{T-dt} \left(\frac{b(\lambda_{t+dt}, t+dt) + b(\lambda_t, t)}{2} d\lambda_t \right).$$

The Ito calculus gives more compact formulas and is better for probabilistic computations. The Stratonovitch calculus has more intrinsic properties with respect to the differential geometry.

Theorem 4.6.1 (Girsanov). *Let x be the diffusion defined by the stochastic differential equation*

$$\begin{cases} dx_t &= b(x_t, t)dt \pm \sqrt{dt} \\ x_0 &= 0 \end{cases}$$

Let $\hat{x}$ be the random walk (equivalent to x) defined in proposition 4.5.1. Let λ be a trajectory of the Wiener walk on $[0, T]$. We have :

$$P(\hat{x} = \lambda) = P(w = \lambda) \exp(M(\lambda))$$

where the function M, defined on the set of the trajectories of the Wiener walk, is such that :

$$M(\lambda) \simeq B(\lambda_T, T) - \int_0^T \left(\frac{\partial B}{\partial t} + \frac{1}{2} b^2 + \frac{1}{2} \frac{\partial b}{\partial x} \right) dt$$

Moreover, the function $M(\lambda)$ is S-continuous on the set of trajectories of the Wiener walk.

Proof. We apply the order 2 Taylor formula to the result of the proposition 4.5.2 :

$$M(\lambda) = \sum_{t=0}^{T-dt} \log(1 + b(\lambda_t, t)d\lambda_t)$$

$$= \sum_{t=0}^{T-dt} \left(b(\lambda_t, t)d\lambda_t - \frac{1}{2}b^2(\lambda_t, t)d\lambda_t^2 + \oslash dt \right)$$

$$= \sum_{t=0}^{T-dt} b(\lambda_t, t)d\lambda_t - \frac{1}{2} \sum_{t=0}^{T-dt} b^2(\lambda_t, t)d\lambda_t^2 + \oslash$$

We now substitute using Ito's formula (proposition 4.6.1) in the last expression. The result is obviously of class S^0 on the set of all bounded functions, because we have Riemann integrals and the functions B, b and their derivatives are standard C^∞.

4.7 The "density" of a diffusion

In this section we want to apply the preceding theorems to compute the probability that a trajectory of a given stochastic process is included in the ε-neighbourhood of a given curve γ. This is a generalization of the proposition 4.2.2.

Theorem 4.7.1 (Density of a diffusion). *Let x be a diffusion given on $[0,T]$ by $x_0 = 0$ and :*

$$x_{t+dt} = x_t + b(x_t, t)dt \pm \sqrt{dt}$$

Let $\gamma(t)$ be a standard C^1 curve defined for t in $[0,T]$, and such that $\gamma(0) = 0$. Let ε be an infinitesimal but large enough real number. We have now the following expression for the "density" of the diffusion x at a point γ :

$$P(\sup_{t \in [0,T]} |x_t - \gamma(t)| < \varepsilon) = (1 + \oslash)\frac{4}{\pi} \exp\left(\frac{-\pi^2 T}{8\varepsilon^2}\right) \exp\left(\mathcal{F}(\gamma)\right)$$

where $\quad \mathcal{F}(\gamma) = -\frac{1}{2} \int_0^T \left([b(\gamma(t), t) - \gamma'(t)]^2 + \frac{\partial b}{\partial x}(\gamma(t), t) \right) dt$

Remark 4.7.1. The reason for the choice of an "infinitesimal but large enough" ε is that we will use a method of averaging around γ. We want to average over all the halo of γ, but we must replace this external halo by an internal set, the ε-neighbourhood of γ.

We could formulate a different, stronger hypothesis on ε, namely, that ε be greater than $f(dt)$ for all standard continuous functions vanishing at zero. We could also use the theory of Peraire (see [95]) of nonstandard analysis with many levels. In that case we would choose dt infinitesimal with respect to ε.

Proof. The first step of the proof consists in proving that the density given in the theorem is the density of a process $\hat{x}$ equivalent to x. In the second step, we will prove that the density (in the sense of the theorem) is independent of the choice of the process equivalent to x.

– The computation is very easy , all the main tools being in the preceding lemmas and propositions. We begin with a change of variable (see 4.4.3) to study the ε-neighbourhood of zero :

$$x_t = \gamma(t) + y_t \quad \text{gives} \quad dy_t = (b(\gamma(t)+y_t,t) - \gamma'(t))\,dt \pm \sqrt{dt}$$

Next, we apply the proposition 4.5.1 to build a process $\hat{y}$ equivalent to y such that its trajectories are the same as the trajectories of the Wiener walk. The process $\hat{x}$ is the inverse image of $\hat{y}$ by the change of variable. The Girsanov's calculus (theorem 4.6.1) gives for any trajectory λ of the Wiener walk :

$$P(\hat{y}=\lambda) = P(w=\lambda)\,\exp(M(\lambda))$$

$$\text{with} \quad M(\lambda) \simeq C(\lambda_T, T) - \int_0^T \left(\frac{\partial C}{\partial t} + \frac{1}{2}c^2 + \frac{1}{2}\frac{\partial c}{\partial y} \right) dt$$

$$\text{where} \quad c(y,t) = b(\gamma(t)+y,t) - \gamma'(t) \quad \text{and} \quad C(y,t) = \int_0^y c(\eta,t)d\eta$$

After some computations, we obtain

$$\lambda \simeq 0 \quad \Rightarrow \quad M(\lambda) \simeq -\frac{1}{2}\int_0^T \left([b(\gamma(t),t) - \gamma'(t)]^2 + \frac{\partial b}{\partial x}(\gamma(t),t) \right) dt$$

which is the expression for $\mathcal{F}(\gamma)$ given in the theorem.
The desired probability can be written :

$$
\begin{aligned}
P(\sup_{t\in[0,T]} |\hat{x}_t - \gamma(t)| < \varepsilon) &= P(\sup_{t\in[0,T]} |\hat{y}_t| < \varepsilon) \\
&= \sum_{||\lambda||<\varepsilon} P(\hat{y}=\lambda) \\
&= \sum_{||\lambda||<\varepsilon} P(w=\lambda)\,\exp(M(\lambda)) \\
&= (1+\oslash)\exp(M(0)) \sum_{||\lambda||<\varepsilon} P(w=\lambda) \\
&= (1+\oslash)\exp(\mathcal{F}(\gamma))\frac{4}{\pi}\exp\left(\frac{-\pi^2 T}{8\varepsilon^2}\right)
\end{aligned}
$$

– Let x be a stochastic process, equivalent to $\hat{x}$. Let $\varphi_x(\varepsilon)$ be the number, defined for all ε by :

$$\varphi_x(\varepsilon) = \frac{P(\sup_{t\in[0,T]} |x_t - \gamma(t)| < \varepsilon)}{\frac{4}{\pi}\exp\left(\frac{-\pi^2 T}{8\varepsilon^2}\right)\exp(\mathcal{F}(\gamma))}$$

We have to prove that for all ε infinitesimal but large enough, we have $\varphi_x(\varepsilon) \simeq \varphi_{\hat{x}}(\varepsilon) \simeq 1$. Let h_ε be the characteristic function of the interval $[0,\varepsilon]$. An elementary relation of probability theory is

$$P(\sup_{t\in[0,T]} |x_t - \gamma(t)| \le \varepsilon) = E(h_\varepsilon(||x - \gamma||))$$

The function $h_\varepsilon(\|x - \gamma\|)$ is not limited and uniformly S-continuous, even for standard ε. Consequently, we cannot apply directly the definition 4.3.2 of equivalent processes. We introduce an "approximation" k_ε of h_ε such that the function $k_\varepsilon(r)$ of the two real positive variables ε and r is standard C^∞. Its value is 1 when $r < \varepsilon$, and 0 when $r > \varepsilon + \varepsilon^4$, it is always in the interval $[0,1]$ (The reader may construct such a function). We have now, for all ε :

$$h_\varepsilon \leq k_\varepsilon \leq h_{\varepsilon + \varepsilon^4}$$

It is now very easy to prove the following inequalities :

$$\varphi_x(\varepsilon) \leq \frac{E(k_\varepsilon(\|x - \gamma\|))}{\frac{4}{\pi} \exp\left(\frac{-\pi^2 T}{8\varepsilon^2}\right) \exp(\mathcal{F}(\gamma))} \leq \varphi_x(\varepsilon + \varepsilon^4) \frac{\exp\left(\frac{-\pi^2 T}{8(\varepsilon + \varepsilon^4)^2}\right)}{\exp\left(\frac{-\pi^2 T}{8\varepsilon^2}\right)}$$

Evidently we have the same inequalities for the process $\hat{x}$.

If ε is appreciable, the function k_ε is limited and uniformly S-continuous. So the expectations $E(k_\varepsilon(\|x - \gamma\|))$ and $E(k_\varepsilon(\|\hat{x} - \gamma\|))$ are equivalent. The denominators above are appreciable. So we obtain, for all appreciable ε, the inequalities :

$$\varphi_x(\varepsilon) \lesssim \varphi_{\hat{x}}(\varepsilon + \varepsilon^4) \frac{\exp\left(\frac{-\pi^2 T}{8(\varepsilon + \varepsilon^4)^2}\right)}{\exp\left(\frac{-\pi^2 T}{8\varepsilon^2}\right)}$$

$$\varphi_{\hat{x}}(\varepsilon) \lesssim \varphi_x(\varepsilon + \varepsilon^4) \frac{\exp\left(\frac{-\pi^2 T}{8(\varepsilon + \varepsilon^4)^2}\right)}{\exp\left(\frac{-\pi^2 T}{8\varepsilon^2}\right)}$$

Using the permanence lemmas, these inequalities are true even for ε infinitesimal large enough. For such an ε the quotient of exponentials is equivalent to 1. Consequently, for all ε infinitesimal large enough, we have:

$$\varphi_x(\varepsilon) \lesssim \varphi_{\hat{x}}(\varepsilon + \varepsilon^4) \qquad \varphi_{\hat{x}}(\varepsilon) \lesssim \varphi_x(\varepsilon + \varepsilon^4)$$

Using the first step of the proof, we may verify that the inequalities of the theorem are correct for the process x.

4.8 Conclusion

The result of this work could be written using the terminology of measure theory. Let Λ be the set of the trajectories of the Wiener walk with step dt on the interval $[0,T]$. This set can be considered as a *sample* (see [36, 11]) of the set $\mathcal{B}$ of all the bounded functions on $[0,T]$:

If $\mathcal{A}$ is a set included in $\mathcal{B}$, we call a *Wiener sample-measure* of A the quotient of the cardinals of

$\mathcal{A} \cap \Lambda$ and Λ. The classical Wiener measure is now the Loeb measure associated with this sample-measure.

Theorem 4.6.1 computes the measure associated with a given diffusion x, by comparing it with the Wiener measure. The sample-measure μ associated to x is defined by

$$\mu(\mathcal{A}) = \sum_{\lambda \in \Lambda} P(x = \lambda)$$

Theorem 4.7.1 shows that the associated Loeb measure is independent of the chosen process in the equivalence class of the diffusion x.

5. Infinitesimal algebra and geometry

Michel Goze

5.1 A natural algebraic calculus

5.1.1 The Leibniz rules

Throughout this chapter, the referential framework is the standard vector space K^n where $K = \mathbb{R}$ *or* $\mathbb{C}$. As K^n is a standard vector space, from the transfer principle there is a standard basis $(e_1, ..., e_n)$ for the space. Except where otherwise noted, this basis will be used below to establish properties of the vector calculus.

Definition 5.1.1. *A vector v of K^n is called infinitesimal (resp. limited) if all its components are infinitesimal (resp. limited). It is called appreciable if it is limited and if at least one of its components is appreciable. If none of these conditions hold, v is said to be infinitely large.*

The aim of this chapter is to show that a limited vector of K^n (or more precisely the corresponding point of the affine plane K^n) hides an intrinsic geometry and an algebraic process that we will illustrate through what we judge to be significant examples. How are this geometry and this algebra to be brought to light? The chosen point of the affine plane is defined by its coordinates. Suppose that it is in the halo of origin so that its coordinates are infinitesimal. Naturally, we may compare them. The first chapter provides us with calculus rules for the orders of magnitude. Some rules are obvious and natural and thus without interest. Others are not universal. The result depends essentially on the hypotheses, and also on the infinitesimals that we want to compare. These are the rules, usually called Leibniz rules corresponding, in fact, to indeterminate forms in classical analysis, that we shall actually apply (the point remaining fixed, the information emerging from the rules) in order to reveal what is hidden by the point.

5.1.2 The algebraic-geometric calculus underlying a point of the plane

Here we examine the simplest case, that in which the vector v is an infinitesimal vector of K^2. One denotes by M the corresponding point of the affine plane and one writes $M = (\varepsilon_1, \varepsilon_2)$. One supposes that this point is in the halo of origin, i.e. $\varepsilon_1 \simeq 0, \varepsilon_2 \simeq 0$. One goes on to compare these two infinitesimals by estimating the quotient $\varepsilon_1/\varepsilon_2$ which one supposes well defined. There are points in the halo $h(0)$ of 0 for which this quotient is infinitesimal or limited, other points for which this quotient is infinitely large. Suppose that for the given point M, the ratio $\varepsilon_1/\varepsilon_2$ is limited. Then we have

$$\varepsilon_1/\varepsilon_2 = a + \varepsilon_3$$

with a standard (possibly zero) and $\varepsilon_3 \simeq 0$. Then

$$\varepsilon_1 = a\varepsilon_2 + \varepsilon_2\varepsilon_3.$$

Then $M = (a\varepsilon_2 + \varepsilon_2\varepsilon_3, \varepsilon_2)$ and this point appears as the image of the point M_1 of the halo of the origin having as coordinates:

$$M_1 = (\varepsilon_3, \varepsilon_2)$$

by the standard algebraic transformation Q_a defined by :

$$Q_a : K^2 \longrightarrow K^2$$

$$Q_a(x, y) = (ay + xy, y).$$

Another way to examine this point M is to write :

$$M = (a\varepsilon_2 + \varepsilon_2\varepsilon_3, \varepsilon_2) = \varepsilon_2(a + \varepsilon_3, 1) = \varepsilon_2(a, 1) + \varepsilon_2\varepsilon_3(1, 0)$$

which suggests a standard frame

$$V_1 = (a, 1), V_2 = (1, 0)$$

of the plane entirely determined by the point M. These two tools, the algebraic transformation and the geometrical frame, are standard and perfectly characterized by the point M. The aim of the following sections is to develop, in a general framework, this approach, this disclosure of standard objects defined by an infinitesimal point and to interpret these.

5.2 A decomposition theorem for a limited point

Let M be a limited point of the standard affine plane K^n, $K = \mathbb{R}$ or $\mathbb{C}$, n being standard. It admits an unique shadow M_0 and the vector $M - M_0$ is infinitesimal. We go on to suppose that $M_0 = O$ and that M is an infinitesimal point.

5.2.1 The decomposition theorem

Theorem 5.2.1. *Let M be an infinitesimal point of K^n, $M \neq 0$. There is an integer k, $k \leq n$, and there are infinitesimal scalars $\varepsilon_1, \cdots, \varepsilon_k$ of K, and a standard frame $(V_1, \cdots, V_k)$ of K^n such that*

$$M = \varepsilon_1 V_1 + \varepsilon_1 \varepsilon_2 V_2 + \cdots + \varepsilon_1 \varepsilon_2 \cdots \varepsilon_k V_k$$

Such an expression will be said to be a decomposition of M. This decomposition is not unique. More precisely :

Proposition 5.2.1. *If $M = \varepsilon_1 V_1 + \varepsilon_1 \varepsilon_2 V_2 + \cdots + \varepsilon_1 \varepsilon_2 \cdots \varepsilon_k V_k$ and $M = \varepsilon_1' W_1 + \varepsilon_1' \varepsilon_2' W_2 + \cdots + \varepsilon_1' \varepsilon_2' \cdots \varepsilon_p' W_p$ are two decompositions of M, then :*

1. *$k = p$*
2. *the flag $D = (V_1, (V_1, V_2), \cdots, (V_1, V_2, \cdots, V_k))$ is equal to the flag $D' = (W_1, (W_1, W_2), \cdots, (W_1, W_2, \cdots, W_k))$*
3. *$\varepsilon_1' \cdots \varepsilon_i' = \sum_{j=1}^{j=i} a_{ij} \varepsilon_1 \cdots \varepsilon_j$, a_{ij} is standard and $a_{ii} \neq 0$.*

Proof of the theorem: Let $M \simeq 0$ be in K^n. We put $M = (\alpha_1, \cdots, \alpha_n)$, the scalars α_i being infinitesimals (this is equivalent to fixing from the outset a standard basis of K^n). Let i_0, $1 \leq i_0 \leq n$ be such that

$$|\alpha_{i_0}| \geq |\alpha_i| \quad \forall i, \ 1 \leq i \leq n.$$

If the point M is not zero, the coordinate α_{i_0} is not zero and the point M/α_{i_0} is appreciable. Its coordinates are $(\alpha_1/\alpha_{i_0}, \cdots, 1, \cdots, \alpha_n/\alpha_{i_0})$. Let W_1 be the vector OM/α_{i_0}. Its shadow V_1 is not zero and has as coordinates $(a_1, \cdots, a_{i_0} = 1, \cdots, a_n)$. We put

$$\alpha_{i_0} = \varepsilon_1, \quad \alpha_i/\alpha_{i_0} = a_i + \alpha_i^1, \quad \alpha_i^1 \simeq 0.$$

Then, we write

$$M = \varepsilon_1 W_1 = \varepsilon_1 V_1 + \varepsilon_1 W_1^1, \quad W_1^1 = (\alpha_1^1, \cdots, \alpha_{i_0-1}^1, 0, \alpha_{i_0+1}^1, \cdots, \alpha_n^1) \simeq 0.$$

If $W_1^1 = 0$, the theorem is proved ($k = 1$). If not, one iterates this process, applying it to the point M_1 of K^n defined by $OM_1 = W_1^1$. This point, close to the origin, admits the decomposition

$$M_1 = \varepsilon_2 V_2 + \varepsilon_2 W_2,$$

$$W_2 = (\alpha_1^2, \cdots, \alpha_{i_0-1}^2, 0, \alpha_{i_0+1}^2, \cdots, \alpha_{i_1-1}^2, 0, \alpha_{i_1+1}^2, \cdots, \alpha_n) \simeq 0$$

where i_1 is defined by $|\alpha_{i_1}^2| \geq |\alpha_i^2|$ $i = 1, \cdots, n$.

So $M = \varepsilon_1 V_1 + \varepsilon_1 \varepsilon_2 V_2 + \varepsilon_1 \varepsilon_2 W_2$. The vectors V_1 and V_2 are standard and independent. If the vector W_2 is zero, then $M = \varepsilon_1 V_1 + \varepsilon_1 \varepsilon_2 V_2$ and the point M admits a decomposition of length 2. If W_2 is not zero, we continue the previous process by factorizing the coordinates by a coordinate with greatest modulus . So, at each step, we construct a standard vector V_i independent of the vectors V_j, $j \neq i$ already constructed. The number of steps is bounded by the dimension of the standard space K^n. In the end, we obtain the decomposition stated in the theorem. ☐

Proof of the proposition: Consider two decompositions of the point M :

$$M = \varepsilon_1 V_1 + \varepsilon_1 \varepsilon_2 V_2 + \cdots + \varepsilon_1 \varepsilon_2 \cdots \varepsilon_k V_k$$

$$M = \varepsilon_1' W_1 + \varepsilon_1' \varepsilon_2' W_2 + \cdots + \varepsilon_1' \varepsilon_2' \cdots \varepsilon_p' W_p$$

Suppose $| \varepsilon_1 | \geq | \varepsilon_1' |$ and $\varepsilon_1 \neq 0$; this is equivalent to supposing that the point M is not the origin. Then we have :

$$V_1 + \varepsilon_2 V_2 + \cdots + \varepsilon_2 \cdots \varepsilon_k V_k = \frac{\varepsilon_1'}{\varepsilon_1} W_1 + \cdots + \frac{\varepsilon_1'}{\varepsilon_1} \varepsilon_2' \cdots \varepsilon_p W_p \qquad (5.1)$$

We take the shadow of each side, each one being limited. We obtain :

$$V_1 = {}^{\circ}\left(\frac{\varepsilon_1'}{\varepsilon_1}\right) W_1.$$

Since they are members of a family of independent vectors, V_1 and W_1 are nonzero and so, necessarily,

$${}^{\circ}\left(\frac{\varepsilon_1'}{\varepsilon_1}\right) \neq 0.$$

Put $\varepsilon_1' = a_{11}\varepsilon_1 + \varepsilon_1 \varepsilon_1^{(2)}$ with a_{11} standard and not zero and $\varepsilon_1^{(2)} \simeq 0$. Then we have $V_1 = a_{11} W_1$ and the previous identity is reduced to

$$\varepsilon_2 V_2 + \cdots + \varepsilon_2 \cdots \varepsilon_k V_k = \varepsilon_1^{(2)} W_1 + (a_{11} + \varepsilon_1^{(2)})\varepsilon_2' W_2 + \cdots + (a_{11} + \varepsilon_1^{(2)})\varepsilon_2' \cdots \varepsilon_p' W_p$$

We put $\varepsilon_2^{(2)} = (a_{11} + \varepsilon_1^{(2)})\varepsilon_2'$. The above equation gives :

$$\varepsilon_2 V_2 + \cdots + \varepsilon_2 \cdots \varepsilon_k V_k = \varepsilon_1^{(2)} W_1 + \varepsilon_2^{(2)} W_2 + \cdots + \varepsilon_2^{(2)} \varepsilon_3' \cdots \varepsilon_p W_p. \qquad (5.2)$$

We are lead to compare the three infinitesimal $\varepsilon_2, \varepsilon_1^{(2)}, \varepsilon_2^{(2)}$. If $\frac{\varepsilon_2}{\varepsilon_1^{(2)}}$ and $\frac{\varepsilon_2^{(2)}}{\varepsilon_1^{(2)}}$ are infinitesimals, the vector W_1 is necessarily equal to zero which is contrary to the hypothesis. Likewise, if the ratios $\frac{\varepsilon_2}{\varepsilon_2^{(2)}}$ and $\frac{\varepsilon_1^{(2)}}{\varepsilon_2^{(2)}}$ are infinitesimal, the vector W_2 is zero, and this is absurd. This shows that the scalars $\frac{\varepsilon_1^{(2)}}{\varepsilon_2}$ and $\frac{\varepsilon_2^{(2)}}{\varepsilon_2}$ are limited. We can divide each member of (1.2) by ε_2 and take the shadow of the two new members. We obtain :

$$V_2 = {}^{\circ}\left(\frac{\varepsilon_1^{(2)}}{\varepsilon_2}\right) W_1 + {}^{\circ}\left(\frac{\varepsilon_2^{(2)}}{\varepsilon_2}\right) W_2.$$

Put

$$\varepsilon_1^{(2)} = a_{12}\varepsilon_2 + \varepsilon_2 \varepsilon_1^{(3)}$$

$$\varepsilon_2^{(2)} = a_{22}\varepsilon_2 + \varepsilon_2 \varepsilon_2^{(3)}.$$

Then, we have:

$$V_2 = a_{12}W_1 + a_{22}W_2, \quad a_{22} \neq 0$$

$$\varepsilon_1' = a_{11}\varepsilon_1 + a_{12}\varepsilon_1\varepsilon_2 + \varepsilon_1\varepsilon_2\varepsilon_1^{(3)}$$

$$\varepsilon_1'\varepsilon_2' = a_{22}\varepsilon_1\varepsilon_2 + \varepsilon_1\varepsilon_2^{(2)}\varepsilon_2^{(3)}$$

and the identity (1.2) is reduced to :

$$\varepsilon_3 V_3 + \cdots + \varepsilon_3\varepsilon_4 \cdots \varepsilon_k V_k = \varepsilon_1^{(3)} W_1 + \varepsilon_2^{(3)} W_2 + \varepsilon_3^{(2)} W_3 + \cdots + \varepsilon_3^{(2)}\varepsilon_4' \cdots \varepsilon_p W_p \tag{5.3}$$

with $\varepsilon_3^{(2)} = \frac{\varepsilon_2^{(2)}}{\varepsilon_2}\varepsilon_3'$. At each of the k steps the identity is reduced by comparing infinitesimals. At the end we obtain :

$$V_i = \sum_{j=1}^{i} a_{ji}W_j, \quad a_{ii} \neq 0, \quad i = 1, \cdots, k$$

$$\varepsilon_1' \cdots \varepsilon_i' = \sum_{j=i}^{k} a_{ij}\varepsilon_1 \cdots \varepsilon_i \cdots \varepsilon_j$$

and the equation (5.3) is reduced to

$$0 = \varepsilon_1^{(k+1)} W_1 + \cdots + \varepsilon_k^{(k+1)} W_k + \varepsilon_{k+1}^{(2)} W_{k+1} + \cdots + \varepsilon_{k+1}^{(2)} \cdots \varepsilon_p' W_p \tag{5.4}$$

This implies $p \geq k$. The vectors (W_i) being standard and linearly independent, each infinitesimal scalar appearing in the linear equation (5.4) is zero. So

$$\varepsilon_1^{(k+1)} = \cdots = \varepsilon_k^{(k+1)} = \varepsilon_{k+1}^{(2)} = 0.$$

As ε_{k+1}' has the same order as $\varepsilon_{k+1}^{(2)}$, $\varepsilon_{k+1}' = 0$. Then, $k = p$ and the proposition follows. $\qquad \square$

5.2.2 Geometrical approach

The theorem above reveals a standard flag tied to an infinitesimal point close to the origin; the proposition shows the unicity of this flag. We will attempt, in this section, to construct directly this geometrical object forgetting for the moment the algebraic calculation which has helped to discover it.

We consider the point M of K^n, $M \simeq 0$ and $M \neq 0$. The straight line OM is not standard unless the ratio of the coordinates of M is standard. This line admits a standard shadow Δ_0. Let V_1 be a standard vector in the direction of Δ_0. If the straight line OM were standard, one would have $\Delta_0 = OM$ and the point M would have a decomposition of length 1. In the contrary case, Δ_0 and OM are distinct and they determine a plane Π. If this plane is not standard, its shadow Π_0 is distinct from Π. The frame V_1, V_2 of the decomposition of M is a basis of Π_0. But if the plane Π is standard, then the point M admits a decomposition of length 2. Otherwise, Π and Π_0 generate a vector subspace of dimension 3 of K^n (obviously if $n \geq 3$) and $\Pi \cap \Pi_0 = \Delta_0$. The frame (V_1, V_2, V_3) is a basis of this space of dimension 3.

Proposition 5.2.2. *The length k of the decomposition of the point M corresponds to the dimension of the smallest standard vector subspace of K^n passing through the point M.*

5.2.3 Algebraic approach

Let $M \simeq 0$ in K^n and

$$M = \varepsilon_1 V_1 + \varepsilon_1 \varepsilon_2 V_2 + \cdots + \varepsilon_1 \varepsilon_2 \cdots \varepsilon_k V_k$$

be a decomposition of this point. We will suppose in this section that $k = n$; in fact this corresponds to the generic case since, in general, the infinitesimal points are not in a proper standard subspace of K^n.

This decomposition permits one to display and to describe a natural transformation of K^n to itself by the correspondence :

$$(\varepsilon_1, \cdots, \varepsilon_n) \longrightarrow M = \varepsilon_1 V_1 + \cdots + \varepsilon_1 \varepsilon_2 \cdots \varepsilon_n V_n.$$

We consider the transformation $\Phi_1 : K^n \longrightarrow K^n$ defined by:

$$\Phi_1(X_1, X_2, \cdots, X_n) = X_1 V_1 + X_1 X_2 V_2 + \cdots + X_1 X_2 \cdots X_n V_n.$$

It is a standard birational transformation of K^n completely determined by the standard point M. Under it the point M corresponds to an unique infinitesimal point M_1 of K^n (recall the hypothesis on the length of M) whose coordinates relative to the canonical basis of K^n, are $\varepsilon_1, \varepsilon_2, \cdots, \varepsilon_n$. This point M_1 being infinitesimal and certainly not zero admits a decomposition of length $k \leq n$. As before, we suppose that this length is maximal and equal to n. The decomposition can be written:

$$M_1 = \alpha_1 W_1 + \alpha_1 \alpha_2 W_2 + \cdots + \alpha_1 \alpha_2 \cdots \alpha_n W_n.$$

We associate with it the standard birational transformation Φ_2 defined by:

$$\Phi_2(X_1, X_2, \cdots, X_n) = X_1 W_1 + X_1 X_2 W_2 + \cdots + X_1 X_2 \cdots X_n W_n$$

which satisfies $\Phi_2(M_2) = M_1$ where $M_2 = (\alpha_1, \alpha_2, \cdots, \alpha_n)$.

Proposition 5.2.3. *The point M defines a sequence (Φ_p) of standard birational transformation of K^n such that $\Phi_p(M_p) = M_{p-1}$, $\Phi_{p-1}(M_{p-1}) = M_{p-2}$, ..., $\Phi_1(M_1) = M$ and such that each point M_i is infinitesimal.*

Remark 5.2.1. If one of the points M_i admits a decomposition of length strictly less than n, the algebraic transformation Φ_{i+1}, restricted to K^p, is not bijective. It is an algebraic transformation of K^p into K^n where p is the length of M_i. But the sequence of algebraic transformations associated to the point M still exists. The simpler and more typical case is probably the case corresponding to the point $M = (\varepsilon, \varepsilon)$. It is an infinitesimal point of K^2. It has a length equal to 1:

$$M = \varepsilon V_1 \quad , \quad V_1 = (1,1)$$

and the associated transformation is $\Phi_1(X) = (X, X)$. We can see that this point is situated on a standard straight line, and this is not a generic situation.

5.3 Infinitesimal riemannian geometry: the case of curves in $\mathbb{R}^3$

5.3.1 Orthonormal decomposition of a point

We restrict ourselves in this section to the case where the points are in the standard space $\mathbb{R}^3$, our goal being to approach the riemannian or euclidean geometry of differentiable curves of $\mathbb{R}^3$.

Theorem 5.3.1. *Let M be an infinitesimal point in $\mathbb{R}^3$, $M \neq 0$. There is an unique orthonormal decomposition of M oriented by the vector $\mathbf{OM}$.*

The decomposition

$$M = \varepsilon_1 V_1 + \varepsilon_1 \varepsilon_2 V_2 + \varepsilon_1 \varepsilon_2 \varepsilon_3 V_3$$

is said to be orthonormal whenever the basis (V_1, V_2, V_3) is orthonormal. Even when the length of M is less than 3, we still speak of an orthonormal decomposition whenever the frame defined by the decomposition is orthonormal.

Proof. Let $M = \varepsilon_1 V_1 + \varepsilon_1 \varepsilon_2 V_2 + \varepsilon_1 \varepsilon_2 \varepsilon_3 V_3$ be a decomposition of M. Another decomposition $M = \varepsilon_1' W_1 + \varepsilon_1' \varepsilon_2' W_2 + \varepsilon_1' \varepsilon_2' \varepsilon_3' W_3$ satisfies:

$$W_i = a_{ii} V_i + \sum_{j<i} a_{ij} V_j \quad , \quad a_{ii} \neq 0.$$

In particular, one has $W_1 = a_{11} V_1$. We take $a_{11} = \frac{\pm 1}{\|V_1\|}$. The vector W_1 has then length equal to 1. One chooses the sign of a_{11} so as to orientate the vector W_1 in the direction of the vector $\mathbf{OM}$. The vector W_1 then is defined uniquely. In the plane (V_1, V_2), we choose W_2 so as to make the orientation of the frame (W_1, W_2) positive. As for the vector W_3, it is defined so as to orthonormalize and orientate the frame (W_1, W_2, W_3) positively.

5.3.2 The Serret-Frenet frame of a differentiable curve in $\mathbb{R}^3$

Let C a standard parametrized curve of $\mathbb{R}^3$ given by

$$C(t) = (x(t), y(t), z(t))$$

where $x(t), y(t), z(t)$ are functions of class C^k, $k \geq 3$. For simplicity, suppose that $C(0) = 0$.

Theorem 5.3.2. *Every point $M \simeq 0$, $M \neq 0$ of the curve C having a positive (and infinitesimal) curvilinear abscissa defines the same frame: the Serret-Frenet frame. A point whose abscissa is negative defines the frame whose orientation is opposite to that of the Serret-Frenet frame.*

Recall that the curvilinear abscissa is given by the arc length calculated from the origin.

Remark 5.3.1. The fact of finding both the Serret-Frenet frame and its opposite is not surprising. Classically, the curvilinear abscissa is defined up to sign: it is only the square of this abscissa which is uniquely defined.

Proof. The proof is based on the geometrical characterization of the Serret-Frenet frame: the first vector W_1 is the unit tangent vector oriented in the direction of the increasing abscissa, the frame (W_1, W_2) is a positively oriented orthonormal basis of the osculating plane. But this plane is defined by the limit of planes turning around the tangent W_1 and containing the straight line OM, as the point M tends to O. It corresponds actually to a plane defined by the decomposition of M. As to the third vector of the Serret-Frenet frame, it is equal to the inner product of the first two vectors. It is the vector W_3.

Remark 5.3.2. Suppose that there exists a point $M \simeq 0$, $M \neq 0$ and $M \in C$ whose decomposition has a length equal to 2. It is contained in a standard vector plane of dimension 2. In this case, the differentiable curve C is locally plane.

5.3.3 The curvature and the torsion

Consider the Taylor expansion of the function $C(s)$, parametrized by the arc length s, in the neighbourhood to the origin:

$$C(s) = C(0) + sC'(0) + \frac{s^2}{2!}C^{(2)}(0) + \frac{s^3}{3!}(C^{(3)}(0) + \varepsilon(s)).$$

If (W_1, W_2, W_3) is the Serret-Frenet frame, we have:

$$C'(0) = W_1$$

$$C^{(2)}(0) = \rho(0)W_2$$

$$C^{(3)}(0) = \rho'(0)W_2 - \rho^2(0)W_1 + \rho(0)\tau(0)W_3$$

where ρ and τ are, respectively, the curvature and the torsion. In order to emphasize the link with the formulae of the classical mechanics, it would have been preferable to denote:

$$W_1 = \boldsymbol{T}(0)$$

$$W_2 = \boldsymbol{N}(0)$$

$$W_3 = B(0).$$

but we prefer to retain our notation. The Taylor formula can be written as :

$$C(s)) = sW_1 \left(1 - \frac{s^2}{3!}\rho^2(0) + \frac{s^2}{3!}\alpha_1(s)\right) + \frac{s^2}{2!}W_2 \left(\rho(0) + \frac{s}{3}\rho'(0) + \frac{s}{3}\alpha_2(s)\right) +$$

$$+ \frac{s^3}{3!}W_3(\rho(0)\tau(0) + \alpha_3(s))$$

where $(\alpha_1(s), \alpha_2(s), \alpha_3(s))$ are the components of the vector $\varepsilon(s)$ with respect to the frame (W_1, W_2, W_3).

Suppose $M = C(s)$. If M is supposed infinitesimal, its abscissa s is infinitesimal. We put $\varepsilon = s$ and we obtain

$$M = \varepsilon W_1(1 - \frac{\varepsilon^2}{3!}\rho^2(0) + \frac{\varepsilon^2}{3!}\alpha_1(\varepsilon)) + \frac{\varepsilon^2}{2!}W_2(\rho(0) + \frac{\varepsilon}{3}\rho'(0) + \frac{\varepsilon}{3}\alpha_2(\varepsilon))$$

$$+ \frac{\varepsilon^3}{3!}W_3(\rho(0)\tau(0) + \alpha_3(\varepsilon)).$$

By comparing this with $M = \varepsilon_1 W_1 + \varepsilon_1\varepsilon_2 W_2 + \varepsilon_1\varepsilon_2\varepsilon_3 W_3$, we obtain

$$\frac{\varepsilon}{\varepsilon_1} = 1 + \varepsilon_1', \quad \varepsilon_1' \simeq 0$$

$$\frac{\varepsilon^2}{2!}\rho(0) = \varepsilon_1\varepsilon_2 + \varepsilon_1\varepsilon_2\varepsilon_2'.$$

Thus, $(\varepsilon_1 + \varepsilon_1\varepsilon_1')^2\rho(0)/2! = \varepsilon_1\varepsilon_2 + \varepsilon_1\varepsilon_2\varepsilon_2'$, and $\varepsilon_1^2\rho(0)/2! = \varepsilon_1\varepsilon_2 + \varepsilon_1\varepsilon_2\varphi$, where φ is an infinitesimal.

As the curvature $\rho(O)$ is standard, we obtain

$$\boxed{\rho(0) = 2^0\left(\frac{\varepsilon_2}{\varepsilon_1}\right).}$$

Likewise, $(\varepsilon^3/3!)\rho(0)\tau(0) = \varepsilon_1\varepsilon_2\varepsilon_3 + \varepsilon_1\varepsilon_2\varepsilon_3\varphi$, so that

$$\rho(0)\tau(0) = \frac{3!\varepsilon_1\varepsilon_2\varepsilon_3}{\varepsilon^3} + \frac{3!\varepsilon_1\varepsilon_2\varepsilon_3\varphi}{\varepsilon^3}$$

which implies that

$$\tau(0) =^0 \left(\frac{3!\varepsilon_1\varepsilon_2\varepsilon_3\varepsilon^2}{2!\varepsilon^3\varepsilon_1\varepsilon_2}\right).$$

Thus we obtain for the torsion

$$\boxed{\tau(0) = 3^0(\varepsilon_3/\varepsilon_1).}$$

Theorem 5.3.3. *Let $M \in C(s)$ be a point close to $0 = c(0)$ whose orthonormal decomposition is:*

$$M = \varepsilon_1 W_1 + \varepsilon_1\varepsilon_2 W_2 + \varepsilon_1\varepsilon_2\varepsilon_3 W_3$$

Then the metric invariants of the curve at the origin are given by

$$\rho(0) = \overset{0}{2}(\varepsilon_2/\varepsilon_1) \quad \tau(0) = \overset{0}{3}(\varepsilon_3/\varepsilon_1).$$

Remarks 5.3.3. *1. If M is infinitely close to a standard point M_0 located on the curve (C), the decomposition of M is $\varepsilon_1 W_1 + \varepsilon_1\varepsilon_2 W_2 + \varepsilon_1\varepsilon_2\varepsilon_3 W_3$, and (W_1, W_2, W_3) is the Serret-Frenet frame at the point M_0 and we also have:*

$$\rho(M_0) = \overset{0}{2}(\varepsilon_2/\varepsilon_1) \quad \tau(M_0) = \overset{0}{3}(\varepsilon_3/\varepsilon_1).$$

2. The well-known Serret-Frenet formulae, which describe the variation of the moving frame, are:

$$\begin{pmatrix} dT/ds \\ dN/ds \\ dB/ds \end{pmatrix} = \begin{pmatrix} 0 & \rho & 0 \\ -\rho & 0 & \tau \\ 0 & -\tau & 0 \end{pmatrix} \begin{pmatrix} T \\ N \\ B \end{pmatrix}$$

Thus there appears an antisymmetric matrix whose components are the curvature and the torsion. This matrix is the particular case, corresponding to dimension 3, of the curvature tensor on a riemannian manifold. Using the decomposition of a point, we can also display the curvature matrix.

If we know the Serret-Frenet frame in each standard point of the standard curve (C), we deduce, using the transfer principle, the expression for this frame at all points on (C). If A is the matrix corresponding to the frame, this matrix is orthogonal. By comparing A with its shadow, we obtain, as a first approximation, an antisymmetric matrix corresponding to the matrix of curvatures.

5.4 The theory of moving frames: the case of surfaces in $\mathbb{R}^3$

5.4.1 The theory of moving frames

This theory was elaborated by Elie Cartan in order to have an efficacious method for the study of submanifolds of homogeneous spaces. The key to the theory is based on the fact that the set of frames is a Lie group G and this group operates naturally on the homogeneous manifold G/H (H being the isotropic group). The structural equations, or Maurer Cartan equations, of G give the variations of the moving frame. But the particular case of plane curves, illustrates the interest of the equations of variation: here the moving frame is the Serret Frenet frame and the famous equations of the same name, which describe the evolution of this frame, give the metric invariants of the curve, that is, the curvature and the torsion. When we know that these are the only metric invariants, we understand better the aim of the Cartan theory , namely, to describe the invariants of the submanifolds of an homogeneous space.

Throughout this section, the basic field will be the field of reals (recall the importance of this field: a nondegenerate holomorphic curve in $P^n C$ is determined, up to a transformation, by its first fundamental form).

The Lie group of the transformations of $\mathbb{R}^n$. Let $E(n)$ be the Lie group of transformations of $\mathbb{R}^n$. It is defined by the following matrices:

$$\begin{pmatrix} a_{11} & \cdots & a_{1n} & x_1 \\ \vdots & \vdots & \vdots & \vdots \\ a_{n1} & \cdots & a_{nn} & x_n \\ 0 & \cdots & 0 & 1 \end{pmatrix}$$

where the matrix $A = (a_{ij})$ is in $SO(n)$. Let $(X_1, \cdots, X_n)$ be a moving frame in $\mathbb{R}^n$. We may consider each vector field X_i as a mapping of $\mathbb{R}^n$ in $\mathbb{R}^n$. Let θ^i be the dual forms of X_i: $\theta^i(X_j) = \delta^i_j$. Then, by putting

$$dX_i = \sum_{i=1}^n \omega^i_j X_j$$

we have:

$$d\theta^i = -\sum \omega^i_k \theta^k$$

$$d\omega^i_j = -\sum \omega^i_k \wedge \omega^k_j.$$

These last equations are the Maurer-Cartan equations of the Lie group of motions $E(n)$. Note that given a matrix of Pfaffian forms (ω^i_j) of $\mathbb{R}^n$ such that

$$d\omega^i_j = -\sum \omega^i_k \wedge \omega^k_j$$

one is assured of the existence, in a neighbourhood $V(0)$ of the origin, of functions A^j_i satisfying $dA^j_i = \sum_k A^k_i \omega^k_j$, and also of the existence in $V(0)$ of a moving frame $(X_1, \cdots, X_n)$ such that the dual forms θ^i satisfy:

$$d\theta^l = -\sum_k \omega^i_k \wedge \omega^k.$$

If the frame (X_i) is orthonormal the matrix (ω^j_i) is antisymmetric. The structural equations so obtained are the equations of the group of the motions of the homogeneous space $\mathbb{R}^n = E(n)/SO(n)$.

The case of a manifold. Let $(X_1, \cdots, X_n)$ be a moving frame and $(\theta_1, \cdots, \theta_n)$ be the dual forms. We can find Pfaffian forms ω_j^i such that

$$\begin{cases} \omega_j^i = -\omega_i^j \\ d\theta^i = \sum_k \theta^k \wedge \omega_k^i \end{cases}$$

but, contrary to the euclidean case where the structural equations are $d\omega_j^i = -\sum \omega_k^i \wedge \omega_j^k$, there now appear some residual terms:

$$d\omega_j^i = -\sum \omega_k^i \wedge \omega_j^k + \Omega_j^i.$$

The 2-forms Ω_j^i are the curvature forms, and the forms (ω_j^i) are the forms of the Levi-Civita connection. The riemannian invariants can be read from the structural equations:

$$\begin{cases} \omega_j^i = \sum_k \Gamma_{kj}^i \theta^k \\ \Omega_j^i = \frac{1}{2} \sum_{k,l} R_{jkl}^i \theta^k \wedge \theta^l \end{cases}$$

which specialists will recognize.

5.4.2 The moving frame: an infinitesimal approach

We have seen, in the case of differentiable curves, that every point infinitely close to a standard point and located on the curve determines, up to orientation, only one frame, the Serret-Frenet frame, which corresponds to the moving frame of Elie Cartan. What do we observe in manifolds of greater dimension, in particular on surfaces ? We are given a standard point on a standard differentiable surface. We consider "the set" of the points infinitely close to this standard point and located on the surface. What are the standard properties related to the surface and determined by this set ? The decomposition of all these points gives a set (in the sense of Set Theory) of frames centred at the standard point of the given surface. It is this set of frames which generalizes to the surface case, the notion of Serret-Frenet frames defined by "the set" of the points infinitely close to a standard point on a standard curve. We propose to describe this object attached to the surface and defined by all the points close to a standard point, and to understand its role in the determination of the riemannian invariants.

Let (S) be a standard surface embedded in $\mathbb{R}^3$ defined by the parametric equations :

$$\begin{cases} x = x(u,v) \\ y = y(u,v) \\ z = z(u,v) \end{cases}$$

where x, y, z are C^∞ functions (or of class at least C^3. Suppose that $x(0,0) = 0, y(0,0) = 0, z(0,0) = 0$ and denote by $h(0)$ the halo of the origin in $\mathbb{R}^3$. We denote also by $h(0) \cap S$ the external set of those points of (S) infinitely close to the origin. We shall examine the standard set defined by the orthonormal frames corresponding to the points of $h(0) \cap S$. If necessary using a change of coordinates, we may suppose that the surface is defined, in a neighbourhood of the origin, by an equation having the following form:

$$z = f(x, y).$$

Let $M \in h(0) \cap S$ be and $M = \varepsilon_1 W_1 + \varepsilon_1 \varepsilon_2 W_2 + \varepsilon_1 \varepsilon_2 \varepsilon_3 W_3$ be its orthonormal decomposition. If the components of the vector W_i in the canonical basis are (a_i^1, a_i^2, a_i^3), then the fact that M is on S implies that:

$$\varepsilon_1 a_1^3 + \varepsilon_1 \varepsilon_2 a_2^3 + \varepsilon_1 \varepsilon_2 \varepsilon_3 a_3^3 = f(\varepsilon_1 a_1^1 + \varepsilon_1 \varepsilon_2 a_2^1 + \varepsilon_1 \varepsilon_2 \varepsilon_3 a_3^1, \varepsilon_1 a_1^2 + \varepsilon_1 \varepsilon_2 a_2^2 + \varepsilon_1 \varepsilon_2 \varepsilon_3 a_3^2).$$

By supposing that f is a differentiable mapping of class at least 3, we have the Taylor expansion:

$$f(x, y) = x f_x'(0) + y f_y'(0) + x^2 f_{x^2}^{(2)}(0) + 2xy f_{xy}^{(2)}(0) + y^2 f_{y^2}^{(2)}(0) + \cdots$$

so that $\varepsilon_1 a_1^3 = \varepsilon_1 a_1^1 A + \varepsilon_1 a_1^2 B$, with $A = f_x'(0)$, $B = f_y'(0)$, which implies that $a_1^3 = A a_1^1 + B a_1^2$, and that

$$\varepsilon_1 \varepsilon_2 a_2^3 + \varepsilon_1 \varepsilon_2 \varepsilon_3 a_3^3 = (\varepsilon_1 \varepsilon_2 a_2^1 + \varepsilon_1 \varepsilon_2 \varepsilon_3 a_3^1) A + (\varepsilon_1 \varepsilon_2 a_2^2 + \varepsilon_1 \varepsilon_2 \varepsilon_3 a_3^2) B$$

$$+ (\varepsilon_1 a_1^1 + \varepsilon_1 \varepsilon_2 a_2^1 + \varepsilon_1 \varepsilon_2 \varepsilon_3 a_3^1)^2 p + 2\varepsilon_1^2 (a_1^1 + \varepsilon_2 a_2^1 + \varepsilon_2 \varepsilon_3 a_3^1)(a_1^2 + \varepsilon_2 a_2^2 + \varepsilon_2 \varepsilon_3 a_3^2) r$$

$$+ \varepsilon_1^2 (a_1^2 + \varepsilon_2 a_2^2 \varepsilon \varepsilon_2 \varepsilon_3 a_3^2)^2 q + \cdots$$

where $p = f_{x^2}^{(2)}(0)$, $q = f_{y^2}^{(2)}(0)$, $r = f_{xy}^{(2)}(0)$. Thus $W_1 = (a_1^1, a_1^2, A a_1^1 + B a_1^2)$, with

$$(a_1^1)^2 + (a_1^2)^2 + (A a_1^1 + B a_1^2)^2 = (a_1^1)^2 (1 + A^2) + (a_1^2)^2 (1 + B^2) + 2 A B a_1^1 a_1^2 = 1$$

Proposition 5.4.1. *Let W^1 be a vector satisfying the preceding relation. Then there are an orthonormal frame (W_1, W_2, W_3) with positive orientation and a point M in $h(0) \cap S$ whose decomposition is associated with this frame.*

In fact, the preceding conditions involve the infinitesimals ε_i and don't establish restrictions on the standard coefficients (a_i^j), except in special cases corresponding to particular structures of S; for example, if S is flat or locally flat. But even in such a situation, this curve shall yield some information concerning the ε_i. So, if S is flat, we will find $\varepsilon_3 = 0$ and the coefficients a_i^j are arbitrary.

Example 5.4.1. We consider the surface S whose equation is $z = x^2 + y^2$. We have $W_1 = (a_1^1, a_1^2, 0)$ and the relation concerning the infinitesimals is:

$$\varepsilon_1 \varepsilon_2 a_2^3 + \varepsilon_1 \varepsilon_2 \varepsilon_3 a_3^3 = \varepsilon_1^2 (a_1^1 + \varepsilon_2 a_2^1 + \varepsilon_2 \varepsilon_3 a_3^1)^2 + \varepsilon_1^2 (a_1^2 + \varepsilon_2 a_2^2 + \varepsilon_2 \varepsilon_3 a_3^2)^2$$

$$= \varepsilon_1^2 ((a_1^1)^2 + (a_1^2)^2) + \varepsilon_1^2 (2 \varepsilon_2 (a_1^1 a_2^1 + a_1^2 a_2^2) + 2 \varepsilon_2 \varepsilon_3 (a_1^1 a_3^1 + a_1^2 a_3^2))$$

$$+\varepsilon_2^2((a_1^2)^2+(a_2^2)^2)+2\varepsilon_2^2\varepsilon_3(a_2^1a_3^1+a_2^2a_3^2)+\varepsilon_2^2\varepsilon_3^2((a_3^1)^2+(a_3^2)^2)$$

Then:

$$\varepsilon_2(a_2^3+\varepsilon_3a_3^3)=\varepsilon_1+\varepsilon_1\varepsilon_2^2((a_2^1)^2+(a_2^2)^2+2\varepsilon_3(a_2^1a_3^1+a_2^2a_3^2)+\varepsilon_3^2((a_1^3)^2+(a_3^2)^2))$$

This relation permits the expression of ε_1 in terms of ε_2 and ε_3 and, thus, allows one to define a parametrization of S.

Proposition 5.4.2. *Let $F(0)$ be the set of the standard frames defined by the points of $h(0)\cap S$. It is a standard set provided with a structure of a submanifold of $SO(3)$ and of dimension equal to 2.*

Proof. One identifies the set of positively oriented orthonormal frames with the Lie group $SO(3)$ of orthogonal matrices having a determinant equal to 1. It is a 3-dimensional Lie group. The elements of $h(0)\cap S$ define frames corresponding to the matrices of $SO(3)$ satisfying some standard linear condition.

5.4.3 The Serret-Frenet fibre bundle

To each standard point M_0 of the standard surface (S), one associates the standard manifold $F(M_0)$. So one defines a differentiable manifold $F(S)$, fibred on (S):

$$F(S)\overset{\pi}{\longrightarrow}S$$

the fibre on M_0 of S being the manifold $F(M_0)$. A similar construction has been carried out in the case of curves, but in this case, the manifold $F(M_0)$ was reduced to one element only and the fibre bundle $F(S)$ was seen to be of dimension 1. The variation of the moving frame can be observed by the variations of $F(M_0)$ in $F(S)$.

The case of a surface is analogous. The riemannian invariants of the surface may be read from the variation of the fibre $F(M_0)$ (in this case the fibres have dimension 2) in the manifold $F(S)$. Note that the Lie group $Gl(3,\mathbb{R})$ acts on $F(S)$, but that the fibre bundle is not principal, the fibre being a 2-dimensional submanifold of $SO(3)$.

5.5 Infinitesimal linear algebra

5.5.1 Nonstandard vector spaces

Let E be a vector subspace of $\mathbb{R}^n$ with n standard. This subspace is not assumed to be standard. Let $^{\circ}E$ be the standard subset of $\mathbb{R}^n$ whose standard elements are the shadows of the limited vectors of E.

Theorem 5.5.1. *If E is a vector subspace of dimension k of $\mathbb{R}^n$, then $^{\circ}E$ is a standard vector subspace of $\mathbb{R}^n$ of the same dimension as E.*

Proof. The continuity of the addition of vectors and multiplication by a scalar implies that $\overset{\circ}{E}$ is a vector subspace of $\mathbb{R}^n$. We shall show that E and its shadow have the same dimension. Let $(e_1, \cdots, e_k)$ be a basis of E. Consider a standard inner product $\prec, \succ$ on $\mathbb{R}^n$. The Gram-Schmidt orthonormalization process permits one to construct an orthonormal frame $(f_1, \cdots, f_k)$ of E. These vectors are limited. Their shadows $(\overset{\circ}{f_1}, \cdots, \overset{\circ}{f_n})$ satisfy

$$\prec \overset{\circ}{f_i}, \overset{\circ}{f_j} \succ = \overset{\circ}{} \prec f_i, f_j \succ = \delta_{ij}.$$

They are orthonormal and linearly independent. So $\dim \overset{\circ}{E} \geq \dim E$. If $(g_1, \cdots, g_p)$ is a basis of $\overset{\circ}{E}$, there are limited vectors $h_1, \cdots, h_p$ of E such that $\overset{\circ}{h_i} = g_i$. The linear independence of the vectors g_i implies the linear independence of the vectors h_i. So $\dim E \geq \dim \overset{\circ}{E}$ and the theorem is proved. $\square$

Corollary 5.5.1. *Let $G(k,n)$ be the Grassmannian of the k-planes of $\mathbb{R}^n$. This manifold is compact.*

In fact, every element of $G(k,n)$ is a k-dimensional subspace of $\mathbb{R}^n$. It admits a shadow which is, from the preceding theorem, an element of $G(k,n)$. This shows that the space is compact.

5.5.2 Perturbation of linear operators

Let V be a standard vector space of standard dimension n on K, $K = \mathbb{R}$ or $\mathbb{C}$. Let f_0 be a standard linear operator on V (by linear operator one means an endomorphism of V).

Definition 5.5.1. *A perturbation f of f_0 is a linear operator of V satisfying $f(e_i) \simeq f_0(e_i)$, where $(e_1, \cdots, e_n)$ is a standard basis of V.*

Remarks 5.5.1. *As V is a standard vector space, it admits, from the transfer principle, a standard basis. The matrix of f relative to the standard basis (e_i) has coefficients infinitely close to the corresponding coefficients of the matrix of f_0.*

By continuity, one can deduce that $f(X) \simeq f_0(X)$ for every limited vector X of V.

Proposition 5.5.1. *Let f be a perturbation of f_0. The eigenvalues of f are limited and their shadows are the eigenvalues of f_0.*

This is a consequence of the following theorem:

Theorem 5.5.2. *Let $P_0(X)$ be a standard polynomial in $\mathbb{C}^n[X]$ of degree n with complex coefficients, n being supposed standard. Let $P(X)$ be a member of $\mathbb{C}^n[X]$ and such that $P(X) \simeq P_0(X)$. Then the roots of $P(X)$ are limited and their shadows are the roots of $P_0(X)$.*

(The proof is given in [65].)

Let us return to the operators f and f_0. Their characteristic polynomials satisfy $C_\lambda(f) \simeq C_\lambda(f_0)$, with $C_\lambda(f) = det(f - \lambda Id)$. The eigenvalues $\lambda_1, \cdots, \lambda_p$ of f are thus limited and their shadows are clearly the eigenvalues of f_0.

Proposition 5.5.2. *If $\lambda_1, \cdots, \lambda_k$ are the eigenvalues of f having the same shadow ρ, then*

$$^\circ(E_{\lambda_i}) \subseteq E_\rho^o \quad 1 \le i \le k$$

where E_{λ_i} denotes the eigenspace of f associated with the eigenvalue λ_i and E_ρ^o the eigenspace of f_0 associated with ρ.

Proof. If v_i is a limited eigenvector associated with λ_i, then $^o v_i$ is an eigenvector of f_0 associated with ρ. We deduce that

$$^\circ E_{\lambda_1} \oplus {}^\circ E_{\lambda_2} \oplus \cdots \oplus {}^\circ E_{\lambda_k} \subseteq E_\rho^o.$$

However, we will not compare, for this moment, the spaces $^\circ(E_{\lambda_1} \oplus \cdots \oplus E_{\lambda_k})$ with E_ρ^o. This is related to the fact that if E and F are subspaces of $\mathbb{C}^n$,

$$^\circ E \oplus {}^\circ F \subset {}^\circ(E \oplus F)$$

and this inequality may be strict. We will give, at the end, a proof of the Jordan reduction theorem which closes this gap.

Now, let $F_{\lambda_i} = Ker(f - \lambda_i I)^{r_i}$ be where r_i is the multiplicity of λ_i. To distinguish the eigenspaces of f and of f_0, we will denote by F_ρ^0 the eigenspace of f_0 associated with the eigenvalue ρ.

Theorem 5.5.3. *If $\lambda_1, \cdots, \lambda_k$ are the eigenvalues of f having the same shadow ρ, then*

$$^\circ F_{\lambda_1} \oplus \cdots \oplus {}^\circ F_{\lambda_k} = F_\rho^0.$$

Proof. Let $F = F_{\lambda_1} \oplus \cdots \oplus F_{\lambda_k}$. Then $^\circ F \subset F_\rho^0$. In fact, if v is a limited vector of F_{λ_i}, then

$$(f - \lambda_i I)^{r_i}(v) = 0 \Longrightarrow {}^\circ [(f - \lambda_i I)^{r_i}(v)] = 0$$

so $(f^0 - \rho I)^r ({}^\circ v) = 0$ and $^\circ v \in F_\rho^0$. Here r denotes the multiplicity of ρ and, as the referential field is $\mathbb{C}$, one has $r = \sum_{i=1}^k r_i$. By comparing the dimensions of F, $^\circ F$ and F_ρ^0, one may deduce the theorem.

5.5.3 The Jordan reduction of a complex linear operator

Let $T_0 : \mathbb{C}^n \longrightarrow \mathbb{C}^n$ be a linear operator. As the Jordan reduction of the matrix of this operator concerns essentially the nilpotent part of T_0, we suppose it nilpotent. Then there is an integer k, $k \leq n$, such that $T_0^k = 0$. Recall that a Jordan block of order s is a s-square matrix of the form:

$$J_s = \begin{pmatrix} 0 & 1 & 0 & \cdots & 0 \\ 0 & 0 & 1 & 0 & 0 \\ \vdots & \vdots & \vdots & \vdots & \vdots \\ 0 & 0 & 0 & 0 & 1 \\ 0 & 0 & 0 & 0 & 0 \end{pmatrix}$$

Theorem 5.5.4. *Every nilpotent operator on $\mathbb{C}^n$ admits a matrix of the form*

$$\begin{pmatrix} J_{s_1} & 0 & 0 \\ \vdots & \vdots & \vdots \\ 0 & 0 & J_{s_p} \end{pmatrix}$$

where J_{s_i} is a Jordan block of order s_i.

Proof. Suppose T_0 nilpotent. The eigenvalues are zero. There is a perturbation T of T_0 whose eigenvalues $\varepsilon_1, \cdots, \varepsilon_n$ are infinitesimal and pairwise distinct. Let $v_1, \cdots, v_n$ be a base of appreciable eigenvectors of T corresponding to the eigenvalues $\varepsilon_1, \cdots, \varepsilon_n$ and let $u_1, \cdots, u_p$ be the linearly independent eigenvectors of T_0, with $p = \dim Ker T_0$ and $s \neq n$. As each vector $\overset{o}{v_i}$ is an eigenvector of T_0, we can suppose that these vectors v_i are such that:

$v_1, \cdots, v_{s_1}$ have as shadow u_1

$v_{s_1+1}, \cdots, v_{s_2}$ have as shadow u_2

$\cdots\cdots$

$v_{s_{p-1}+1}, \cdots, v_{s_p}$ have as shadow u_p.

Let E_i be the linear space generated by the vectors $v_{s_{i-1}+1}, \cdots, v_{s_i}$. Its dimension is equal to s_i. Its shadow, which has the same dimension, contains, as eigenvectors of T_0, only the multiples of u_i. We have

Proposition 5.5.3. *The linear space $\overset{o}{E_i}$ is invariant under T_0.*

Proof. Indeed E_i is invariant for T.

Let us return to the proof of the theorem. As each $\overset{o}{E_i}$ is invariant, we consider only the restriction of T_0 to these linear subspaces. The dimension of $\overset{o}{E_i}$ corresponds to the length of the ith Jordan block. To simplify the notation, we consider only the first Jordan block ($i = 1$). Let $E_{1,j}$ be the linear subspace of E_1 generated by the vectors $v_1, \cdots, v_j$. We have the filtration,

$$E_{1,1} \subset E_{1,2} \subset \cdots \subset E_{1,s_1} = E_1$$

and we deduce the filtration,

$$\overset{\circ}{E}_{1,1} \subset \overset{\circ}{E}_{1,2} \subset \cdots \subset \overset{\circ}{E}_{1,s_1}.$$

We consider a basis $(w_1 = u_1, \cdots, w_{s_1})$ of $\overset{\circ}{E}_{s_1}$ adapted to this filtration. We verify that:

$$T_0(w_1) = 0$$

$$T_0(w_2) = T_0(^{\circ}(av_1 + bv_2)) \simeq a\varepsilon_1 v_1 + b\varepsilon_2 v_2.$$

As this last vector is in $E_{1,2}$, $T_0(w_2) = a'w_1 + b'w_2$, a simple change of basis permits one to put:

$$T_0(w_2) = w_1.$$

We continue this process for the successive elements of the basis W_i. This gives the Jordan basis sought.

Application Consider a linear problem, or a problem giving a linear situation shaped by a nondiagonalizable matrix. Instead of working with the Jordanizable matrix, we perturb the problem so as to have a diagonalizable model, whose shadow defines a Jordan basis as above.

The reader may find further readings in [63, 66, 64, 65].

6. General topology

Tewfik Sari

Let X be a set. Let $\mathcal{O}$ be a set of subsets of X having the following properties: (i) $\emptyset$ and X are in $\mathcal{O}$, (ii) the intersection of any two elements of $\mathcal{O}$ is in $\mathcal{O}$ and (iii) the union of any family of elements of $\mathcal{O}$ is in $\mathcal{O}$. We say that $\mathcal{O}$ defines a *topology* on X and we call the pair $(X, \mathcal{O})$ a *topological space* and the sets of $\mathcal{O}$ *open sets*. For brevity we often denote the topological space by X and the topology by τ. The reader is expected to be familiar with the elements of the theory of topological spaces.

The main purpose of this chapter is to give (Section 6.3) an external definition of a topology and to show that it may be useful when the usual definition and study of a topology is complicated. This is, in particular, the case for the topologies on the power set of a topological space (Section 6.4) and for the study of the semicontinuity properties of set-valued mappings and the Painlevé-Kuratowsky convergence of families of subsets in a topological space (Section 6.5). The uniform structures will be described in the same way (Section 6.6). In Sections 6.1 and 6.2 we recall the notion of a halo in topological spaces and show how to use it in the discussion of the usual topological notions.

6.1 Halos in topological spaces

6.1.1 Topological proximity

Let X be a standard topological space. Let x and y be points in X. We write $y \simeq x$, which is read as 'y is *infinitely close* to x', to mean that y is in any standard open set which contains x. The relation $y \simeq x$ may be considered as an abbreviation of the *external formula* $\forall^{st} U \in \mathcal{O}(x \in U \Rightarrow y \in U)$, so we write:

$$y \simeq x \Leftrightarrow \forall^{st} U \in \mathcal{O}(x \in U \Rightarrow y \in U)$$

We call this external relation on X, *topological proximity* on X: it is a reflexive and transitive relation on X. In general, topological proximity is not a symmetric relation. For example, for the usual topology on R, we have $\varepsilon \simeq 0$ for any infinitesimal ε, but 0 *is not infinitely close* to ε since $R \setminus \{0\}$ is a standard open set containing ε but not 0. The topological proximity $y \simeq x$ in

R must not be confused with the *metric proximity* defined by requiring that $y - x$ be infinitesimal.

6.1.2 The halo of a point

We want to consider the set of all points in X that are infinitely close to a given point. But in *Internal Set Theory* (I.S.T. see [91]), if X is a set and A is an external formula the notation $Y = \{y \in X : A\}$ is not allowed; it is an instance of *illegal set formation* because only internal formulas can be used to define subsets in this manner. However, for convenience, we adopt such a notation and we call Y an *external- set*[1]. We consider external-sets only as abbreviations of external formulas: indeed the formula $y \in Y$ will be considered as an abbreviation of the formula $y \in X \& A$. In the same way we consider inclusions, intersections, unions, images by mappings of external-sets and more generally all the usual set constructions, wherever this will economize writing or thinking. For example, for any set X, we denote by $^\sigma X = \{y \in X : \exists^{st} x \ \ y = x\}$ the external-subset of the standard elements in X; the formula $y \in {^\sigma X}$ is nothing other than "y is standard in X".

Definition 6.1.1. *Let X be a standard topological space. The* topological halo *(we abbreviate this as* monad *or* halo *) of a point x in X is the external-subset of X, denoted by $hal(x) = \{y \in X : y \simeq x\}$, whose elements are those elements of X which are infinitely close to x.*

The properties of reflexivity and transitivity of topological proximity translate immediately into the following properties of the halos.

Proposition 6.1.1. *Let X be a standard topological space. For any $x \in X$, we have $x \in hal(x)$. For any $x \in X$ and any $y \in hal(x)$, we have $hal(y) \subset hal(x)$.*

Examples: Let X be a standard set.

– For the trivial topology on X we have $hal(x) = X$ for all x in X.
– For the discrete topology on X we have $hal(x) = \{x\}$ for all standard x in X. When the set X is infinite the equality $hal(x) = \{x\}$ does not hold if x is not standard.

---

[1] The hyphen between "external" and "set" should be noticed. The adjective "external" is not qualifying the noun "set". It may happen that there exists a subset Z of X such that $\forall y \in X(y \in Z \Leftrightarrow y \in X \& A)$, that is, external-sets may be sets. If there does not exist such a set Z, the external-set $Y = \{y \in X : A\}$ should be called a *strictly external- set*. Our terminology differs slightly from the terminology used in Section 1 where we had *internal* and *external* sets at our disposal. Of course we can also refer to some axiom systems involving internal and external sets (for example the Hrbacek's theory [74] or the I.S.T.E. theory of Lutz and Goze [85]); we refrain from doing it since the axiom system of I.S.T. is sufficient for our proofs.

– Let d be a standard metric on X. For the metric topology on X we have $\mathrm{hal}\,(x) = \{y \in X : d(x,y) \simeq 0\}$ for all standard x in X. The equality does not hold in general for nonstandard points.

The halo of a point x is the intersection of all standard open sets which contain x. It is also the intersection of all standard neighbourhoods of x (Exercise 6.1.1), so we have:

$$\mathrm{hal}\,(x) = \bigcap_{V \in {}^\circ \mathcal{V}(x)} V$$

where $\mathcal{V}(x)$ is the set of neighbourhoods of x.

A *base* of the topology of a topological space X is a set $\mathcal{B}$ of open sets of X such that any open set of X is an union of elements of $\mathcal{B}$. A *subbase* of the topology of X is a set $\mathcal{B}$ of open sets of X such that the set of finite intersections of elements of $\mathcal{B}$ is a base of the topology of X. Let $\mathcal{B}$ be a standard base or subbase of a standard topological space X. The $\mathcal{B}$-*halo* of a point x of X, denoted by $\mathrm{hal}\,_\mathcal{B}(x)$, is the intersection of all standard elements of $\mathcal{B}$ that contain x. For any standard x in X we have $\mathrm{hal}\,_\mathcal{B}(x) = \mathrm{hal}\,(x)$. This property is not true in general for a nonstandard point (Exercise 6.1.2).

Proposition 6.1.2. *Let E be a subset of the standard topological space X. Let $x \in E$. Then we have:*

$$hal(x) \cap E \neq \emptyset \Leftrightarrow \forall^{st} V \in \mathcal{V}(x) \ \ E \cap V \neq \emptyset$$

Proof. To say $\mathrm{hal}\,(x) \cap E \neq \emptyset$ is the same as saying $\exists y \in E \ \forall^{st} U \in \mathcal{V}(x) \ y \in U$, so by idealization we have $\forall^{st\ fin} U' \subset \mathcal{V}(x) \ \exists y \in E \ \forall U \in U' \ y \in U$. To say that y is in every element of U' is the same as saying that $y \in V = \bigcap_{U \in U'} U$. Thus the last formula is equivalent to $\forall^{st} V \in \mathcal{V}(x) \ \exists y \in E \ y \in V$. $\square$

6.1.3 The shadow of a subset

Let E be a subset (standard or not) of the standard topological space X . Then we define the *shadow* of E, denoted by $\mathcal{E}$, as follows:

$$\mathcal{E} = {}^S\{x \in X : \mathrm{hal}\,(x) \cap E \neq \emptyset\} = {}^S\{x \in X : \exists y \in E \ y \simeq x\}$$

By Proposition 6.1.2 the shadow of E is given also by:

$$\mathcal{E} = {}^S\{x \in X : \forall^{st} V \in \mathcal{V}(x) \ E \cap V \neq \emptyset\}$$

Theorem 6.1.1. *The shadow of a subset E of the standard topological space X is closed.*

Proof. (See also [100] p. 101, [108] p. 201, [91] p. 1178 or [50] p. 63) Let x be standard in the closure of $\mathcal{E}$. Then every open neighbourhood of x contains a point of $\mathcal{E}$, so by transfer every standard open neighbourhood V of x contains a standard point y of $\mathcal{E}$. For a standard point y of $\mathcal{E}$, there is z in E with $z \simeq y$, so z is in V^2. That is to say $\forall^{st}V \in \mathcal{V}(x)$ $E \cap V \neq \emptyset$. Thus x is in $\mathcal{E}$. By transfer every point in the closure of $\mathcal{E}$ is in $\mathcal{E}$. Thus $\mathcal{E}$ is closed.   □

6.1.4 The halo of a subset

Let A be a subset of a standard topological space X. The *halo* of A, denoted by hal (A) is the intersection of all standard open sets that contain A. It is also the intersection of all standard neighbourhoods of A. Since $A \subset B \Rightarrow$ hal $(A) \subset$ hal (B) for all subsets A and B of X and hal $(x) =$ hal $(\{x\})$ for all $x \in X$, we have $\bigcup_{x \in A}$ hal $(x) \subset$ hal (A). In fact the converse inclusion is true also, as shown in the following proposition (see also [108] p. 199).

Proposition 6.1.3. *The halo of a subset A of the standard topological space X is the union of the halos of the elements of A.*

Proof. Let y be such that $y \notin$ hal (x) for all $x \in A$. Then we have:

$$\forall x \in A \; \exists^{st}U \in \mathcal{O} \; (x \in U \; \& \; y \notin U)$$

Using the idealization principle we get that:

$$\exists^{st \; fin}\alpha \subset \mathcal{O} \; \forall x \in A \; \exists U \in \alpha \; (x \in U \; \& \; y \notin U)$$

Let $\beta = \{U \in \alpha : y \notin U\}$. The set β being standard and finite, $V = \bigcup_{U \in \beta} U$ is a standard open set of X such that $A \subset V$ and $y \notin V$. Thus $y \notin$ hal (A).
□

Proposition 6.1.4. *Any subset A of the standard topological space X has an infinitesimal neighbourhood, that is to say a neighbourhood V of A such that $V \subset$ hal(A).*

Proof. Let $\mathcal{V}(A)$ be the set of neighbourhoods of A. To say $\exists V \in \mathcal{V}(A) \; V \subset$ hal (A) is the same as saying $\exists V \in \mathcal{V}(A) \; \forall^{st}U \in \mathcal{V}(A) \; V \subset U$. By idealization this is equivalent to

$$\forall^{st \; fin}U' \subset \mathcal{V}(A) \; \exists V \in \mathcal{V}(A) \; \forall U \in U' \; V \subset U$$

---

[2] We use here the following characterization of a standard open set V (Section 6.2.3):

$$\forall^{st}y \in V \; \forall z \in X \; (z \simeq y \Rightarrow z \in V)$$

which is obviously satisfied: take $V = \bigcap_{U \in U'} U$.   □

Exercises for Section 6.1

1. Let X be a standard topological space and let $\mathcal{V}(x)$ be the set of all neighbourhoods of a point x. Prove that $y \simeq x \Leftrightarrow \forall^{st} V \in \mathcal{V}(x) \ y \in V$.

2. Let $\mathcal{B}$ be a base (or a subbase) of the standard topological space X. Prove that hal $(x) \subset$ hal $_\mathcal{B}(x)$ for any $x \in X$. Prove that hal $(x) =$ hal $_\mathcal{B}(x)$ for any standard $x \in X$. Can you give an example that shows that the equality does not hold for nonstandard points ?

3. Let A and B be subsets of the standard topological space X. Let $\mathcal{C}$ be the set of closed sets of X. Prove that:

$$A \subset \mathcal{B} \Rightarrow \forall^{st} F \in \mathcal{C} \ (B \subset F \Rightarrow A \subset F)$$

Prove the converse when A is standard.

4. Let A and B be subsets of the standard topological space X. Let $\mathcal{K}$ be the set of compact subsets of X. Prove that

$$\mathcal{B} \subset A \Rightarrow \forall^{st} K \in \mathcal{K} \ (B \cap K \neq \emptyset \Rightarrow A \cap K \neq \emptyset)$$

Prove the converse when the space X is locally compact and A is a standard closed subset of X.

5. Let X be a standard compact topological space. Let A and B be subsets of X. Prove that $^{o}B \subset A \Rightarrow B \subset$ hal (A).

6.2 What purpose do halos serve ?

Let X be a standard set. Our aim is to show (Section 6.3) that a standard topology on X is completely defined by the halos of the standard points in X. More precisely, given a *halic preorder* $\mathcal{P}$ (see Section 6.3.1 for the definition of this notion) on a standard set X, there is a unique standard topology on X whose associated topological proximity $y \simeq x$ coincides with the preorder $\mathcal{P}$, when restricted to the standard points x in X (in a sense which will be made precise in Theorem 6.3.1). Before giving this result let us recall some useful facts about the nonstandard approach to topological spaces and the use of halos. Indeed the topological proximity relation permits a very simple approach to the usual topological notions such as the continuity of mappings, the closeness and compactness, the separation properties, the comparison of topologies... The reader may consider the following nonstandard characterizations of the usual topological notions as "definitions" of the corresponding notions. In some of the exercises we propose to prove that the nonstandard "definitions" are equivalent the the standard ones. These exercises are intended to the reader who is familiar with I.S.T.

6.2.1 Comparison of topologies

Let τ_1 and τ_2 be two standard topologies on a standard set X. Then τ_1 is finer than τ_2 (or τ_2 is coarser than τ_1) if and only if for any standard x in X we have $\mathrm{hal}_1(x) \subset \mathrm{hal}_2(x)$ where $\mathrm{hal}_i(x)$ is the halo of x for the topology τ_i. Hence two standard topologies are equal if and only if they have the same halos for standard points.

6.2.2 Continuity

Let $f : X \longrightarrow Y$ be a standard mapping of the standard topological space X into the standard topological space Y. Let x be standard in X.

- f is continuous at x if and only if $\forall y \in X (y \simeq x \Rightarrow f(y) \simeq f(x))$, which may be written also $f(\mathrm{hal}\,(x)) \subset \mathrm{hal}\,(f(x))$
- f is continuous on X if and only if for all standard x in X we have $f(\mathrm{hal}\,(x)) \subset \mathrm{hal}\,(f(x))$.

Hence the set of continuous mappings of a standard topological space X into a standard topological space Y is the standard set $\mathcal{C}(X,Y)$ defined as follows:

$$\mathcal{C}(X,Y) = {}^S\{f \in Y^X : \forall^{st}x \in X \;\; f(\mathrm{hal}\,(x)) \subset \mathrm{hal}\,(f(x))\}$$

where Y^X is the set of mappings of X into Y.

6.2.3 Neighbourhoods, open sets and closed sets

Let X be a standard topological space. A standard subset E of X is a neighbourhood of a standard subset A of X if and only if E contains the halo of A. The set $\mathcal{V}(x)$ of neighbourhoods of a standard point x of X is given by:

$$\mathcal{V}(x) = {}^S\{E \subset X : \mathrm{hal}\,(x) \subset E\}$$

A standard subset E of X is an open set if and only if E contains the halos of all its standard points. The set $\mathcal{O}$ of open sets of a standard topological space X is given by:

$$\mathcal{O} = {}^S\{E \subset X : \forall^{st}x \in E \;\; \mathrm{hal}\,(x) \subset E\}$$

A standard point x is in the closure of a standard subset E if and only if its halo intersects E. Therefore, the closure of a standard subset E of X is equal to its shadow. A standard subset E of X is closed if and only if E contains any standard point whose halo intersects E. The set $\mathcal{C}$ of closed subsets of X is given by:

$$\mathcal{C} = {}^S\{E \subset X : \forall^{st}x \in X(\mathrm{hal}\,(x) \cap E \neq \emptyset \Rightarrow x \in E)\}$$

6.2.4 Separation and compactness

Let X be a standard topological space.

- X is a T_0-space (or Kolmogorov space) if and only if two standard points x and y are equal whenever $x \simeq y$ and $y \simeq x$, that is,

$$\forall^{st} x \in X \ \forall^{st} y \in X \ (\mathrm{hal}\,(x) = \mathrm{hal}\,(y) \Rightarrow x = y)$$

- X is a T_1-space (or Kuratowski space) if and only if two standard points x and y are equal whenever $x \simeq y$, that is,

$$\forall^{st} x \in X \ \forall^{st} y \in X \ (\mathrm{hal}\,(x) \subset \mathrm{hal}\,(y) \Rightarrow x = y)$$

- X is a T_2-space (or Hausdorff space or separated space) if and only if the halos of two standard distinct points are disjoint, that is,

$$\forall^{st} x \in X \ \forall^{st} y \in X \ (\mathrm{hal}\,(x) \cap \mathrm{hal}\,(y) \neq \emptyset \Rightarrow x = y)$$

- X is a regular space if and only if a standard point x whose halo intersects a standard subset A belongs to the closure of A, that is,

$$\forall^{st} x \in X \ \forall^{st} A \subset X \ (\mathrm{hal}\,(x) \cap \mathrm{hal}\,(A) \neq \emptyset \Rightarrow x \in \overline{A})$$

- X is a T_3-space if it is regular and T_0.
- X is a normal space if and only if the halos of two standard sets A and B are disjoint whenever the closures of A and B are disjoint, that is,

$$\forall^{st} A \subset X \ \forall^{st} B \subset X \ (\mathrm{hal}\,(A) \cap \mathrm{hal}\,(B) \neq \emptyset \Rightarrow \overline{A} \cap \overline{B} \neq \emptyset)$$

- X is a T_4-space if it is normal and T_1.

Axiom T_2 is equivalent the following: each singleton $\{x\}$ is closed. Indeed, for standard points x and y in X, $y \in \overline{\{x\}}$ is equivalent to $y \simeq x$. The implications $T_2 \Rightarrow T_1 \Rightarrow T_0$ are trivial. We have also $T_4 \Rightarrow T_3 \Rightarrow T_2$ (Exercise 6.2.6).

The space X is compact[3] if and only if the halos of the standard points of X cover X. If the space X is compact and separated then the halos of its standard points form a partition of X. Hence any topology which is strictly finer than a compact and separated topology is not compact, but separated and any topology which is strictly coarser than a compact and separated topology is not separated, but compact (Exercise 6.2.3).

A standard subset E of X is compact if and only if it is covered by the halos of its standard points. The set $\mathcal{K}$ of compact subsets of the standard topological space X is given by:

$$\mathcal{K} = {}^S\{E \subset X : \forall y \in E \ \exists^{st} x \in E \ y \simeq x\}$$

---

[3] The term "compact" is used here in the meaning that any open covering of X has a finite subcovering. Bourbaki [24] calls such a space quasicompact and reserves the term compact to a quasicompact and separated space.

Exercises for Section 6.2

1. Show that the external characterization of continuous, open, closed, compact, separated... in this section are equivalent to the usual ones.

2. Let A be a subset of the standard topological space X. We denote by $A_{NS} = \{y \in X : \exists^{st} x \in A \ \ y \simeq x\}$ the external-subset of X of *near-standard points in* A. Assume A is standard. What can you say about the set A in case $A \subset A_{NS}$ and in case $A_{NS} \subset A$?

3. Let τ be a topology on the set X. Prove that if τ is separated then any topology that is strictly finer than τ is not compact. Prove that if τ is compact then any topology which is strictly coarser than τ is not separated.

4. Let X be a standard topological space. Let x be standard in X. Prove that $y \simeq x \Leftrightarrow \forall^{st} A \subset X(y \in A \Rightarrow x \in \overline{A})$.

5. Prove that a standard space X is regular if and only if it satisfies the following axiom:

$$\forall^{st} x \in X \ \ \forall y \in X \ \ (\mathrm{hal}\,(x) \cap \mathrm{hal}\,(y) \neq \emptyset \Rightarrow y \simeq x)$$

6. Prove that $T_4 \Rightarrow T_3 \Rightarrow T_2$.

7. Prove that each compact Hausdorff space is normal. Prove that each compact regular space is normal.

8. Let $X \times Y$ be the product space of two standard topological spaces. Prove that $\mathrm{hal}\,(x,y) \subset \mathrm{hal}\,(x) \times \mathrm{hal}\,(y)$ for any $(x,y) \in X \times Y$. Prove that $\mathrm{hal}\,(x,y) = \mathrm{hal}\,(x) \times \mathrm{hal}\,(y)$ for any standard $(x,y) \in X \times Y$.

9. Let $X \times Y$ be the product space of two standard topological spaces. Prove that $\mathrm{hal}\,(A \times B) \subset \mathrm{hal}\,(A) \times \mathrm{hal}\,(B)$ for any $A \subset X$ and $B \subset Y$. Prove that $\mathrm{hal}\,(A \times B) = \mathrm{hal}\,(A) \times \mathrm{hal}\,(B)$ when the subsets A and B are standard and compact.

10. Prove the theorem of Wallace: for A and B compact subsets of the topological spaces X and Y, respectively, and any neighbourhood W of $A \times B$ in $X \times Y$, there exist neighbourhoods U of A in X and V of B in Y such that $U \times V \subset W$.

6.3 The external definition of a topology

The results in Section 6.2 illustrate the fact that a standard topological space is characterized by the halos of its standard points. The main purpose of this section is to show that once a halo is given for each standard point of a standard set X, with the properties of reflexivity and transitivity pointed out in Proposition 6.1.1, there is a unique standard topology on X whose topological halos for the standard points are the given halos.

Proposition 6.3.1. *Let* $(\mu(x))_{x \in X}$ *be a family of subsets (or external-subsets) of a standard set* X. *Then the set* $\mathcal{O} = {}^S\{U \subset X : \forall^{st} x \in U \ \ \mu(x) \subset U\}$ *is the set of open sets of a standard topology on* X, *and the associated*

topological halo, hal(x), of a standard point x of X, satisfies $\mu(x) \subset hal(x)$
(equality does not hold in general).

Proof. The set $\mathcal{O}$ is standard and satisfies the axioms for a topology. More-over for any standard x in X, $\mu(x)$ is included in any standard element of $\mathcal{O}$ which contains x. Thus $\mu(x) \subset hal(x)$. If x is not standard we have nei-ther the inclusion $hal(x) \subset \mu(x)$ — take for example $\mu(x) = \{x\}$ for all x in N from which one gets the discrete topology on N — nor the inclusion $hal(x) \supset \mu(x)$ — take for example $\mu(x) = \{y \in R : y - x \simeq 0\}$ for all x in R from which one gets the usual metric topology on R.   □

The topology constructed in Proposition 6.3.1 was considered by Hurd and Loeb ([75] p. 114). They observed also that the topological halos of this topology are strictly larger in general than the given sets $\mu(x)$, since they did not impose any property on the sets $\mu(x)$. When the sets $\mu(x)$ are the $\mathcal{P}$-halos associated to a halic preorder $\mathcal{P}$ on X (see Section 6.3.1 for the definition of this notion) we obtain the equality $hal(x) = \mu(x)$ for all standard x in X.

6.3.1 Halic preorders and $\mathcal{P}$-halos

Let us introduce the following notions.

- A formula of I.S.T. is said to be *halic (or monadic)* whenever it is of the form $\forall^{st}r\,A$ where A is an internal formula.
- A binary formula $\mathcal{P}$ is said to be *quasi-reflexive* on a set X whenever $\forall^{st}x \in X \;\; \mathcal{P}[x,x]$ holds.
- A binary formula $\mathcal{P}$ is said to be *quasi-transitive* on a set X whenever $\forall^{st}x \in X \;\; \forall y \in X \;\; \forall z \in X \;\; (\mathcal{P}[z,y]\&\mathcal{P}[y,x] \Rightarrow \mathcal{P}[z,x])$ holds.
- A *halic preorder* on a standard set X is a halic binary formula $\mathcal{P}$ which is quasi-reflexive and quasi-transitive on X.
- Let $\mathcal{P}$ be a halic preorder X. For any x in X, the $\mathcal{P}$-halo of x in X is the external-set $p(x) = \{y \in X : \mathcal{P}[y,x]\}$.

Remark 6.3.1. *Let $\mathcal{P}[x,y]$ be a halic binary formula with free variables x and y all of whose parameters (the other free variables of $\mathcal{P}$) take standard values. Assume that $\mathcal{P}$ is quasi-reflexive on a standard set X. By transfer this formula is in fact reflexive since $\forall x \in X \;\; \mathcal{P}[x,x]$ holds. However, we cannot infer transitivity from quasi-transitivity.*

The properties of quasi-reflexivity and quasi-transitivity of $\mathcal{P}$ translate into the following properties of the $\mathcal{P}$-halos.

Proposition 6.3.2. *Let $p(x)$ be the $\mathcal{P}$-halo of x associated with the halic preorder $\mathcal{P}$ on X. Then we have:*
$\forall^{st}x \in X \;\; x \in p(x)$
$\forall^{st}x \in X \;\; \forall y \in p(x) \;\; p(y) \subset p(x)$

Example: In a standard topological space X topological proximity is a halic preorder. Here, the formula $\mathcal{P}[y, x]$ is the external formula $y \simeq x$ which is an abbreviation for the halic formula $\forall^{st} U \in \mathcal{O}(x \in U \Rightarrow y \in U)$. The $\mathcal{P}$-halos associated with this halic preorder are the topological halos. In this example the formula $\mathcal{P}$ is transitive on X and not merely quasi- transitive.

The main result of this section (Theorem 6.3.1) asserts that a halic preorder $\mathcal{P}$ on a standard set X defines a unique standard topology on X such that the topological halo of any standard point x is equal to its $\mathcal{P}$-halo[4]. This result was also obtained independently by Vakil [113]. His approach differs slightly from ours in the following sense. Firstly the paper [113] was mainly motivated by the study of the Wattenberg [116] extension of monad systems, and there is no applications to the definition of standard topologies. Secondly his formulation of the crucial notion of halos is somewhat more restrictive than ours: his halos are the intersections of standard elements of some standard filter when our halos are defined by halic formulas. His results are presented in the model theoretic form of nonstandard analysis [100, 75, 108]. Our approach is both a generalization of the result of Vakil and a translation of his result into the axiom system of I.S.T. It should be noticed that Haddad [67] obtained also, in 1978, some results related to the present problem.

Theorem 6.3.1. *Let $\mathcal{P}$ be a halic preorder on the standard set X. There is a unique standard topology on X such that:*

$$\forall^{st} x \in X \ \ hal(x) = p(x)$$

where $hal(x)$ is the topological halo of x and $p(x)$ its $\mathcal{P}$- halo.

Proof. Since two standard topologies are equal whenever they have the same halos of standard points, there is at most one topology whose halos for the standard points x in X are the $\mathcal{P}$-halos $p(x)$. Indeed the set $\mathcal{O}$ of open sets for such a topology must satisfy (see Section 6.2.3) the property:

$$\mathcal{O} = {}^{S}\{U \subset X : \forall^{st} x \in U \ \ p(x) \in U\}$$

By proposition 6.3.1 we have $p(x) \subset hal(x)$ for any standard x in X. The proof of the inclusion $hal(x) \subset p(x)$ is more subtle and requires the quasi-reflexivity and the quasi-transitivity of $\mathcal{P}$ which was not used until now. In fact we shall give an equivalent description of the topology by defining its neighbourhood system. The rest of the proof is given in Theorem 6.3.2 and its proof. $\square$

---

[4] This result was obtained in the spring of 1988, during a stay in the Econometric Institute of the Groningen University (see [103]). I thank this institute for its hospitality. I thank I.P. van den Berg for many fruitful conversations which were at the origin of my interest on this subject. I thank A. Fruchard for many interesting remarks during the writing of this chapter.

The standard topology on X whose existence is asserted by Theorem 6.3.1 is called the *topology defined by the formula:*

$$\forall^{st} x \in X \ \forall y \in X \ (y \simeq x \Leftrightarrow P[y,x])$$

To be brief we shall omit the quantifiers $\forall^{st} x \in X$ and $\forall y \in X$ and speak about the "topology defined by the formula $y \simeq x \Leftrightarrow P[y,x]$", but it is to be understood that the equivalence holds only when x is standard in X. For example[5] let X be a standard set.

– The *trivial topology* on X is the topology defined by the formula:

$$y \simeq x \Leftrightarrow (y,x) \in X \times X$$

– The *discrete topology* on X is the topology defined by the formula:

$$y \simeq x \Leftrightarrow y = x$$

– Let d be a standard metric on X. The *metric topology* on X is the topology defined by the formula:

$$y \simeq x \Leftrightarrow d(x,y) \simeq 0$$

The two last examples show that the halic preorder P does not coincide with topological proximity, for all x in X. The possibility of defining the topology even if we describe its topological proximity $y \simeq x$ approximately only— i.e. for the standard x only and not for all x — illustrates how handy is Theorem 6.3.1 in defining new topologies.

6.3.2 The ball of centre x and radius α

Definition 6.3.1. *Let X be a standard set. Let $P[y,x] \equiv \forall^{st} r A[y,x,r]$ be a halic formula. For any set α and any x in X the* ball of centre x and radius α *associated with P, denoted by $B(x,\alpha)$, is the set:*

$$B(x,\alpha) = \{y \in X : \forall r \in \alpha \ A[y,x,r]\}$$

Proposition 6.3.3. *Let P be a halic preorder on the standard set X. Let x be in X. Then the P-halo $p(x)$ of x is the intersection of all the balls $B(x,\{r\})$ of centre x and radius $\{r\}$ for all standard r. Let F be a finite set that contains every standard set. Then $B(x,F) \subset p(x)$.*

---

[5] In the examples of Section 6.1.2 we have described the topological proximity for topologies which were supposed to be already defined. Here we are going to use Theorem 6.3.1 to "define" topologies. These first examples are rudimentary. For more elaborate constructions see Sections 6.3.3 and 6.4.

Proof. This is an immediate consequence of the definitions.   □

Recall that a topology on a set X may be defined by a *neighbourhood system* on X, that is, a mapping $\mathcal{V}$ that associates with each point x in X a set $\mathcal{V}(x)$ of subsets of X, called the *neighbourhoods* of x, such that:

(V_1) $x \in U$ for all U in $\mathcal{V}(x)$.

(V_2) If U_1 and U_2 are in $\mathcal{V}(x)$ then $U_1 \cap U_2 \in \mathcal{V}(x)$.

(V_3) If $U \in \mathcal{V}(x)$ and $U \subset V$ then $V \in \mathcal{V}(x)$.

(V_4) For any $U \in \mathcal{V}(x)$ there is $W \in \mathcal{V}(x)$ such that $U \in \mathcal{V}(y)$ for all $y \in W$.

Let $\mathcal{V}$ be a neighbourhood system on a set X. Then there is a unique topology on X such that for all x in X, $\mathcal{V}(x)$ is the set of neighbourhoods of x for this topology. The set of open sets for this topology is the set $\mathcal{O}$ of subsets U of X such that $U \in \mathcal{V}(x)$ for all x in U.

Proposition 6.3.4. *Let $\mathcal{P}$ be a halic preorder on the standard set X. Let $\mathcal{V}$ be the standard mapping of X into its power set that is defined, for x standard in X, by:*

$$\mathcal{V}(x) = {}^S\{U \subset X : p(x) \subset U\}$$

where $p(x)$ is the $\mathcal{P}$-halo of x. Then, for all x in X we have:

$$\mathcal{V}(x) = \{U \subset X : \exists^{fin}\alpha\ B(x,\alpha) \subset U\}$$

where $B(x,\alpha)$ is the ball of centre x and radius α associated with $\mathcal{P}$. Moreover $\mathcal{V}$ is a neighbourhood system on X[6].

Proof. By definition of $\mathcal{V}$ we have:

$$\forall^{st}x\forall^{st}U(U \in \mathcal{V}(x) \Leftrightarrow \forall y \in X(\forall^{st}r\ A[y,x,r] \Rightarrow y \in U))$$

The reduction algorithm of external formulas of Nelson [91] transforms this formula into:

$$\forall x\forall U(U \in \mathcal{V}(x) \Leftrightarrow \exists^{fin}\alpha\forall y \in X(\forall r \in \alpha\ A[y,x,r] \Rightarrow y \in U))$$

Let us prove now that $\mathcal{V}$ is a neighbourhood system on X. By transfer it suffices to demonstrate the properties ($V_1 - V_4$), characterizing a neighbourhood system, for x standard in X. Moreover by transfer also it suffices to do this for their standard data. The proofs of the properties ($V_1 - V_3$) are straightforward and require only the external definition of $\mathcal{V}$. Indeed let U be standard in $\mathcal{V}(x)$. By the quasi-reflexivity of the preorder $\mathcal{P}$ we have $x \in p(x)$. Since $p(x) \subset U$ we have $x \in U$ and the property (V_1) is satisfied. Let U_1 and U_2 be standard in $\mathcal{V}(x)$. Since $p(x) \subset U_1$ and $p(x) \subset U_2$ we have $p(x) \subset U_1 \cap U_2$.

---

[6] It is called the neighbourhood system defined by $\mathcal{P}$. Hence the set U is a neighbourhood of x if and only if *it contains a ball of centre x and finite radius α.* It may be noticed that this sentence is strongly similar to the usual description of the neighbourhoods in the case of metric spaces.

Then $U_1 \cap U_2 \in \mathcal{V}(x)$ and the property (V_2) is satisfied. Let U be standard in $\mathcal{V}(x)$. Let V be standard and such that $U \subset V$. Since $p(x) \subset U \subset V$ we have $V \in \mathcal{V}(x)$ and the property (V_3) is satisfied. The proof of the property (V_4) requires the internal description of $\mathcal{V}$ given above. Let U be standard in $\mathcal{V}(x)$. Let F be a finite set that contains every standard set. Then $W = B(x, F)$ is an element of $\mathcal{V}(x)$. By Proposition 6.3.3 $W \subset p(x)$. By the quasi-transitivity of $\mathcal{P}$, we have $p(y) \subset p(x)$ for all y in W. Using $B(y, F) \subset p(y)$ and $p(x) \subset U$ we obtain $B(y, F) \subset U$. Thus $U \in \mathcal{V}(y)$ and the property (V_4) is satisfied.

$\square$

Theorem 6.3.2. *Let $\mathcal{P}$ be a halic preorder on the standard set X. Let $\mathcal{V}$ be the neighbourhood system defined by $\mathcal{P}$. Then, for any standard x in X, the halo of x for the topology defined by $\mathcal{V}$ is equal to its $\mathcal{P}$-halo.*

Proof. The set of open sets for the topology defined by $\mathcal{V}$ is given by $\mathcal{O} = {}^S\{U \subset X : \forall^{st} x \in U \;\; p(x) \in U\}$. We have already proved (see the proof of Theorem 6.3.1) the inclusion $\mathrm{hal}(x) \supset p(x)$. Conversely, let x be standard in X and let $y \simeq x$. For any standard r, the set $V_r = B(x, \{r\})$ is a standard neighbourhood of x, so $y \in V_r$. By Proposition 6.3.3 $y \in p(x)$. Thus $\mathrm{hal}(x) \subset p(x)$.

$\square$

6.3.3 Product spaces and function spaces

In this section we are going to define topologies by using Theorem 6.3.1, that is, by describing the halos of the standard points. Since we know the neighbourhood system of the topology (Proposition 6.3.4) it is easy to show that our definitions are equivalent to the usual ones. However we shall work only with the external definition, our aim being to convince you that this approach permits us to do many things in the study of topological spaces. Some topological notions — such as regularity or normality — require the knowledge of all the open sets (that is, the knowledge of all the halos), and not merely the knowledge of the halos of the standard points. For the study of such notions it suffices to construct the neighbourhood system as explained in Proposition 6.3.4.

Let $(X_a)_{a \in A}$ be a standard family of topological spaces. The *product topology* on the product space $Y = \prod_{a \in A} X_a$ is the topology defined by the formula:

$$y \simeq x \Leftrightarrow \forall^{st} a \in A \;\; y_a \simeq x_a$$

The *strong topology* on the product space Y is the topology defined by the formula:

$$y \simeq x \Leftrightarrow \forall^{st} (U_a)_{a \in A} \forall a \in A (U_a \in \mathcal{O}_a \,\&\, x_a \in U_a \Rightarrow y_a \in U_a)$$

where $\mathcal{O}_a$ is the set of open sets of the factor space X_a. If A is a finite set, the strong topology on Y coincides with the product topology. It should be noted that the definition of the strong topology on Y requires the definition of the set of open sets of each factor space X_a of Y, whereas, the definition of the product topology on Y requires only the definition of the halos of the standard points of the standard factor spaces X_a of Y. Hence we can define the product topology on Y by a halic formula even when the topology of each standard factor space X_a is defined itself by a halic preorder $\mathcal{P}_a$. We do so as follows:

$$y \simeq x \Leftrightarrow \forall^{st} a \in A \ \mathcal{P}_a[y_a, x_a]$$

Let us give an elementary proof of the Tychonoff's theorem (The product of compact spaces is compact), that shows that we can obtain deep results, by working only with the external definition of the product topology (see also [100] p. 95, [108] p. 202 or [91] p. 1185). Assume that each factor space X_a is compact. Let $y = (y_a)_{a \in A}$ be an element of Y. For any standard a in A, there is a standard x_a in X_a such that $y_a \simeq x_a$. By the standardization principle, there is a standard $x = (x_a)_{a \in A}$ in X that satisfies $\forall^{st} a \in A \ y_a \simeq x_a$. Hence $y \simeq x$, that is, Y is compact.

When all the factor spaces X_a of the family $(X_a)_{a \in A}$ are equal to the same topological space X, the product space Y is the set of mappings $f : A \to X$ of A into X, denoted by $Y = X^A$, and the product topology on X^A is called the *pointwise convergence topology*. This is the topology defined by the formula:

$$g \simeq f \Leftrightarrow \forall^{st} a \in A \ g(a) \simeq f(a)$$

More generally, if $\mathcal{K}$ is a standard set of subsets of A the $\mathcal{K}$-*topology* on X^A is the topology defined by the formula

$$g \simeq f \Leftrightarrow \forall^{st} K \in \mathcal{K} \ g(K) \subset \mathrm{hal}(f(K))$$

If $\mathcal{K}$ is the set of singletons $\{a\}$ with $a \in A$, then the $\mathcal{K}$-topology on X^A is the pointwise convergence topology. If A is itself a topological space and $\mathcal{K}$ is the set of compact subsets of A, then the $\mathcal{K}$-topology on X^A is called the *compact-open topology* on X^A. This topology is finer than the pointwise convergence topology since all the singletons are compact.

Exercises for Section 6.3

1. Show that the product topology and the compact-open topology defined in Section 6.3.3 are the same as the usual ones.

2. Let X and Y be topological spaces. Let F be a subset of Y^X. A topology on F is said to be *jointly continuous* on a subset A of X if and only if the mapping P of $F \times A$ into Y which carries (f, x) into $f(x)$ is continuous (see [77] p. 223). Prove that each topology which is jointly continuous on compact subsets of X is finer than the compact open topology on F.

3. Let $(Y_i)_{i \in I}$ be a standard family of topological spaces. Let X be a standard set. Let $(f_i)_{i \in I}$ be a standard family of mappings $f_i : X \to Y_i$ of

X into Y_i. Prove that the coarsest topology on X such that the mappings f_i are continuous is the topology defined by the formula:

$$y \simeq x \Leftrightarrow \forall^{st} i \in I \ \ f_i(y) \simeq f_i(x)$$

6.4 The power set of a topological space

The study of the power set of a topological space illustrates much better than the previous examples how handy is the external definition of a topology. Many topologies have been defined on the power set of a topological space (see [33, 57, 58, 59, 87, 88, 96]): the *lower semi-finite topology* of Michael [87] is called the *λ-topology* by Feichtinger [58] and the *local topology* by Effros [57], the *upper semi-finite topology* of Michael [87] is called the *κ-topology* by Ponomarev [96], the *global topology* of Effros [57] had been proposed a long time before by Choquet [33] for the "topologization" of the notion of *convergence of a family of subsets* in a topological space (Section 6.5.3). Wattenberg [117] used also Nonstandard Analysis for the study of some topologies on the set of closed subsets.

Let $\mathcal{P}(X)$ be the power set of a standard topological space X. Let $\mathcal{A}$ be a standard set of subsets of X. We define the *lower $\mathcal{A}$-topology*[7] on $\mathcal{P}(X)$ as being the topology defined by the formula:

$$B \simeq A \Leftrightarrow \forall^{st} U \in \mathcal{A}(B \subset U \Rightarrow A \subset U)$$

We define also the *upper $\mathcal{A}$-topology*[8] on $\mathcal{P}(X)$ as being the topology defined by the formula:

$$B \simeq A \Leftrightarrow \forall^{st} U \in \mathcal{A}(A \subset U \Rightarrow B \subset U)$$

6.4.1 The Vietoris topology

Let X be a standard topological space. Let $\mathcal{O}$ be the set of open sets of X. Let $\mathcal{C}$ be the set of closed sets of X. The lower $\mathcal{C}$-topology on $\mathcal{P}(X)$ is the lower semi-finite topology of Michael. The upper $\mathcal{O}$-topology on $\mathcal{P}(X)$ is the upper semi-finite topology of Michael (Exercise 6.4.1). The intersection of the lower and upper semi-finite topologies of Michael is the *Vietoris topology*.

---

[7] For this topology, if a subset B is infinitely close to a standard subset A, then B may be "smaller" than A but just a little since A is contained in every standard subset $U \in \mathcal{A}$ that contains B. This is the meaning of the adjective "lower".

[8] For this topology, if a subset B is infinitely close to a standard subset A, then B may be "larger" than A but just a little since B is contained in every standard subset $U \in \mathcal{A}$ that contains A. This is the meaning of the adjective "upper".

Proposition 6.4.1. *Let X be a standard topological space:*
i) For the lower $\mathcal{C}$-topology on $\mathcal{P}(X)$ we have $B \simeq A \Leftrightarrow A \subset {}^\circ B$ for any standard $A \subset X$ and any $B \subset X$.
ii) For the upper $\mathcal{O}$-topology on $\mathcal{P}(X)$ we have $B \simeq A \Leftrightarrow B \subset \mathrm{hal}(A)$ for any standard $A \subset X$ and any $B \subset X$.
iii) For the Vietoris topology on $\mathcal{P}(X)$ we have

$$B \simeq A \Leftrightarrow B \subset \mathrm{hal}(A) \ \& \ A \subset {}^\circ B$$

for any standard $A \subset X$ and any $B \subset X$.

Proof. *i)* By Exercise 6.1.3. *ii)* By definition of the halo of the subset A. *iii)* By definition of the Vietoris topology.   □

Theorem 6.4.1. *The set $\mathcal{H}(X)$ of non-empty closed subsets of a compact topological space endowed with the Vietoris topology is compact.*

Proof. By transfer we may assume that X is standard. Let $B \in \mathcal{H}(X)$ and $A = {}^\circ B$. Then A is closed. Since X is compact, A is non-empty. Using $\mathcal{B} \subset A$ and Exercise 5 of Section 6.1 we obtain $B \subset \mathrm{hal}(A)$. We have also $A \subset {}^\circ B$. Thus $B \simeq A$ for the Vietoris topology. We have shown that every element in $\mathcal{H}(X)$ is nearstandard in $\mathcal{H}(X)$, so $\mathcal{H}(X)$ is compact.   □

6.4.2 The Choquet topology

Let X be a standard topological space. Let $\mathcal{K}$ be the set of compact subsets of X. Let $\mathcal{K}' = \{X \setminus K : K \in \mathcal{K}\}$ be the set of the complementary subsets of compact subsets of X. Taking negations, the upper $\mathcal{K}'$- topology on $\mathcal{P}(X)$ is the topology defined by the formula $B \simeq A \Leftrightarrow \forall^{st} K \in \mathcal{K} \ (B \cap K \neq \emptyset \Rightarrow A \cap K \neq \emptyset)$, that is, the subset B is infinitely close to the standard subset A if A intersects any standard compact subset that intersects B. The upper $\mathcal{K}'$-topology on $\mathcal{P}(X)$ is the global topology of Effros (Exercise 6.4.1). Since the lower $\mathcal{C}$-topology is the local topology of Effros, the intersection of the lower $\mathcal{C}$-topology and the upper $\mathcal{K}'$-topology is the Choquet topology named also the Fell topology [57].

Proposition 6.4.2. *Let X be a standard topological space. Then we have ${}^\circ B \subset A \Rightarrow B \simeq A$ for the upper $\mathcal{K}'$-topology on $\mathcal{P}(X)$*

Proof. By Exercise 6.1.4.   □

Theorem 6.4.2. *The set $\mathcal{C}(X)$ of closed subsets of the topological space X, endowed with the Choquet topology is compact.*

Proof. By transfer we may assume that X is standard. Let $B \in \mathcal{C}(X)$ and $A = {}^{\circ}B$. Then we have $A \in \mathcal{C}(X)$. Using $A \subset {}^{\circ}B$ and Proposition 6.4.1 we have $B \simeq A$ for the lower $\mathcal{C}$-topology. Using ${}^{\circ}B \subset A$ and Proposition 6.4.2 we have $B \simeq A$ for the upper $\mathcal{K}'$- topology. Thus $B \simeq A$ for the Choquet topology. We have shown that every element in $\mathcal{C}(X)$ is nearstandard in $\mathcal{C}(X)$, so $\mathcal{C}(X)$ is compact. $\qquad\square$

Exercises for Section 6.4

1. Show that the topologies of Vietoris and Choquet defined in this section are the same as the usual ones.

2. Let E be a closed subset of X. Show that $\mathcal{P}(E)$ is closed in $\mathcal{P}(X)$, endowed with the lower $\mathcal{C}$-topology.

3. Let E be an open subset of X. Show that $\mathcal{P}(E)$ is open in $\mathcal{P}(X)$, endowed with the upper $\mathcal{O}$-topology.

4. Show that the space $\mathcal{C}(X)$ of closed subsets of X, endowed with the lower $\mathcal{C}$-topology, is T_0.

6.5 Set-valued mappings and limits of sets

6.5.1 Semicontinuous set-valued mappings

Let Y be a standard topological space. For every standard $a \in Y$ and every $b \in Y$ we have

$$b \simeq a \Leftrightarrow \{a\} \subset {}^{\circ}\{b\} \Leftrightarrow \{b\} \subset \operatorname{hal}(\{a\})$$

Then, a standard mapping $f : X \to Y$ is continuous at the standard point $x \in X$, if and only if it satisfies one of the two equivalent properties: $\forall y \in X(y \simeq x \Rightarrow \{a\} \subset {}^{\circ}\{b\})$ or $\forall y \in X(y \simeq x \Rightarrow \{b\} \subset \operatorname{hal}(\{a\}))$ where $a = f(x)$ and $b = f(y)$. For set-valued mappings these properties are not equivalent, so they define two kinds of *semicontinuities*.

Definition 6.5.1. *Let $f : X \to \mathcal{P}(Y)$ be a standard set-valued mapping of the standard topological space X into the power set $\mathcal{P}(Y)$ of the standard topological space Y. Let x be standard in X.*

– *The mapping f is said to be* lower semicontinuous *(l.s.c.) at x if*

$$\forall y \in X(y \simeq x \Rightarrow f(x) \subset {}^{\circ}f(y))$$

– *The mapping f is said to be* upper semi-continuous *(u.s.c.) at x if*

$$\forall y \in X(y \simeq x \Rightarrow f(y) \subset \operatorname{hal}(f(x)))$$

– *The mapping f is said to be* continuous *at x if it is both l.s.c. and u.s.c. at x.*

These definitions are the same as the usual ones (Exercise 6.5.1).

Examples: Orbits of vector fields Let $\mathcal{X}(\Omega)$ be the set of continuous vector fields $f : \Omega \to R^n$ on the open set $\Omega \subset R^n$, such that, for any $a \in \Omega$ the solution $\pi_f(t,a)$ of the differential equation $x'(t) = f(x(t))$, with the initial condition $\pi_f(0,a) = a$ is unique. Let $I_f(a) = (\alpha_f(a), \omega_f(a))$, $-\infty \leq \alpha_f(a) < \omega_f(a) \leq \infty$, be its maximal interval of definition. Let $\gamma_f(a) = \{\pi_f(t,a) \in R^n : t \in I_f(a)\}$ be the orbit through a. The mappings I_f (resp. γ_f) of Ω into $\mathcal{P}(R)$ (resp. $\mathcal{P}(R^n)$) are l.s.c. on Ω. Indeed, assume that f and a are standard then, by the Short Shadow Lemma[9], for every $b \simeq a$ and every standard $t \in I_f(a)$, we have $t \in I_f(b)$ and $\pi_f(t,b) \simeq \pi_f(t,a)$. Hence $I_f(a) \subset {}^S I_f(b) \subset \mathbf{I}_f(b)$ and $\gamma_f(a) \subset {}^\diamond \gamma_f(b)$. In fact the result of the Short Shadow Lemma is much stronger than these classical semicontinuity results. Indeed let I (resp. γ) be the mapping of $\mathcal{X}(\Omega) \times \Omega$ into $\mathcal{P}(R)$ (resp. $\mathcal{P}(R^n)$) which carries (f,a) into $I_f(a)$ (resp. $\gamma_f(a)$). Then the mappings I and γ are l.s.c. when $\mathcal{X}(\Omega)$ is endowed with the topology of uniform convergence on compacta.

Julia sets Let c_0 be a standard complex number and let $c \simeq c_0$. Suppose $c_o \neq \frac{1}{4}$. It was proved (Theorem 2.3.3) that the halo of the Julia set $J(c)$ contains the halo of the Julia set $J(c_o)$. We deduce easily from this that $J(c_o)$ is contained in the shadow of $J(c)$. Thus the mapping J that associates to each complex number c its Julia set $J(c)$ is l.s.c. on $C \setminus \{\frac{1}{4}\}$.

6.5.2 The topologization of semi-continuities

These external characterizations of semi-continuities permit one to topologize very readily these notions, that is, to define a topology on $\mathcal{P}(Y)$ such that the semi-continuity of the mapping $f : X \to \mathcal{P}(Y)$ at x is the same as the usual continuity of f at x when considered as a mapping between two topological spaces. This is a positive answer to a question asked by Berge who wrote, in 1956, in his book *Espaces topologiques, fonctions multivoques* ([20] p. 135):

"Peut-on associer à $\mathcal{F}'$ [la famille des fermés non vides d'un espace topologique X] une structure topologique telle que l'étude de la continuité d'une application Γ [multivoque] à valeurs dans X, se ramène à celle d'une application univoque à valeurs dans $\mathcal{F}'$. En fait, il existe plusieurs procédés de topologisations, parmi lesquels nous citerons celui de Vietoris et celui de Bourbaki; on notera néanmoins [que] ce point de vue ne peut apporter, à l'étude des applications Γ, que des résultats très partiels; en effet, il suppose que ... Γ possède à la fois les deux semi-continuités..."

---

[9] **The Short Shadow Lemma:** Let Ω be standard and let f be standard in $\mathcal{X}(\Omega)$. Let g be a continuous vector field on Ω such that $g(x) \simeq f(x)$ (this is the metric proximity on R^n) for any nearstandard x in Ω. Let $a \in \Omega$ be standard, let $b \simeq a$ and let t be standard in $I_f(a)$. Then, any maximal solution $\varphi(t)$ of the differential equation $x'(t) = g(x(t))$ with the initial condition $\varphi(0) = b$, is defined at least on $[0,t]$ and satisfies $\varphi(s) \simeq \pi_f(s,a)$ for any $s \in [0,t]$ ([104] p. 6, see also [50] p. 137 for another version of this result).

The same remark was made later, in 1990, by Aubin and Frankowska in their book *Set-Valued analysis* ([3] p. 6):

> "For instance, when we regard a set-valued map as a single- valued map from one set to the power set of the other (supplied with any one of the topologies we can think of), we arrive at continuity concepts which are stronger than both lower and upper semicontinuity..."

The topologies on the power set of a topological space that topologize the semi-continuities have been well known for a long time, in particular among mathematical economists [72]. It is surprising to see that such questions are not so popular, even among specialists of considerable standing in questions connected with set-valued mappings. Meanwhile, the nonstandard approach gives one the answer almost immediately. One has nothing to do but read the definitions!

Theorem 6.5.1. *Let $f : X \to \mathcal{P}(Y)$ be a set-valued mapping. Then*
1. f is l.s.c. at x if and only if it is continuous at x as a mapping of the topological space X into the topological space $\mathcal{P}(Y)$ endowed with the lower $\mathcal{C}$-topology (or lower semi-finite topology of Michael),
2. f is u.s.c. at x if and only if it is continuous at x as a mapping of the topological space X into the topological space $\mathcal{P}(Y)$ endowed with the upper $\mathcal{O}$-topology (or upper semi-finite topology of Michael).

Proof. By transfer we may assume that f and x are standard. The lower semi-continuity of f at x is equivalent to the property $\forall y \in X (y \simeq x \Rightarrow f(x) \subset {}^{o}f(y))$. By Proposition 6.4.1 we know that $f(x) \subset {}^{o}f(y)$ is equivalent to $f(y) \simeq f(x)$ for the lower $\mathcal{C}$-topology on $\mathcal{P}(Y)$. Finally we have:

$$\forall y \in X (y \simeq x \Rightarrow f(y) \simeq f(x))$$

This is the nonstandard characterization of the continuity of f at x when $\mathcal{P}(Y)$ is endowed with the lower $\mathcal{C}$-topology. The same proof holds for upper semicontinuity. $\square$

This result has the following corollary.

Proposition 6.5.1. *A set-valued mapping $f : X \to \mathcal{P}(Y)$ is continuous at $x \in X$ if and only if it is continuous as a mapping of the topological space X into the topological space $\mathcal{P}(Y)$ endowed with the Vietoris topology.*

6.5.3 Limits of Sets

The concepts of upper and lower limit of sets are due to Painlevé [94] in 1902. They were popularized by Kuratowski [78] and Berge [20] (see [3] for many interesting historical and bibliographical comments). They were also studied by Choquet [33] who solved the problem to topologizing them.

Definition 6.5.2. *Let I be a set. A filter on I is a family $\mathcal{F}$ of subsets of I such that $\emptyset \notin \mathcal{F}$, $I \in \mathcal{F}$, and for all subsets S and T of X we have $S \in \mathcal{F} \& T \in \mathcal{F} \Rightarrow S \cap T \in \mathcal{F}$ and $S \in \mathcal{F} \& S \subset T \Rightarrow T \in \mathcal{F}$. Let $(I, \mathcal{F})$ be a standard filtered set. The* halo *(or monad) of $\mathcal{F}$ is the external-set*

$$hal(\mathcal{F}) = \bigcap_{S \in {}^{\sigma}\mathcal{F}} S$$

Definition 6.5.3. *Let $(A_i)_{i \in I}$ be a standard family of subsets of the standard topological space Y indexed by the elements of the standard filtered set $(I, \mathcal{F})$. The* lower limit *of the family (A_i) is the subset*

$$\liminf_{\mathcal{F}} A_i = {}^S\{y \in Y : \forall i \in hal(\mathcal{F}) \ y \in {}^{\circ}A_i\}$$

The upper limit *of the family (A_i) is the subset*

$$\limsup_{\mathcal{F}} A_i = {}^S\{y \in Y : \exists i \in hal(\mathcal{F}) \ y \in {}^{\circ}A_i\}$$

The family (A_i) is said to converge *to the subset A, which is denoted by $A = \lim_{\mathcal{F}} A_i$, if ${}^{\circ}A_i = A$ for any $i \in hal(\mathcal{F})$.*

This is equivalent to the usual definitions (Exercise 6.5.2).

Examples: **The discrete case** We consider the Fréchet base $\mathcal{F}$ on N defined by

$$\mathcal{F} = \{S_n \subset N : S_n = \{k \in N : k > n\} \ n \in N\}$$

Then $hal(\mathcal{F})$ is the external-set of unlimited integers. Let $(A_n)_{n \in N}$ be a standard sequence of subsets of the topological space Y. The lower limit of the sequence (A_n) is the subset

$$\liminf_{n \to \infty} A_n = {}^S\{y \in Y : \forall n \simeq \infty \ y \in {}^{\circ}A_n\}$$

The upper limit of the sequence (A_n) is the subset

$$\limsup_{n \to \infty} A_n = {}^S\{y \in Y : \exists n \simeq \infty \ y \in {}^{\circ}A_n\}$$

The sequence (A_n) converges to the subset A, which is denoted by $A = \lim_{n \to \infty} A_n$ if ${}^{\circ}A_n = A$ for any unlimited n.

The continuous case We consider the filtered space $(X, \mathcal{F})$ where $\mathcal{F}$ is the set of neighbourhoods (or a base of this set) of a standard point x in the standard topological space X. Then $hal(\mathcal{F}) = hal(x)$. Let $(A_{x'})_{x' \in X}$ be a standard family of subsets of the standard topological space Y. The lower limit of the family $(A_{x'})$, when x' tends to x, is the subset

$$\liminf_{x' \to x} A_{x'} = {}^S\{y \in Y : \forall x' \simeq x \ y \in {}^{\circ}A_{x'}\}$$

The upper limit of the family $(A_{x'})$ when x' tends to x is the subset

$$\limsup_{x' \to x} A_{x'} = {}^S\{y \in Y : \exists x' \simeq x \ y \in {}^{\circ}A_{x'}\}$$

The family $(A_{x'})$ converges to the subset A when x' tends to x, which is denoted by $A = \lim_{x' \to x} A_x$, if ${}^{\circ}A_{x'} = A$ for any $x' \simeq x$.

6.5.4 The topologization of the notion of the limit of sets

The natural question in this context is to ask whether there exits a topology on the set of closed subsets of the topological space Y such that the convergence of the filtered family $(A_i)_{i \in I}$ is the usual notion of convergence in a topological space. This question was answered by Choquet [33] who introduced the notion of pseudo-topology and showed that the Painlevé-Kuratowski is topological when the space Y is locally compact (in fact this condition is also necessary). The same question was considered later by Effros [57] who gave the same answer without referring to the previous work by Choquet. In fact, the subject seems to be very difficult and is not popular even among specialists. See Berge ([20] p. 135) who wrote on this question:

> "Quant à l'etude de la convergence d'une suite (F_n), nous avons déjà signalé qu'elle ne possède pas, en général, les caractéristiques d'une suite d'éléments dans un espace topologique véritable; et l'introduction d'une "pseudo-topologie" conduit alors à des complications inextricables."

By the external characterization of the convergence of a standard filtered family in a standard topological space (Exercise 6.5.3), we obtain that the Painlevé-Kuratowski convergence in the standard topological space Y is topological, if and only if there is a standard topology on the set $\mathcal{C}$ of closed subsets of Y such that, for any standard $A \in \mathcal{C}$ and any $B \in \mathcal{C}$

$$B \simeq A \Leftrightarrow \mathcal{B} = A$$

Remark 6.5.1. We have $\mathcal{B} = A \Leftrightarrow \mathcal{B} \subset A \& A \subset \mathcal{B}$. The formula $A \subset \mathcal{B}$ is halic (Proposition 6.4.1), but the formula $\mathcal{B} \subset A$ is not halic in general. Hence the formula $B \simeq A \Leftrightarrow \mathcal{B} = A$ does not define a topology on $\mathcal{C}$. The problem is then to find conditions on the topological space Y such that the formula $\mathcal{B} \subset A$ is halic.

Proposition 6.5.2. Let Y be a standard locally compact topological space. Then for any standard closed subset A of Y and any subset B of Y we have $\mathcal{B} = A$ if an only if $B \simeq A$ for the Choquet topology on $\mathcal{P}(Y)$.

Proof. Let $\mathcal{C}$ be the set of closed subsets of Y. Let $\mathcal{K}$ be the set of compact subsets of Y. We have already seen (Exercises 6.1.3 and 6.1.4) that

$$A \subset \mathcal{B} \Leftrightarrow \forall^{st} F \in \mathcal{C}(B \subset F \Rightarrow A \subset F)$$

for any standard $A \subset Y$ and any $B \subset Y$ and

$$\mathcal{B} \subset A \Leftrightarrow \forall^{st} K \in \mathcal{K}(B \cap K \neq \emptyset \Rightarrow A \cap K \neq \emptyset)$$

for any standard closed $A \subset Y$ and any $B \subset Y$. Consequently $\mathcal{B} = A \Leftrightarrow B \simeq A$ for the Choquet topology.   □

This result has the following corollary.

Theorem 6.5.2. *Let Y be a locally compact space. The Painlevé-Kuratowski convergence on the set C of closed subsets of Y is the topological convergence for the Choquet topology on C.*

Exercises for Section 6.5

1. Let $f : X \to \mathcal{P}(Y)$ be a mapping of the topological space X into the power set $\mathcal{P}(X)$ of the topological space Y. Let $x \in X$. Prove that the mapping f is l.s.c. at x if for every open set U of Y with $f(x) \cap U \neq \emptyset$ there is a neighbourhood V of x such that $f(y) \cap U \neq \emptyset$ for every $y \in V$. Prove that the mapping f is u.s.c. at x if for every open set U of Y with $f(x) \subset U$ there is a neighbourhood V of x such that $f(y) \subset U$ for every $y \in V$.

2. Let $(A_i)_{i \in I}$ be a family of subsets of the topological space Y indexed by the elements of the filtered set $(I, \mathcal{F})$. Show that the lower limit of the family (A_i) is the subset

$$\liminf_{\mathcal{F}} A_i = \{y \in Y : \forall U \in \mathcal{V}(y) \; \exists S \in \mathcal{F} \; \forall i \in S \; A_i \cap U \neq \emptyset\}$$

Show that the upper limit of the family (A_i) is the subset

$$\limsup_{\mathcal{F}} A_i = \{y \in Y : \forall U \in \mathcal{V}(y) \; \forall S \in \mathcal{F} \; \exists i \in S \; A_i \cap U \neq \emptyset\}$$

Show that the family (A_i) converges to the subset A if and only if

$$A = \liminf_{\mathcal{F}} A_i = \limsup_{\mathcal{F}} A_i.$$

3. Let $(x_i)_{i \in I}$ be a filtered family of elements of the topological space X. This family is said to converge to the element $x \in X$, which is denoted by $x = \lim_{\mathcal{F}} x_i$ if $\forall U \in \mathcal{V}(x) \; \exists T \in \mathcal{F} \; \forall i \in T \; x_i \in U$. Prove that, for standard data, we have $x = \lim_{\mathcal{F}} x_i \Leftrightarrow \forall i \in \text{hal}(\mathcal{F}) \; x_i \simeq x$.

4. Let $f : X \to \mathcal{P}(Y)$ be a set-valued map. The *graph* of f is the set $\text{Graph}(f) = \{(x, y) \in X \times Y : y \in f(x)\}$. Show that a point (x, y) belongs to the closure of the graph of f if and only if $y \in \limsup_{x' \to x} f(x')$. Show that f is l.s.c. at x if and only if $f(x) \subset \liminf_{x' \to x} f(x')$.

5. Prove that the set $C(X)$ of closed subset of a locally compact topological space, endowed with the Choquet topology, is Hausdorff.

6. Prove that the lower limit and the upper limit of a filtered family of subsets of a topological space are closed.

6.6 Uniform spaces

There are certain notions as such as "completeness" or "uniform continuity" that are defined on metric spaces but not on general topological spaces. The concept of uniform space was introduced by A. Weil [118] to generalize the notion of metric space.

Definition 6.6.1. *Let X be a set. For $U, V \subset X \times X$ and $u \in X$ we define:*
$U^{-1} = \{(x, y) \in X \times X : (y, x) \in U\}$
$V \circ U = \{(x, y) \in X \times X : \exists z \in X \ (x, z) \in U \& (z, y) \in V\}$
$U[u] = \{x \in X : (x, u) \in U\}.$

Definition 6.6.2. *A* uniformity *on the set X is a set $\mathcal{U}$ of subsets of $X \times X$, said to be the* entourages, *such that:*
(a) each member of $\mathcal{U}$ contains the diagonal $\Delta = \{(x, x) : x \in X\}$,
(b) if $U \in \mathcal{U}$ and $U \subset V$ then $V \in \mathcal{U}$,
(c) if $U, V \in \mathcal{U}$ then $U \cap V \in \mathcal{U}$,
(d) if $V \in \mathcal{U}$ then $V^{-1} \in \mathcal{U}$ and,
(e) if $U \in \mathcal{U}$ then there is $V \in \mathcal{U}$ such that $V \circ V \subset U$.

Examples: A metric (and more generally a pseudo-metric) d on a set X determines the uniformity

$$\mathcal{U}_d = \{U \subset X \times X : \exists \varepsilon > 0 \ \{(x, y) \in X \times X : d(x, y) < \varepsilon\} \subset U\}$$

The usual *metric proximity* defined by $d(x, y) \simeq 0$ in a standard metric space may be generalized to standard uniform spaces as follows.

6.6.1 Uniform proximity

Let X be a standard uniform space. The points x and y in X are said to be *infinitely close* when (x, y) is in any standard entourage, which is denoted by $x \simeq y$ (or $y \simeq x$ since this relation is symmetric). This is an external equivalence relation on X. It is called the *uniform proximity* on X. The formula $x \simeq y$ may be considered as an abbreviation of the external halic formula $\forall^{st} U \in \mathcal{U} \ (x, y) \in U$, so we write:

$$y \simeq x \Leftrightarrow (x, y) \in \text{hal}(\mathcal{U})$$

where $\text{hal}(\mathcal{U}) = \bigcap_{U \in {}^{\sigma}\mathcal{U}} U$ is the halo of the filter $\mathcal{U}$ on $X \times X$. The external set $\text{hal}_{\mathcal{U}}(x) = \{y \in X : y \simeq x\}$ is called the *uniform halo* (we say also abbreviated as *monad* or *halo*) of x.

Let X be a standard uniform space. The uniform proximity on X is an equivalence relation. Thus it is a halic preorder on X. By theorem 6.3.1 it defines a standard topology on X called the *uniform topology* of X. Unless specified otherwise a uniform space will always be assumed to carry its uniform topology.

Remark 6.6.1. *For any standard x in X, we have, by definition of the uniform topology, $y \simeq x$ for the topological proximity of the uniform topology of X if and only if $x \simeq y$ for the uniform proximity on X. In other words the topological halo and the uniform halo of a standard point are equal, but the property is not true in general for nonstandard points (Exercise 6.6.4).*

6.6.2 Limited, accessible and nearstandard points

Definition 6.6.3. *Let $(X, \mathcal{U})$ be a standard uniform space. A point y in X is said to be* limited *if there is a standard $x \in X$ and a standard $U \in \mathcal{U}$ such that $(x, y) \in U$,* accessible *if for any standard $U \in \mathcal{U}$ there is a standard $x \in X$ such that $(x, y) \in U$ and* nearstandard *if there is a standard $x \in X$ such that $y \simeq x$*[10].

Let $X_L = \{y \in X : \exists^{st} x \in X \ \exists^{st} U \in \mathcal{U} \ (x, y) \in U\}$ be the external-set of limited points in X. Let $X_A = \{y \in X : \forall^{st} U \in \mathcal{U} \ \exists^{st} x \in X \ (x, y) \in U\}$ be the external-set of accessible points in X. Let $X_{NS} = \{y \in X : \exists^{st} x \in X \ \forall^{st} U \in \mathcal{U} \ (x, y) \in U\}$ be the external-set of nearstandard points in X. We have the following inclusions

$$X_{NS} \subset X_A \subset X_L \subset X$$

Example: For a standard metric (or pseudo-metric) space (X, d) the external-sets of limited, accessible and nearstandard points are
$$X_L \quad = \{y \in X : \exists^{st} x \in X \ \exists^{st} r > 0 \ d(x, y) < r\}$$
$$X_A \quad = \{y \in X : \forall^{st} r > 0 \ \exists^{st} x \in X \ d(x, y) < r\}$$
$$X_{NS} \ = \{y \in X : \exists^{st} x \in X \ \forall^{st} r > 0 \ d(x, y) < r\}$$

Definition 6.6.4. *The space X is said to be:* bounded *if every point in X is limited, that is, $X = X_L$,* precompact *if every point in X is accessible, that is, $X = X_A$,* compact *if every point in X is nearstandard, that is, $X = X_{NS}$,* preproper *if every limited point in X is accessible, that is, $X_L = X_A$,* proper *if every limited point in X is nearstandard, that is, $X_L = X_{NS}$,* complete *if every accessible point in X is nearstandard, that is, $X_A = X_{NS}$.*

See Exercise 6.6.1 for the equivalent internal definitions.

Theorem 6.6.1. *The topological space X is*
1. compact if and only if it is bounded and proper,
2. precompact if and only if it is bounded and preproper,
3. proper if and only if it is preproper and complete,
4. compact if and only if it is precompact and complete.

Proof. There is nothing to prove, just read the definitions[11].   □

---

[10] This terminology differs from the terminology of Lutz and Goze ([85] p. 94) where our *nearstandard* (resp. *inaccessible*) points was called *approachable* (resp. *uniformly unapproachable*).

[11] It sounds like a miracle. In the usual approach the last property is not trivial at all and needs some work (see for instance the proof in the simpler case of metric spaces given by Dieudonné [56] p. 58). It should be noticed that the usual definitions of these notions make use of the different concepts of coverings, Cauchy filters, convergent filters... while the external definitions proposed here use only some properties of the points of the space.

6.6.3 The external definition of a uniformity

The purpose of this section is to show that, a *halic equivalence relation* $\mathcal{E}$ on the standard set X gives rise to a unique standard uniformity on X whose associated uniform proximity coincides with the relation $\mathcal{E}$. This fact had been already noticed by Stroyan and Luxembourg ([108] p. 211).

Definition 6.6.5. *A* halic equivalence relation *on the standard set X is a halic binary formula $\mathcal{E}$ that is reflexive, symmetric and transitive on X.*

The external-subset $e = \{(x, y) \in X \times X : \mathcal{E}[x, y]\}$ contains the diagonal Δ; it is symmetric and transitive, that is, $e^{-1} = e$ and $e \circ e \subset e$.

Theorem 6.6.2. *Let $\mathcal{E}$ be a halic equivalence relation on the standard set X. Then there is a unique standard uniformity $\mathcal{U}$ on X such that for any $x, y \in X$ we have $x \simeq y \Leftrightarrow \mathcal{E}[x, y]$, that is, $hal(\mathcal{U}) = e$.*

Proof. The proof is similar to that of Theorem 6.3.2. We define the uniformity by $\mathcal{U} = {}^S\{U \subset X \times X : e \subset U\}$. Then $\mathcal{U}$ satisfies the required condition (Exercise 6.6.3). □

The standard uniformity on X whose existence is asserted by Theorem 6.6.2 is called the *uniformity defined by the formula* $x \simeq y \Leftrightarrow \mathcal{E}[x, y]$.

Let X be a standard topological space. By the symmetrization of its topological proximity we obtain a halic equivalence relation on X. By Theorem 6.6.2 this relation defines a standard uniformity on X. The uniform topology associated to this uniformity is in general finer than the given topology on X (Exercise 6.6.2). A topological space (X, τ) is said to be *uniformizable* if there is a uniformity on X whose uniform topology is the topology τ of X. Consequently a standard topological space is uniformizable if and only if its topology may be defined by a halic equivalence relation. Weil [118] (see also [77] p. 188) gave the characterization of the uniformizable topological spaces: a topological space is uniformizable if and only if it is *completely regular*. I don't know if the external approach presented here may be of valuable help in the proof of Weil's theorem. However this approach enables us to give a nice description of the uniformity of a compact Hausdorff space ([24] p. II.27)

Theorem 6.6.3. *A compact Hausdorff topological space is uniformizable.*

Proof. By transfer we may assume that the space is standard. In a standard compact Hausdorff topological space X the halos of the standard points form a partition of X. This partition defines an external equivalence relation, denoted by c, on X. It remains to show that c is a halic relation. We have:

$$(x, y) \in c \Leftrightarrow \exists^{st} z \in X \ \ x, y \in \mathrm{hal}\,(z) \Leftrightarrow \exists z \in X \ \ x, y \in \mathrm{hal}\,(z)$$

The first equivalence is the definition of c. The second equivalence is a consequence of the compactness of X. The external formula on the right may be written as $\exists z \in X \ \ \forall^{st}V \in \mathcal{V}(z) \ \ x,y \in V$. Moreover a formula of type $\exists z \ \forall^{st}v \ \ A$, where A is internal, is a halic formula since it is equivalent by idealization to the formula $\forall^{st} \ {}^{fin}v' \ \exists z \ \forall v \in v' \ A$.   □

Exercises for Section 6.6

1. Give the internal definition of bounded, proper and prepoper. Show that the definition of compact, precompact and complete in this section are equivalent to the usual ones.

2. What is the topology on R obtained by symmetrization of the topological proximity associated to the usual topology on R ?

3. Give the details of the proof of Theorem 6.6.2.

4. Prove that the uniform topology defined in this section is the same as the usual one. Show that the uniform halo and the topological halo of a point x are not equal unless x is standard.

5. Show that the separation axioms T_0, T_1 and T_2 are equivalent for uniform spaces.

6. A mapping $f : X \to Y$ of the standard uniform space X into the standard uniform space Y is said to be uniformly continuous on X if $f(x) \simeq f(y)$ for any $x \simeq y$. Prove that this is equivalent to the usual definition. Prove that a continuous mapping on a compact uniform space is uniformly continuous.

6.7 Answers to the exercises

Exercises for Section 6.1, page 113

1. Let y be in any standard element of $\mathcal{V}(x)$. Then y is in any standard open set which contains x, since such an open set is a standard neighbourhood of x. Thus the implication $\forall^{st}V \in \mathcal{V}(x) \ \ y \in V \Rightarrow y \simeq x$ is true. For the converse, let $y \simeq x$ and let V be standard in $\mathcal{V}(x)$. Then x belongs to the interior V^o of V which is a standard open set. Since $y \simeq x$, we have that $y \in V^o$ and hence $y \in V$.

2. Let $\mathcal{B}$ be a standard base (the proof for a subbase is similar). Let $y \in \mathrm{hal}\,(x)$. Let B be standard in $\mathcal{B}$ such that $x \in B$. Since B is a standard open set containing x we have $y \in B$. So, $y \in \mathrm{hal}\,_\mathcal{B}(x)$. Assume now that x is standard. Let $y \in \mathrm{hal}\,_\mathcal{B}(x)$. Let U be a standard open set such that $x \in U$. By transfer, there is a standard $B \in \mathcal{B}$ such that $x \in B \subset U$. Thus $y \in B$, so $y \in U$. Hence $y \in \mathrm{hal}\,(x)$.

The open intervals form a base $\mathcal{B}$ of the usual topology of R. We have $\mathrm{hal}\,_\mathcal{B}(\frac{1}{\omega}) = \mathrm{hal}\,(0) \cap (0,1)$ for any unlimited odd integer ω. However $V = \bigcup_{n=1}^{\infty}(\frac{1}{2n+2}, \frac{1}{2n})$ is a standard open set that contains $\frac{1}{\omega}$, hence $\mathrm{hal}\,(\frac{1}{\omega}) \subset V$, so $\mathrm{hal}\,_\mathcal{B}(\frac{1}{\omega}) \neq \mathrm{hal}\,(\frac{1}{\omega})$.

3. Let A, B be such that $A \subset \mathcal{B}$. Let F be a standard closed set such that $B \subset F$. Then $A \subset \mathcal{B} \subset \mathcal{F} = F$. Assume now that A is standard. Let

$B \subset X$ be such that $\forall^{st} F \in \mathcal{C}(B \subset F \Rightarrow A \subset F)$. Taking negations we have $\forall^{st} U \in \mathcal{O}(A \cap U \neq \emptyset \Rightarrow B \cap U \neq \emptyset)$. Let x be standard in A. For any standard open neighbourhood U of x, we have $A \cap U \neq \emptyset$ and hence $B \cap U \neq \emptyset$. By Proposition 6.1.2 x is in $\mathcal{B}$. We have shown that every standard point in A is in $\mathcal{B}$, so by transfer every point in A is in $\mathcal{B}$. Thus $A \subset \mathcal{B}$.

4. Let A, B be such that $\mathcal{B} \subset A$. Let K be a standard compact set that intersects B. Let $y \in B \cap K$. Since K is compact there is a standard $x \in K$ such that $y \simeq x$. Thus $x \in {}^{o}B$. Since ${}^{o}B \subset A$ we have $x \in A$, so $A \cap K \neq \emptyset$. Assume now that X is locally compact and A is a standard closed subset of X. Let $B \subset X$ be such that $\forall^{st} K \in \mathcal{K}(B \cap K \neq \emptyset \Rightarrow A \cap K \neq \emptyset)$. Let x be standard in $\mathcal{B}$ and let U be a standard neighbourhood of x. Since X is locally compact there is a standard compact neighbourhood K of x such that $K \subset U$. We have $B \cap K \neq \emptyset$. Then $A \cap K \neq \emptyset$, so $A \cap U \neq \emptyset$. Thus $x \in {}^{o}A = A$, because A is a closed standard set. We have shown that every standard point in $\mathcal{B}$ is in A, so by transfer every point in $\mathcal{B}$ is in A. Thus $A \subset \mathcal{B}$.

5. Let A, B be such that $\mathcal{B} \subset A$. Let U be a standard open set such that $A \subset U$. Let $y \in B$. Since X is compact, there is a standard x in X such that $y \simeq x$. By definition of the shadow of B we have $x \in {}^{o}B$. Since ${}^{o}B \subset A$ we have $x \in A$. Hence $x \in U$. Since U is a standard open set, x is standard and $y \simeq x$, we have $y \in U$. Thus $B \subset U$.

Exercises for Section 6.2, page 116

1. See your favourite treatise on nonstandard analysis !

We shall use frequently the reduction algorithm of external formulas of Nelson [91]. Let us illustrate this algorithm with the external definitions of "f is continuous at x" and "X is compact". The first formula in question is $\forall y \ (y \simeq x \Rightarrow f(y) \simeq f(x))$, that is,

$$\forall y \ (\forall^{st} W \in \mathcal{V}(x) \ y \in W \Rightarrow \forall^{st} V \in \mathcal{V}(f(x)) \ f(y) \in V)$$

By idealization we obtain

$$\forall^{st} V \in \mathcal{V}(f(x)) \ \exists^{st \ fin} W' \subset \mathcal{V}(x) \ \forall y \ (\forall W \in W' \ y \in W \Rightarrow f(y) \in V)$$

To say $\forall W \in W' \ y \in W$ is the same as saying $y \in U$ where $U = \bigcap_{W \in W'} W$, and so our formula is equivalent to

$$\forall^{st} V \in \mathcal{V}(f(x)) \ \exists^{st} U \in \mathcal{V}(x) \ \forall y \ (y \in U \Rightarrow f(y) \in V)$$

Now, since f and x are standard, we can apply transfer; this simply removes the superscripts "st", and we obtain the usual definition of the continuity of the mapping f at x.

Now consider the second formula $\forall y \in X \ \exists^{st} x \in X \ y \simeq x$, that is,

$$\forall y \in X \ \exists^{st} x \in X \ \forall^{st} U \in \mathcal{V}(x) \ y \in x$$

By standardization, introducing the standard function $\tilde{U} : X \to \mathcal{P}(X)$ that associates to each $x \in X$ a neighbourhood $\tilde{U}(x) \in \mathcal{V}(x)$, the formula becomes

$$\forall^{st}\tilde{U} \ \forall y \in X \ \exists^{st} x \in X \ \ y \in \tilde{U}(x)$$

By idealization we obtain

$$\forall^{st}\tilde{U} \ \exists^{st \ fin} x' \subset X \ \forall y \in X \ \exists x \in x' \ \ y \in \tilde{U}(x)$$

Now, since X is standard, we can apply transfer; this simply removes the superscripts "st", and we obtain the usual definition of the compactness of the topological space X.

2. $A \subset A_{NS}$ means A is compact and $A_{NS} \subset A$ means A is open.

3. By transfer we may assume that all the data are standard. Let τ_1 be a topology that is strictly finer τ. Then $\forall^{st} x \in X \ \ \mathrm{hal}_1(x) \subset \mathrm{hal}(x)$ and $\exists^{st} y \in X \ \mathrm{hal}(y) \setminus \mathrm{hal}_1(y) \neq \emptyset$. Let $z \in \mathrm{hal}(y) \setminus \mathrm{hal}_1(y)$. If τ_1 were compact then $\exists^{st} x \in X \ \ z \in \mathrm{hal}_1(x) \subset \mathrm{hal}(x)$. Since τ is separated, $x = y$. Thus $z \in \mathrm{hal}_1(y)$, which is a contradiction. Let τ_2 be a topology that is strictly coarser than τ. Then $\forall^{st} x \in X \ \ \mathrm{hal}(x) \subset \mathrm{hal}_2(x)$ and $\exists^{st} y \in X \ \mathrm{hal}_2(y) \setminus \mathrm{hal}(y) \neq \emptyset$. Let $z \in \mathrm{hal}_2(y) \setminus \mathrm{hal}(y)$. Since τ is compact $\exists^{st} x \in X \ \ z \in \mathrm{hal}(x) \subset \mathrm{hal}_2(x)$. Thus $\mathrm{hal}_2(x) \cap \mathrm{hal}_2(y) \neq \emptyset$ and $x \neq y$, that is, the topology τ_2 is not separated.

4. Let $y \simeq x$ and let A be a standard subset of X. If $y \in A$ then $x \in \overline{A}$. For the converse direction, assume that y is not infinitely close to x, then there is a standard open set U such that $x \in U$ and $y \notin U$. From $y \in X \setminus U$ we deduce $x \in \overline{X \setminus U} = X \setminus U$, a contradiction.

5. Assume that X satisfies this axiom. Let x and A be standard such that $\mathrm{hal}(x) \cap \mathrm{hal}(A) \neq \emptyset$. Since $\mathrm{hal}(A) = \bigcup_{y \in A} \mathrm{hal}(y)$, there is y in A such that $\mathrm{hal}(x) \cap \mathrm{hal}(y) \neq \emptyset$, hence $y \simeq x$. Thus $x \in \overline{A}$. Consequently X is regular. For the converse let x be standard and let y be such that $\mathrm{hal}(x) \cap \mathrm{hal}(y) \neq \emptyset$. Let A be a standard subset of X such that $y \in A$. Then $\mathrm{hal}(x) \cap \mathrm{hal}(A) \neq \emptyset$, that is, $x \in \overline{A}$. By the previous exercise we obtain that $y \simeq x$.

6. Assume that X is regular and T_0. Let x and y be standard in X such that $\mathrm{hal}(x) \cap \mathrm{hal}(y) \neq \emptyset$. By the previous exercise we have $y \simeq x$ and $x \simeq y$, hence $x = y$. Thus the space is separated. Assume now that X is T_1 and normal. Let $x \in X$ and $A \subset X$ be standard such that $\mathrm{hal}(x) \cap \mathrm{hal}(A) \neq \emptyset$. Hence $\overline{\{x\}} \cap \overline{A} \neq \emptyset$. By T_1 we have $\overline{\{x\}} = \{x\}$, so $x \in \overline{A}$. Thus the space is regular.

7. By transfer we assume that the compact Hausdorff space X is standard. Let A and B be standard subsets of X such that $\mathrm{hal}(A) \cap \mathrm{hal}(B) \neq \emptyset$. There is $x \in A$ and $y \in B$ such that $\mathrm{hal}(x) \cap \mathrm{hal}(y) \neq \emptyset$. Since X is compact there is a standard $x_0 \in \overline{A}$ and a standard $y_0 \in \overline{B}$ such that $x \simeq x_0$ and $y \simeq y_0$. Hence $\mathrm{hal}(x_0) \cap \mathrm{hal}(y_0) \neq \emptyset$. Since X is Hausdorff $x_0 = y_0$, so $\overline{A} \cap \overline{B} \neq \emptyset$. Thus the space is normal. Assume now that X is a regular compact space. Let A and B be standard subsets of X such that $\mathrm{hal}(A) \cap \mathrm{hal}(B) \neq \emptyset$. There is $x \in A$ such that $\mathrm{hal}(x) \cap \mathrm{hal}(B) \neq \emptyset$. Since X is compact there is a standard

$x_0 \in \overline{A}$ such that $x \simeq x_0$. Hence $\operatorname{hal}(x_0) \cap \operatorname{hal}(B) \neq \emptyset$. Since X is regular $x_0 \in \overline{B}$, so $\overline{A} \cap \overline{B} \neq \emptyset$. Thus the space is normal.

8. Let $(x', y') \in \operatorname{hal}(x, y)$. Let U be a standard open set of X containing x. Let V be a standard open set of Y containing y. Then $U \times V$ is a standard open set of $X \times Y$ containing (x, y). Thus $(x', y') \in U \times V$, that is to say $x' \in U$ and $y' \in V$. Hence $x' \in \operatorname{hal}(x)$ and $y' \in \operatorname{hal}(y)$. Assume now that x and y are standard and let $(x', y') \in \operatorname{hal}(x) \times \operatorname{hal}(y)$. A base $\mathcal{B}$ of the product topology on $X \times Y$ is given by the products $U \times V$ where U is an open set of X and V is an open set of Y. Let $U \times V$ be any standard element of $\mathcal{B}$. Then $(x', y') \in U \times V$. Hence $(x', y') \in \operatorname{hal}_{\mathcal{B}}(x, y) = \operatorname{hal}(x, y)$.

9. Using the previous exercise and Proposition 6.1.3 we have

$$
\operatorname{hal}(A \times B) = \bigcup_{(x,y) \in A \times B} \operatorname{hal}(x, y) \subset \bigcup_{(x,y) \in A \times B} \operatorname{hal}(x) \times \operatorname{hal}(y)
$$

$$
= \bigcup_{x \in A} \operatorname{hal}(x) \times \bigcup_{y \in B} \operatorname{hal}(y) = \operatorname{hal}(A) \times \operatorname{hal}(B)
$$

for all $A \subset X$ and $B \subset Y$. Assume now that the subsets A and B are standard and compact. Let $x' \in \operatorname{hal}(A)$ and $y' \in \operatorname{hal}(B)$. By Proposition 6.1.3 there exist $x \in A$ and $y \in B$ such that $x' \in \operatorname{hal}(x)$ and $y' \in \operatorname{hal}(y)$. Since A and B are standard and compact there exist a standard u in A and a standard v in B such that $x \in \operatorname{hal}(u)$ and $y \in \operatorname{hal}(v)$. By the transitivity of topological proximity we have that $x' \in \operatorname{hal}(u)$ and $y' \in \operatorname{hal}(v)$. Thus $(x', y') \in \operatorname{hal}(u) \times \operatorname{hal}(v) = \operatorname{hal}(u, v)$. Hence $(x', y') \in \operatorname{hal}(A \times B)$.

10. By transfer we may assume that all the data X, Y, A, B and W are standard. By the previous exercise $\operatorname{hal}(A) \times \operatorname{hal}(B) = \operatorname{hal}(A \times B) \subset W$. By Proposition 6.1.4 there exist infinitesimal neighbourhoods U and V of A and B such that $U \subset \operatorname{hal}(A)$ and $V \subset \operatorname{hal}(B)$. Thus $U \times V$ is a neighbourhood of $A \times B$ and $U \times V \subset W$.

Exercises for Section 6.3, page 122

1. The family of sets of the form $\{x \in Y : x_a \in U\}$ where U is an open subset of X_a is a subbase of the product topology ([77] p. 90). Thus, for any standard x in Y and any y in Y, we have $y \simeq x$ for the product topology, if and only if (Exercise 6.1.2) $\forall^{st} a \in A \ \forall^{st} U \in \mathcal{O}_a (x_a \in U \Rightarrow y_a \in U)$. The family of sets of the form $\{f \in X^A : f(K) \subset U\}$ where K is a compact subset of A and U is an open subset of X is a subbase for the compact-open topology ([77] p. 221). We may also use Exercise 6.1.2.

2. By transfer we may assume that all the data are standard. Let τ be a standard topology on F that is jointly continuous on compact subsets of X. Let f be standard in F. Let $g \simeq f$ for the topology τ. Let K be a standard compact subset of X. Let U be a standard open subset of Y such that $f(K) \subset U$. For any $y \in K$ there is a standard $x \in K$ such that $y \simeq x$. By the continuity of the mapping P we have $g(y) \simeq f(x)$. Since $f(x) \in U$

and U is open we have $g(y) \in U$, so $g(K) \subset U$. Thus $g \simeq f$ for the compact open topology. Hence τ is finer than the compact-open topology.

3. Let τ_0 be the topology defined in this exercise. By transfer the mappings f_i are all continuous for the topology τ_0 since they are continuous for standard $i \in I$. Let τ be a standard topology on X such that all the mappings f_i are continuous. Let x be standard in X. Let $y \simeq x$ for the topology τ. We have $f_i(y) \simeq f_i(x)$ for any standard $i \in I$, so $y \simeq x$ for the topology τ_0. Hence τ is finer than τ_0.

Exercises for Section 6.4, page 125

1. The family of sets of the form $\{A \in \mathcal{P}(X) : A \subset U\}$ where U is an open subset of X is a subbase of the upper semi-finite topology of Michael ([87] p. 179). The family of sets of the form $\{A \in \mathcal{P}(X) : A \cap U \neq \emptyset\}$ where U is an open subset of X is a subbase of the lower semi-finite topology of Michael ([87] p. 179) or the local topology of Effros ([57] p. 931). The family of sets of the form $\{A \in \mathcal{P}(X) : A \subset K\}$ where K is a the complement of a compact subset of X is a subbase for the global topology of Effros ([57] p. 931). Finally, we use Exercise 6.1.2.

2. By transfer we may assume that X and E are standard. Let A be standard in the closure of $\mathcal{P}(E)$. Then there is $B \in \mathcal{P}(E)$ such that $B \simeq A$ for the lower $\mathcal{C}$-topology. Hence $A \subset \mathcal{B} \subset \mathcal{E} = E$, because E is closed, so $A \in \mathcal{P}(E)$. By transfer every point in the closure of $\mathcal{P}(E)$ is in $\mathcal{P}(E)$. Thus $\mathcal{P}(E)$ is closed.

3. By transfer we may assume that X and E are standard. Let A be standard in $\mathcal{P}(E)$. Let $B \in \mathcal{P}(X)$ be such that $B \simeq A$ for the upper $\mathcal{O}$-topology. Hence $B \subset \mathrm{hal}\,(A) \subset \mathrm{hal}\,(E) = E$, because E is open, so $B \in \mathcal{P}(E)$. Thus $\mathcal{P}(E)$ is open.

4. By transfer, we may assume that X is standard. Let A and B be standard closed subsets of X such that $B \simeq A$ and $A \simeq B$ for the lower $\mathcal{C}$-topology. By Proposition 6.4.1 $A \subset \mathcal{B} = B$ and $B \subset \mathcal{A} = A$, so $A = B$. Thus the space $\mathcal{C}(X)$ is T_0.

Exercises for Section 6.5, page 130

1. By transfer, we may assume that f and x are standard. Let us denote by $\mathcal{O}$ the set of open sets of Y and by $\mathcal{V}(x)$ the set of neighbourhoods of x. We have proved (Exercise 6.1.3) that $f(x) \subset {}^o f(y)$ is equivalent to $\forall^{st} U \in \mathcal{O}(f(x) \cap U \neq \emptyset \Rightarrow f(y) \cap U \neq \emptyset)$. Then the formula $\forall y \in X[y \simeq x \Rightarrow f(x) \subset {}^o f(y)]$ is equivalent to:

$$\forall y \in X[\forall^{st} V \in \mathcal{V}(x)\ \ y \in V \Rightarrow \forall^{st} U \in \mathcal{O}(f(x) \cap U \neq \emptyset \Rightarrow f(y) \cap U \neq \emptyset)]$$

The reduction algorithm of external formulas transforms this into:

$$\forall U \in \mathcal{O}\ \exists V \in \mathcal{V}(x)\ \forall y \in X[y \in V \Rightarrow (f(x) \cap U \neq \emptyset \Rightarrow f(y) \cap U \neq \emptyset)]$$

The formula in the brackets is equivalent to $f(x) \cap U \neq \emptyset \Rightarrow (y \in V \Rightarrow f(y) \cap U \neq \emptyset)$. Finally we have:

$$\forall U \in \mathcal{O}[f(x) \cap U \neq \emptyset \Rightarrow \exists V \in \mathcal{V}(x) \ \forall y \in V \ f(y) \cap U \neq \emptyset]$$

This is the internal definition of lower semi-continuity of f at x.

By definition of the halo of a subset, $f(y) \subset \mathrm{hal}\,(f(x))$ is equivalent to $\forall^{st} U \in \mathcal{O}(f(x) \subset U \Rightarrow f(y) \subset U)$. Then the formula $\forall y \in X \ [y \simeq x \Rightarrow f(y) \subset \mathrm{hal}\,(f(x))]$ is equivalent to:

$$\forall y \in X[\forall^{st} V \in \mathcal{V}(x) \ y \in V \Rightarrow \forall^{st} U \in \mathcal{O} \ (f(x) \subset U \Rightarrow f(y) \subset U)]$$

The reduction algorithm of external formulas transforms this into:

$$\forall U \in \mathcal{O} \ \exists V \in \mathcal{V}(x) \ \forall y \in X[y \in V \Rightarrow (f(x) \subset U \Rightarrow f(y) \subset U)]$$

Finally we have:

$$\forall U \in \mathcal{O}[f(x) \subset U \Rightarrow \exists V \in \mathcal{V}(x) \ \forall y \in V \ f(y) \subset U]$$

This is the internal definition of the upper semi continuity of f at x.

2. Let y be standard in Y. To say $\forall i \in \mathrm{hal}\,(\mathcal{F}) \ y \in {}^{\circ}A_i$ is the same as saying

$$\forall i \in I \ (\forall^{st} S \in \mathcal{F} \ i \in S \Rightarrow \forall^{st} U \in \mathcal{V}(y) \ U \cap A_i \neq \emptyset)$$

The reduction algorithm of external formulas transforms this formula into

$$\forall U \in \mathcal{V}(y) \ \exists S \in \mathcal{F} \ \forall i \in I \ (i \in S \Rightarrow U \cap A_i \neq \emptyset)$$

which is the internal definition of y being in the lower limit of the family (A_i). To say $\exists i \in \mathrm{hal}\,(\mathcal{F}) \ y \in {}^{\circ}A_i$ is the same as saying

$$\exists i \in I \ \forall^{st} S \in \mathcal{F} \ i \in S \ \& \ \forall^{st} U \in \mathcal{V}(y) \ U \cap A_i \neq \emptyset)$$

The reduction algorithm of external formulas transforms this formula into

$$\forall U \in \mathcal{V}(y) \ \forall S \in \mathcal{F} \ \exists i \in I \ (i \in S \& U \cap A_i \neq \emptyset)$$

which is the internal definition of y being in the upper limit of the family (A_i). The last property is an immediate consequence of the two first.

3. To say $\forall i \in \mathrm{hal}\,(\mathcal{F}) \ x_i \simeq x$ is the same as saying $\forall i \in I \ (\forall^{st} S \in \mathcal{F} \ i \in S \Rightarrow \forall^{st} U \in \mathcal{V}(x) \ x_i \in U)$. The reduction algorithm of external formulas transforms this formula into $\forall U \in \mathcal{V}(x) \ \exists^{fin} S' \subset \mathcal{F} \ \forall i \in I(\forall S \in S' \ i \in S \Rightarrow x_i \in U)$. To say $\forall S \in S' \ i \in S$ is the same as saying $i \in T$ for $T = \bigcap_{S \in S'} S$. Thus the formula is equivalent to $\forall U \in \mathcal{V}(x) \ \exists T \in \mathcal{F} \ \forall i \in I(i \in T \Rightarrow x_i \in U)$.

4. By transfer we may assume that all the data are standard. Let (x, y) be standard in $X \times Y$. Then (x, y) is in the closure of the graph of f if and only if there is x' in X and $z \in f(x')$ such that $x' \simeq x$ and $z \simeq y$, so $y \in {}^{\circ}f(x')$ for some $x' \simeq x$, that is, $y \in \limsup_{x' \to x} f(x')$. Moreover f is

lower semicontinuous at x if and only if $f(x) \subset \mathcal{G}(x')$ for all $x' \simeq x$, that is, $f(x) \subset \liminf_{x' \to x} f(x')$.

5. By transfer we may assume that X is standard. Let A and B be standard closed subsets of X such that there is a subset C satisfying $C \simeq A$ and $C \simeq B$ for the Choquet topology. By Proposition 6.5.2 we obtain $A = C = B$. Thus the space $\mathcal{C}(X)$ is Hausdorff.

6. Let x be standard in the closure of $\liminf_{\mathcal{F}} A_i$. Then every open neighbourhood of x contains a point of $\liminf_{\mathcal{F}} A_i$, so by transfer every standard open neighbourhood V of x contains a standard point y of $\liminf_{\mathcal{F}} A_i$. For a standard point y of $\liminf_{\mathcal{F}} A_i$, for any $i \in \mathrm{hal}\,(\mathcal{F})$ there is z_i in A_i with $z_i \simeq y$, so z_i is in V. That is to say $\forall i \in \mathrm{hal}\,(\mathcal{F})$ $\forall^{st} V \in \mathcal{V}(x)$ $A_i \cap V \neq \emptyset$. Hence, by Proposition 6.1.2, $\forall i \in \mathrm{hal}\,(\mathcal{F})$ $x \in {}^{\alpha}A_i$. Thus x is in $\liminf_{\mathcal{F}} A_i$. We have shown that every standard point x in the closure of $\liminf_{\mathcal{F}} A_i$ is in $\liminf_{\mathcal{F}} A_i$. By transfer every point in the closure of $\mathcal{E}$ is in $\mathcal{E}$. Thus $\liminf_{\mathcal{F}} A_i$ is closed. For the upper limit the proof is similar[12].

Exercises for Section 6.6, page 134

1. Let $(X, \mathcal{U})$ be a standard uniform space. The external definition of "precompact" and "compact" are respectively:

$$\forall x \ \forall^{st}U \ \exists^{st}y \ x \in U[y]$$

$$\forall x \ \exists^{st}y \ \forall^{st}U \ x \in U[y]$$

It is understood that x and y range over X and that U ranges over $\mathcal{U}$. The reduction algorithm of external formulas transforms these formulas into:

$$\forall U \ \exists^{fin}y' \subset X \ \forall x \ \exists y \in y' \ x \in U[y]$$

$$\forall \tilde{U} \ \exists^{fin}y' \subset X \ \forall x \ \exists y \in y' \ x \in \tilde{U}_y[y]$$

Where $\tilde{U}$ is a mapping that associates with each $y \in X$ an entourage $\tilde{U}_y \in \mathcal{U}$. Thus the space X is:

- precompact if and only if it is covered by a finite number of sets $U[y_1] \ldots U[y_n]$ for any given $U \in \mathcal{U}$. This is equivalent to the usual precompactness (see [77] p. 198 or [24] Theorem 3 p.II.29.),
- compact if and only if any covering of X by sets of type $U[y]$ has a finite subcovering. This is equivalent to the compactness of the space X for the uniform topology associated with its uniformity.

---

[12] It should be noticed that, when we use the internal definitions, the proofs of the closeness of the upper and lower limits are trivial. I gave this "long" proof since I want to avoid giving you the impression that the nonstandard approach to topological spaces is intended to replace the usual one. It is only an addition, so in some cases the standard proof is better than the nonstandard one.

The external definitions of "bounded", "preproper" and "proper" are respectively:

$$\forall x \; \exists^{st} U \; \exists^{st} y \; x \in U[y]$$

$$\forall x (\exists^{st} U \; \exists^{st} y \; x \in U[y] \Rightarrow \forall^{st} V \; \exists^{st} z \; x \in V[z])$$

$$\forall x (\exists^{st} U \; \exists^{st} y \; x \in U[y] \Rightarrow \exists^{st} z \; \forall^{st} V \; x \in V[z])$$

It is understood that x, y and z range over the standard set X and that U and V range over $\mathcal{U}$. The reduction algorithm of external formulas transforms these formulas into:

$$\exists^{fin} U' \subset \mathcal{U} \; \exists^{fin} y' \subset X \; \forall x \; \exists U \in U' \; \exists y \in y' \; x \in U[y]$$

$$\forall U \; \forall y \; \forall V \; \exists^{fin} z' \subset X \; \forall x (x \in U[y] \Rightarrow \exists z \in z' \; x \in V[z])$$

$$\forall U \; \forall y \; \forall \tilde{V} \; \exists^{fin} z' \subset X \; \forall x (x \in U[y] \Rightarrow \exists z \in z' \; x \in \tilde{V}_z[z])$$

Where $\tilde{V}$ is a mapping that associates with each $x \in X$ an entourage $\tilde{V}_z \in \mathcal{U}$. Thus the space X is:

- bounded if and only if it is covered by a finite number of sets $U_1[y_1]$ ... $U_n[y_n]$,
- preproper if and only if every set $U[y]$ is covered by a finite number of sets $V[z_1]$ ... $V[z_n]$ for any given $V \in \mathcal{U}$, that is, every set $U[y]$ is *relatively precompact*,
- proper if and only if for any set $U[y]$ and any covering of X by sets of type $V[z]$ there is a finite subcovering of $U[y]$, that is, every set $U[y]$ is *relatively compact*.

Let us prove now that our definition of completeness is equivalent to, the usual one. Recall that a *Cauchy filter* on a uniform space $(X, \mathcal{U})$ is a filter $\mathcal{F}$ on X such that for any $U \in \mathcal{U}$ there is $A \in \mathcal{F}$ such that $A \times A \subset U$. Recall that a filter $\mathcal{F}$ on a topological space X is said to *converge to a point* $x \in X$ if for any neighbourhood V of x there is $A \in \mathcal{F}$ such that $A \subset V$. The uniform space X is complete if every Cauchy filter on X converges to a point of X.

Proposition 6.7.1. *a) A standard filter $\mathcal{F}$ on the standard uniform space X is a Cauchy filter if and only if $hal(\mathcal{F}) \times hal(\mathcal{F}) \subset hal(\mathcal{U})$.*
b) A standard filter $\mathcal{F}$ on the standard topological space X converges to a standard point $x \in X$ if and only if $hal(\mathcal{F}) \subset hal(x)$.
c) Let $\mathcal{F}$ be a standard filter. A set S contains $hal(\mathcal{F})$ if and only if it contains a standard element A of $\mathcal{F}$.
d) Let $\mathcal{F}$ be a standard Cauchy filter on the standard uniform space X and let $z \in X$ be standard such that there is $y \in hal(\mathcal{F})$ satisfying $y \simeq z$. Then $\mathcal{F}$ converges to z.

Proof.   a. To say $\mathrm{hal}\,(\mathcal{F}) \times \mathrm{hal}\,(\mathcal{F}) \subset \mathrm{hal}\,(\mathcal{U})$ is the same as saying $\forall (x,y)(\forall^{st} B \in \mathcal{F}\ (x,y) \in B \times B \Rightarrow \forall^{st} U \in \mathcal{U}\ (x,y) \in U)$. The reduction algorithm of external formulas transforms this formula into $\forall U \in \mathcal{U}\ \exists^{fin} B' \subset \mathcal{F}\ \forall (x,y)(\forall B \in B'\ (x,y) \in B \times B \Rightarrow (x,y) \in U)$. To say $\forall B \in B'\ (x,y) \in B \times B$ is the same as saying $(x,y) \in A \times A$ where $A = \bigcap_{B \in B'} B$ and so our formula is equivalent to $\forall U \in \mathcal{U}\ \exists A \in \mathcal{F}\ \forall(x,y)((x,y) \in A \times A \Rightarrow (x,y) \in U)$, that is, $\forall U \in \mathcal{U}\ \exists A \in \mathcal{F}\ A \times A \subset U$.

b. To say $\mathrm{hal}\,(\mathcal{F}) \subset \mathrm{hal}\,(x)$ is the same as saying $\forall y (\forall^{st} B \in \mathcal{F}\ y \in B \Rightarrow \forall^{st} V \in \mathcal{V}(x)\ y \in V)$. The reduction algorithm of external formulas transforms this formula into $\forall V \in \mathcal{V}(x)\ \exists^{fin} B' \subset \mathcal{F}\ \forall y(\forall B \in B'\ y \in B \Rightarrow y \in V)$. To say $\forall B \in B'\ y \in B$ is the same as saying $y \in A = \bigcap_{B \in B'} B$ and so our formula is equivalent to $\forall V \in \mathcal{V}(x)\ \exists A \in \mathcal{F}\ \forall y(y \in A \Rightarrow y \in V)$, that is, $\forall V \in \mathcal{V}(x)\ \exists A \in \mathcal{F}\ A \subset V$.

c. To say $\mathrm{hal}\,(\mathcal{F}) \subset S$ is the same as saying $\forall y(\forall^{st} B \in \mathcal{F}\ y \in B \Rightarrow y \in S)$. By idealization we obtain $\exists^{st\ fin} B' \subset \mathcal{F}\ \forall y(\forall B \in B'\ y \in B \Rightarrow y \in S)$. To say $\forall B \in B'\ y \in B$ is the same as saying $y \in A = \bigcap_{B \in B'} B$ and so our formula is equivalent to $\exists^{st} A \in \mathcal{F}\ \forall y(y \in A \Rightarrow y \in S)$, that is, $A \subset S$.

d. Let $x \in \mathrm{hal}\,(\mathcal{F})$. Then $(x,y) \in \mathrm{hal}\,(\mathcal{F}) \times \mathrm{hal}\,(\mathcal{F}) \subset \mathrm{hal}\,(\mathcal{U})$, so $x \simeq y$. Since $y \simeq z$, we have $x \simeq z$, so $x \in \mathrm{hal}\,(z)$.   □

Let $(X, \mathcal{U})$ be a standard uniform space such that every accessible point is nearstandard. Let $\mathcal{F}$ be standard Cauchy filter on X. By Proposition 6.7.1.a we have $\mathrm{hal}\,(\mathcal{F}) \times \mathrm{hal}\,(\mathcal{F}) \subset \mathrm{hal}\,(\mathcal{U})$. Let $y \in \mathrm{hal}\,(\mathcal{F})$ and let U be standard in $\mathcal{U}$. The set $S = \{x \in X : (x,y) \subset U\}$ contains the external-set $\mathrm{hal}\,(\mathcal{F})$, so by Proposition 6.7.1.c there is a standard $A \in \mathcal{F}$ such that $A \subset S$. Let x be standard in A, we have $(x,y) \in U$. Thus y is accessible, so there is a standard z such that $y \simeq z$. By Proposition 6.7.1.d $\mathcal{F}$ converges to z. We have shown that every standard Cauchy filter on X converges to a standard point of X. By transfer every Cauchy filter on X converges, that is, X is complete. Conversely let $(X, \mathcal{U})$ be a standard complete uniform space and let x be a accessible point in X. Then for every standard $U \in \mathcal{U}$ there is a standard $y_U \in X$ such that $(x, y_U) \in U$, so by standardization, there is a standard mapping $y : \mathcal{U} \to X$ such that $(x, y_U) \in U$ for every standard $U \in \mathcal{U}$. Let $A_U = \{y_V \in X : V \subset U\ V \in \mathcal{U}\}$. Then $(A_U)_{U \in \mathcal{U}}$ is a base of a standard Cauchy filter $\mathcal{F}$ on X. Indeed, let $U \in \mathcal{U}$ be standard. By axiom (d) of a uniformity there is a standard $V \in \mathcal{U}$ such that $V \circ V \subset U$. Let $V_1 \subset V$, $V_2 \subset V$ be standard. Then $(x, y_{V_1}) \in V_1$ and $(x, y_{V_2}) \in V_2$. Hence $(y_{V_1}, y_{V_2}) \in V_1 \circ V_2 \subset V \circ V \subset U$. By transfer this holds for all $V_1 \subset V$ and $V_2 \subset V$, that is, $A_V \times A_V \subset U$. We have shown that for any standard $U \in \mathcal{U}$ there is a standard $V \in \mathcal{U}$ such that $A_V \times A_V \subset V$, so by transfer for any $U \in \mathcal{U}$ there is $V \in \mathcal{U}$ such that $A_V \times A_V \subset V$, that is, the filter $\mathcal{F}$ generated by $(A_U)_{U \in \mathcal{U}}$ is a standard Cauchy filter and so has a standard limit z. Let us prove that $x \simeq z$. Let $U \in \mathcal{U}$ be standard. There is a standard $V \in \mathcal{U}$ such that $V \circ V \subset U$. Since $\mathcal{F}$ converges to z there is a standard $W \in \mathcal{U}$, which can be chosen such that $W \subset V$, satisfying $(y_W, z) \in V$. Since $(x, y_W) \in W \subset V$, we have $(x, z) \in V \circ V \subset U$. We have shown that (x, z) is in any standard entourage U so $x \simeq z$. Thus every accessible point x in X is nearstandard.

Remarks 6.7.1.

1. *The external definition of "complete" is:*

$$\forall x (\forall^{st} U \; \exists^{st} y \; (x, y) \in U \Rightarrow \exists^{st} z \; \forall^{st} V \; (x, z) \in V)$$

It is understood that x, y and z range over the standard set X and that U and V range over $\mathcal{U}$. The reduction algorithm of external formulas transforms this formula into:

$$\forall \tilde{y} \forall \tilde{V} \exists^{fin} U' \exists^{fin} z' \subset X \forall x (\forall U \in U' (x, \tilde{y}(U)) \in U \Rightarrow \exists z \in z' (x, z) \in \tilde{V}_z)$$

where $\tilde{V}$ is a mapping that associates with each $z \in X$ an entourage $\tilde{V}_z \in \mathcal{U}$ and $\tilde{y}$ is a mapping that associates with each entourage $U \in \mathcal{U}$ a point $\tilde{y}(U) \in X$. This is an internal (but unknown ?) formulation of completeness.

2. It is the discrete topology. Indeed, let x be standard in R. Let $y \in R$ be such that $y \simeq x$ and $x \simeq y$. Then $y = x$ because if y were different from x than $R \setminus \{x\}$ would be an open set containing y but not x.

3. Let $\mathcal{E}[x, y] \equiv \forall^{st} r \; A[x, y, r]$ be an equivalent halic formula. For any set α let us denote by $E_\alpha = \{(x, y) \in X \times X : \forall r \in \alpha \; A[x, y, r]\}$. Then we have $E_F \subset e$ where F is a finite set that contains every standard set and e is the intersection of all $E_{\{r\}}$ for standard r. Moreover we have

$$\mathcal{U} = \{U \subset X \times X : \exists^{fin} \alpha \; E_\alpha \subset U\}$$

Indeed, by definition of $\mathcal{U}$ we have

$$\forall^{st} U \subset X \times X (U \in \mathcal{U} \Leftrightarrow \forall x \forall y (\forall^{st} r \; A[x, y, r] \Rightarrow (x, y) \in U))$$

The reduction algorithm of external formulas transforms this formula into:

$$\forall U \subset X \times X (U \in \mathcal{U} \Leftrightarrow \exists^{fin} \alpha \; \forall x \forall y (\forall r \in \alpha \; A[x, y, r] \Rightarrow (x, y) \in U))$$

Let us prove now that $\mathcal{U}$ is a uniformity on X. By transfer it suffices to show the characterizing properties of a uniformity for their standard data. Properties $(a - d)$ are immediate. The proof of (e) runs as follows. Let U be standard in $\mathcal{U}$. Let $V = E_F$ where F is a finite set that contains every standard element. Then $V \in \mathcal{U}$ and $V \circ V \subset e \circ e = e \subset U$.

Let us prove now that $\text{hal}(\mathcal{U}) = e$. Let (x, y) be in e. Then, by definition of $\mathcal{U}$, (x, y) is in any standard U in $\mathcal{U}$, so (x, y) is in $\text{hal}(\mathcal{U})$. Let (x, y) be in $\text{hal}(\mathcal{U})$. Then (x, y) is in any $E_{\{r\}}$ with r standard, so (x, y) is in e.

4. Let $(X, \mathcal{U})$ be a standard uniform space. The uniform topology is the topology defined by the formula $y \simeq x \Leftrightarrow \forall^{st} U \in \mathcal{U} \; (x, y) \in U$. By Proposition 6.3.4 a subset O of X is open for this topology if and only if for any $x \in O$ there is a finite subset α of $\mathcal{U}$ such that the ball $B(x, \alpha) = \{y \in X : \forall U \in \alpha \; (y, x) \in U\}$ is included in O. But $V = \bigcap_{U \in \alpha} U$

is an entourage, so that $B(x, \alpha) = V[x]$. We have shown that O is open for the uniform topology if and only if for each $x \in O$ there is $V \in \mathcal{U}$ such that $V[x] \subset O$. This is the usual definition of the uniform topology ([77] p. 178). Consider now the usual metric uniformity on R. Let $\varepsilon > 0$ be an infinitesimal real number. Then $\mathrm{hal}\,(\varepsilon) \subset \mathrm{hal}\,(0) \cap (0, 1)$ but $\mathrm{hal}_\mathcal{U}(\varepsilon) = \mathrm{hal}\,(0)$.

5. We have only to prove that $T_0 \Rightarrow T_2$. By transfer we may assume that the uniform T_0 space X is standard. Let x and y be standard in X such that there is z in X satisfying $z \simeq x$ and $z \simeq y$. By the symmetry and the transitivity of the uniform proximity we have $x \simeq y$ and $y \simeq x$, so by T_0 $x = y$, thus the space is Hausdorff.

6. To say $\forall x \forall y (y \simeq x \Rightarrow f(y) \simeq f(x))$, is the same as saying,

$$\forall x \forall y \;\; (\forall^{st} W \in \mathcal{U}_X \;\; (x, y) \in W \Rightarrow \forall^{st} V \in \mathcal{U}_Y \;\; (f(x), f(y)) \in V)$$

The reduction algorithm of external formulas transforms this formula into

$$\forall V \in \mathcal{U}_Y \;\; \exists^{fin} W' \subset \mathcal{U}_X \;\; \forall x \forall y \;\; (\forall W \in W' \;\; (x, y) \in W \Rightarrow (f(x), f(y)) \in V)$$

To say $\forall W \in W' \;\; (x, y) \in W$ is the same as saying $(x, y) \in U$ where $U = \bigcap_{W \in W'} W$, and so our formula is equivalent to

$$\forall V \in \mathcal{U}_Y \;\; \exists U \in \mathcal{U}_X \;\; \forall x \forall y \;\; ((x, y) \in U \Rightarrow (f(x), f(y)) \in V)$$

This is the usual definition of the uniform continuity of f.

Let us prove now that any continuous mapping f on a compact uniform space X is uniformly continuous on X. By transfer we may assume that X and f are standard. Let x and y be such that $x \simeq y$. Since X is compact there exist standard x_0 and y_0 in X such that $x \simeq x_0$ and $y \simeq y_0$. By the continuity of f on X we have $f(x) \simeq f(x_0)$ and $f(y) \simeq f(y_0)$. By the symmetry and the transitivity of the uniform proximity we obtain $f(x) \simeq f(y)$. Thus f is uniformly continuous on X.

7. Neutrices, external numbers, and external calculus

Fouad Koudjeti and Imme van den Berg

7.1 Introduction

In some branches of pure mathematics such as asymptotics, or mathematics applied to astronomy, physics or economics for example, we are often confronted with laborious calculations where we are asked to handle mathematical entities of different orders of magnitude. In many cases if we decide to push the calculations to the last detail , these same calculations become, if not impenetrable, at least unreadable. In such cases, the classical procedure is to replace the expressions which are too detailed by approximations and to give the "order of magnitude" of the error made. Physicians, astronomers, economists, computer scientists and others have some heuristic methods for the calculation of this order of magnitude. Asymptoticians, on their side, have developed formalisms which start by defining the notion of order of magnitude of a function and then give rules which allow some elementary calculations using this notion. The notation of Landau is perhaps a good example. (See [93].) We also mention the "Infinitärcalcül" of Du Bois–Reymond. (See [68].)

A formalism which is particularly pertinent for us is the Theory of Neutrices of J. G. van der Corput developed in the beginning of the sixties (see [34]). The goal of this theory was, among others, to replace calculations involving functions which are over–specified and too difficult to deal with by calculations with different *orders of magnitude of functions*. The aim of this paper is the same, but our approach is different in the sense that we aim to replace numbers in calculations which are overly detailed and laborious by *orders of magnitude of numbers*. Nonstandard Analysis is the right environment for the achievement of such a goal, because it recognizes many different orders of magnitude of numbers.

Calculations based on different orders of magnitude of numbers regularly appear in publications of people working within the framework of Nonstandard Analysis. Such publications concern for instance the notions of slow-fast systems, canards, jumps of trajectories and expansions in ε-shadows , using implicitly the richness of the external sets, and especially the vagueness of their borders. This use is essentially based on logical procedures like the idealisation principle and on permanence principles such as the "Fehrele principle" and the "Cauchy principle". A rich bibliography concerning these principles can be found in [53] and [15]. Our goal is, among others, to systemize these calculations through the introduction of a symbolism for external sets and the development of a calculus on these sets. Our calculations are like those which have been made during the last fifteen years, but have the advantage of being more direct, effective and formally justified.

Let us recall here that the external sets in $\mathbb{R}$ (see [14], [15], [50] and [91]) are not sets in the normal sense of the word, but collections of real numbers obeying an external formula and disobeying at least one theorem of classical mathematics. The external sets play a primordial role in our work, especially those external sets not satisfying the *least upper bound theorem* (every bounded set of $\mathbb{R}$ has a least upper bound). Such sets have fuzzy "bounds", a characteristic which is adequate for modelling the intuitive notion of the *order of magnitude*.

In the next section we will introduce some symbols related to external notions and we will define the notion of *order of magnitude of a number*. In that section we will also treat the approximation of the exponential integral from a didactic point of view, using the notion of order of magnitude. From the third to the fifth section we formally establish the external calculus. In the third section we introduce two classes of external sets of real numbers: the **neutrices** and the **external numbers**. The first part of the fourth section deals with elementary operations concerning neutrices and external numbers, and in the remaining part of it we derive some of the most remarkable properties of neutrices. This will help us to give a canonical form for the neutrices and to calculate in a direct manner the product of two neutrices, and thus of two external numbers. In the fifth section we fix the notion of external distance and we define the least upper bound and the greatest lower bound of collections of external numbers; notions which are necessary if we aim to build an analysis using external numbers. We then move from the consideration of external sets of real numbers to the consideration of external sets of functions. We will speak about external functions and the integration of external functions. In the final section we derive Stirling's formula using the external tools which we have developed throughout this paper.

7.2 Conventions; an example

In our calculations we use the following symbols for external sets of $\mathbb{R}$

- the symbol £ represents the external set of limited real numbers.
- the symbol @ represents the external set of positive appreciable real numbers.
- the symbol $\oslash$ represents the external set of infinitesimal real numbers.
- the symbol ϕ represents the external set of positive unlimited real numbers.

We allow the use of the common algebraic and analytical operations on these symbols. Later on we will come back to these operations in more detail.

Thus, in particular, we have $-\pounds = \pounds$, $\frac{1}{@} = @$, $\log @ = \pounds$, $\exp \pounds = @$, $\pounds + \pounds = \pounds$ and $\oslash\pounds = \oslash$.

The above–mentioned relations are, in fact, set equalities. However, we now wish to make one important convention, concerning the equality symbol. **To indicate strict set identity we reserve the symbol "≡". The symbol "=" represents from now on inclusion.** Although this convention is unusual, it is associated with a certain tradition in classical asymptotics (see for instance [25]). As an example, if ω is an unlimited positive real number then we write $\omega + \sqrt{\omega} = (1 + \oslash)\omega$ just as we are used to writing $x + \sqrt{x} = (1 + o(1))x$ if $x \longrightarrow \infty$.

We say that **two real numbers** a **and** b **are of the same order of magnitude** if

$$a = @b.$$

Thus **the order of magnitude of a real number** a is the external set $@a$.

For introductory and didactic reasons, we present below a simple, direct and, for the moment, heuristic argument applied to the exponential integral. We prove that for any positive infinitesimal ε we have

$$E_1(\varepsilon) = \varepsilon(1 + \oslash).$$

This is the nonstandard analogue of the fact that in the neighbourhood of 0

$$E_1(x) = \int_0^\infty \frac{e^{-\frac{t}{x}}}{1+t}dt \sim x \cdot (1 + o(1)). \qquad (7.1)$$

In the following sections, we will build a formalism in which this same reasoning will be correct.

The integrand

$$I(t) = \frac{e^{-\frac{t}{\varepsilon}}}{1+t}$$

takes its maximum in 0, and $I(0) = 1 = @$. To approach the value of such an integral, we make use of the fact that the integrand has a unique maximum. A first approximation is to integrate on the domain where the integrand is of the order of its maximum. To determine this domain we solve the **external equation**

$$I(t) = @.$$

We proceed as follows:

$$\frac{e^{-\frac{t}{\varepsilon}}}{1+t} = @ \quad \Leftrightarrow \quad -\frac{t}{\varepsilon} - \log(1 + t) = \pounds \qquad (7.2)$$

At this point, one should remark that $\frac{t}{\varepsilon}$ and $\log(1+t)$ are of different orders of magnitude, because for any positive t we have

$$\frac{t}{\varepsilon} + \log(1+t) = \frac{t}{\varepsilon}(1+\oslash).$$

This means that if t is appreciable or unlimited the expression $-\frac{t}{\varepsilon} - \log(1+t)$ is unlimited. In both cases equation (7.2) has no solution.

If t is infinitesimal then $\log(1+t)$ is infinitesimal, so that

$$-\frac{t}{\varepsilon} - \log(1+t) = \pounds \quad \Leftrightarrow \quad -\frac{t}{\varepsilon} = \pounds$$
$$\Leftrightarrow \quad t = \varepsilon\pounds$$

which means that the integrand is appreciable on $\varepsilon\pounds$. The change of variables $t = \varepsilon s$ expands the domain $\varepsilon\pounds$ to the whole of $\pounds$. This suggests the separation of the integration interval into two parts: that of the limited positive real numbers, and that of the positive unlimited real numbers. We get

$$\int_0^\infty \frac{e^{-\frac{t}{\varepsilon}}}{1+t}dt = \varepsilon\left[\int_0^\pounds \frac{e^{-s}}{1+\varepsilon s}ds + \int_{\oslash}^\infty \frac{e^{-s}}{1+\varepsilon s}ds\right].$$

For any positive limited s, the quotient $\frac{1}{1+\varepsilon s}$ is infinitely close to 1, and thus $\frac{e^{-s}}{1+\varepsilon s} = e^{-s} + \oslash$, and for any unlimited s the quotient $\frac{1}{1+\varepsilon s}$ is limited. Thus

$$\int_0^\pounds \frac{e^{-s}}{1+\varepsilon s}ds + \int_{\oslash}^\infty \frac{e^{-s}}{1+\varepsilon s}ds = \int_0^\pounds (e^{-s}+\oslash)ds + \int_{\oslash}^\infty \pounds e^{-s}ds,$$

$$= \int_0^\pounds e^{-s}ds + \int_0^\pounds \oslash ds + \pounds\int_{\oslash}^\infty e^{-s}ds,$$

$$= (1+\oslash) + \oslash\pounds + \pounds\oslash = 1 + \oslash.$$

Hence $E_1(\varepsilon) = \varepsilon(1+\oslash)$.

In these calculations we have replaced $\frac{e^{-s}}{1+\varepsilon s}$ by $e^{-s} + \oslash$ for limited s, and by $\pounds e^{-s}$ for unlimited s. Through these (rough) approximations we have succeeded in replacing complicated functions by functions which are easy to handle, *i.e.* functions which are integrable by quadrature.

The above reasoning is natural and systematic, yet it is legitimate to wonder if all this external symbolism is not too heavy a method for treating the exponential integral whose approximation is not after all very difficult to obtain. In fact, this same symbolism will allow us to deal, with the same ease and simplicity, with approximation problems which are much more difficult, and with calculations which are more laborious, such as problems where many parameters appear, or problems where many orders of magnitude are involved. Later on in this paper we will give a relevant example of such a case when we derive Stirling's formula.

7.3 Neutrices and external numbers

Definition 7.3.1. *A neutrix is a convex, additive subgroup of* $\mathbb{R}$, *which may be internal or external.*
An external number is the algebraic sum of a real number and a neutrix.

The word *neutrix* has already been used by van der Corput in [34]. For him, a neutrix is an additive subgroup of functions with only one constant function, *i.e.* the function identically equal to zero. Consequently, the neutrices of van der Corput and our neutrices, which concern real numbers, are not the same. Nevertheless, the algebraic and analytic similarities as well as the purpose of the introduction of such a notion have lead us to adopt the same terminology.

We agree to denote a neutrix by an upper–case calligraphic character and an external number by a Greek character.

Let α be an external number. Then

$$\alpha \equiv a + \mathcal{A} \equiv \{a + x \mid x \in \mathcal{A}\}$$

where a is a real number and $\mathcal{A}$ is a neutrix. The neutrix $\mathcal{A}$ is called the *neutrix part* of α.

We see that an external number is obtained by translating a neutrix by a real number, and we may easily verify that the neutrix part of an external number is the set of all real numbers which leave it invariant by translation.

The unique internal neutrices of $\mathbb{R}$ are 0 and $\mathbb{R}$ itself, but there appear to be many external neutrices. To begin with, the principal galaxy

$$\pounds \equiv \bigcup_{st(n)} \,]{-}n,\, n[\,,$$

and the halo of 0

$$\oslash \equiv \bigcap_{st(n)} \,]{-}\frac{1}{n},\, \frac{1}{n}[\,.$$

are neutrices.

If t is a positive, non–appreciable real number, then the *t–halo of* 0 denoted by $t\oslash$ and given by

$$t\oslash \equiv \bigcap_{st(n)} \,]{-}\frac{t}{n},\, \frac{t}{n}[$$

and the *t–galaxy of* 0 denoted by $t\pounds$ and given by

$$t\pounds \equiv \bigcup_{st(n)} \,]{-}tn,\, tn[$$

are neutrices different from $\oslash$ and $\pounds$. If ε is a positive infinitesimal, the ε–*microhalo of* 0 denoted by $M(\varepsilon)$ and given by

$$M(\varepsilon) \equiv \bigcap_{st(n)}]-\varepsilon^n,\ \varepsilon^n[,$$

and the ε–*microgalaxy of* 0 denoted by $m(\varepsilon)$ and given by

$$m(\varepsilon) \equiv \bigcup_{st(n)}]-e^{-\frac{1}{n\varepsilon}},\ e^{-\frac{1}{n\varepsilon}}[$$

are two neutrices different from $\varepsilon\oslash$ and $\varepsilon\pounds$. In fact, they are different from $\eta\pounds$ and $\eta\oslash$ for any real η. For a proof see [15].

The ε–microhalos and ε–microgalaxies appear in a natural way in the context of expansions in ε–shadows and canard values of singularly perturbed differential equations. See [53], [12] et [15].

While real numbers and neutrices are all external numbers here, on the other hand, are three examples of external numbers which are neither real numbers nor neutrices.

Let ω be an unlimited positive real number. The following external sets

1. $1 + \oslash \equiv \{1 + t \mid t = \oslash\}$,
2. $\omega + \pounds \equiv \{\omega + t \mid t = \pounds\}$,
3. $\frac{1}{\omega} + m(\frac{1}{\omega}) \equiv \{\frac{1}{\omega} + s \mid s = m(\frac{1}{\omega})\}$

are all external numbers.

Notice that @, the set of positive appreciable numbers, and $\oslash\!\!\!/$, the external set of positive unlimited real numbers, are not external numbers.

The collection of all external numbers will be written $\mathbb{E}$.

7.4 Basic algebraic properties

Convention: If not explicitly stated otherwise, it will be understood in the remainder of this paper that whenever external numbers or sets appear as arguments in a function, the value of the function consists of all values of the function on the elements of these sets.

For example, if α and β are two given external numbers, then, under the condition that $\alpha^2 + \beta^2$ does not contain 0,

$$\log \frac{|\alpha\beta|}{\alpha^2 + \beta^2} \equiv \left\{ \log \frac{|xy|}{x'^2 + y'^2} \mid x = \alpha,\ x' = \alpha,\ y = \beta,\ y' = \beta \right\}.$$

7.4.1 Elementary operations

Convention 7.4 gives us the following convenient notation for the ε–microhalo and the ε–microgalaxy of 0 (as defined in the Section 7.3 above)

$$M(\varepsilon) \;\equiv\; \pounds\varepsilon^{\oslash}$$

$$m(\varepsilon) \;\equiv\; \pounds e^{-\frac{\oslash}{\varepsilon}}.$$

We notice some elementary properties of the neutrices.

1. A convex subset $\mathcal{A}$ of $\mathbb{R}$ is a neutrix if and only if it is symmetric with respect to 0 and $2\mathcal{A} \equiv \mathcal{A}$.
2. The neutrices are ordered by inclusion.
3. The sum, $\mathcal{A} + \mathcal{B}$, of two neutrices $\mathcal{A}$ and $\mathcal{B}$, is equal to the larger of the two.
4. The product of a neutrix $\mathcal{A}$ with any appreciable real number leaves $\mathcal{A}$ invariant.
5. If a neutrix does not contain 1 then, by external induction, it does not contain $\left(\frac{1}{n}\right)$ for any standard integer n and is thus included in $\oslash$. If a neutrix contains 1 then, by external induction, it contains any standard integer n, and thus it contains the whole of $\pounds$. This proves that the smallest neutrix containing $\oslash$ is $\pounds$.

If the real number a does not belong to $\mathcal{A}$ then $\frac{\mathcal{A}}{a}$ is a neutrix included in $\oslash$, called the *relative uncertainty of* α. The external number α can then be written in the following form: $\alpha \equiv a\left(1 + \frac{\mathcal{A}}{a}\right)$, and the external number $\left(1 + \frac{\mathcal{A}}{a}\right)$ is called the *modulus of* α. It is exactly the collection of all real numbers which leave α invariant under multiplication.

Let now $\alpha \equiv a + \mathcal{A}$ and $\beta \equiv b + \mathcal{B}$ be two external numbers. Using Convention 7.4, we see that

$$\alpha + \beta \;\equiv\; \{x + y \mid x = \alpha,\; y = \beta\} \tag{7.3}$$

$$\equiv\; a + b + \mathcal{A} + \mathcal{B} \tag{7.4}$$

$$\equiv\; a + b + \max(\mathcal{A},\, \mathcal{B}) \tag{7.5}$$

$$\alpha - \beta \;\equiv\; a - b + \max(\mathcal{A},\, \mathcal{B}) \tag{7.6}$$

$$\alpha\beta \;\equiv\; ab + \max(a\mathcal{B},\, b\mathcal{A},\, \mathcal{A}\mathcal{B}). \tag{7.7}$$

If β is not a neutrix then

$$\frac{1}{\beta} \equiv \frac{1}{b + \mathcal{B}} \equiv \frac{1}{b} \cdot \frac{1}{\left(1 + \frac{\mathcal{B}}{b}\right)} \equiv \frac{1}{b}\left(1 + \frac{\mathcal{B}}{b}\right)$$

so

$$\frac{\alpha}{\beta} \equiv \frac{a}{b} + \frac{1}{b^2} \cdot \max(a\mathcal{B},\, b\mathcal{A}). \tag{7.8}$$

Recall that $(\alpha - \alpha)$ is not equal to 0 but to $\mathcal{A}$, *i.e.* to the neutrix part of α which contains 0, and that if α is not a neutrix then $\frac{\alpha}{\alpha}$ is not equal to 1 but to the modulus of α, *i.e.* to $\left(1 + \frac{A}{a}\right)$, which contains 1. If now $\beta \equiv \mathcal{B}$ then $\alpha\beta \equiv a\mathcal{B}$ is a neutrix. The treatment of the product of two neutrices is presented in Section 7.4.3.

Let $\alpha \equiv a + \mathcal{A}$ and $\beta \equiv b + \mathcal{B}$ be two external numbers and let us suppose that $\mathcal{A}$ contains $\mathcal{B}$ and that there exists a real x belonging to both α and β. Since α is invariant if translated by an element of $\mathcal{A}$ and since β is invariant if translated by an element of $\mathcal{B}$, it follows that

$$\alpha \equiv x + \mathcal{A} \ \wedge \ \beta \equiv x + \mathcal{B}.$$

This means that α contains β. So we have the following proposition

Proposition 7.4.1. *Let α and β be two external numbers. Then either α and β are disjoint, or one contains the other.*

Definition 7.4.1. *Let $\alpha \equiv a + \mathcal{A}$ and $\beta \equiv b + \mathcal{B}$ be two external numbers. Then*

$$\alpha < \beta \ \Leftrightarrow \ \forall x \in \alpha, \forall y \in \beta \ \ x < y. \tag{7.9}$$

The expression "$\alpha \leq \beta$" corresponds literally to "$\alpha < \beta$ or $\alpha = \beta$." From Proposition 7.4.1 and Definition 7.4.1, we deduce that

$$\alpha \leq \beta \ \Leftrightarrow \ \forall x \in \alpha, \exists y \in \beta \mid x \leq y. \tag{7.10}$$

One may verify that the external numbers are totally ordered by the relation "$\leq$".

In the same way, we define

$$\alpha > \beta \quad \Leftrightarrow \quad \forall x \in \alpha, \forall y \in \beta \ \ x > y, \tag{7.11}$$

$$\alpha \geq \beta \quad \Leftrightarrow \quad \forall x \in \alpha, \exists y \in \beta \mid x \geq y. \tag{7.12}$$

Recall that

$$\alpha < \beta \ \Leftrightarrow \ \beta > \alpha \tag{7.13}$$

but that $\alpha \leq \beta$ and $\beta \geq \alpha$ are not necessarily equivalent. For instance $1 + \oslash < 2 + \oslash$ and $2 + \oslash > 1 + \oslash$ are both true, but $1 \leq 1 + \oslash$ is true while $1 + \oslash \geq 1$ is false, and $1 \geq 1 + \oslash$ is true while $1 + \oslash \leq 1$ is false.

In general, if $\alpha \subset \beta$ then $\alpha \leq \beta$ and $\alpha \geq \beta$ are both true, while $\beta \leq \alpha$ and $\beta \geq \alpha$ are both false.

Note that the relations "$<$" and "$\leq$" are compatible with addition and multiplication by positive real numbers, and that if moreover α and β are two given external numbers then

$$\alpha < \beta \quad \Leftrightarrow \quad -\alpha > -\beta, \tag{7.14}$$
$$\alpha \leq \beta \quad \Leftrightarrow \quad -\alpha \geq -\beta. \tag{7.15}$$

Note also that the relations "<" and "≤" possess the property of *(simple) dense ordering*. Indeed, if α and β are two given, non identical external numbers then

$$\alpha \leq \beta \quad \Rightarrow \quad \exists \eta \mid \alpha \leq \eta \leq \beta \tag{7.16}$$
$$\alpha < \beta \quad \Rightarrow \quad \exists \eta \mid \alpha < \eta < \beta \tag{7.17}$$

where η is an external number *different* from α and β.

Definition 7.4.2. *Let α and β be two external numbers such that $\alpha \leq \beta$. Then $[\alpha, \beta]$ is the set of all real numbers r satisfying*

$$r \geq \alpha \ \wedge \ r \leq \beta$$

and $]\alpha, \beta]$ is the set of all real numbers r satisfying

$$\alpha < r \leq \beta.$$

If $\alpha < \beta$ then $[\alpha, \beta[$ is the set of all real numbers r satisfying

$$r \geq \alpha \ \wedge \ r < \beta$$

and $]\alpha, \beta[$ is the set of all real numbers r satisfying

$$\alpha < r < \beta.$$

The collections of real numbers $[\alpha, \beta]$, $[\alpha, \beta[$, $]\alpha, \beta]$ and $]\alpha, \beta[$, when well-defined, are called external intervals with upper boundary β and with lower boundary α.

The interval $[\alpha, \beta]$ is said to be closed, *the intervals $[\alpha, \beta[$ and $]\alpha, \beta]$ are said to be* semi–open *and the interval $]\alpha, \beta[$ is said to be* open.

7.4.2 On the shape of a neutrix

For convenience, in the remainder of this section, we suppose that the neutrices considered are external.

Logarithm of a neutrix and external cuts of $\mathbb{R}$**.** We know that it is possible to multiply two real numbers by considering the logarithm of the product and then the exponential of the logarithm. This detour by way of the logarithm has the great advantage of transforming a *product* into a *sum*.

In the same spirit, since multiplying two neutrices involves calculating the set of products of their elements, we define the logarithm of a neutrix in such a manner that

- up to elementary transformations, the exponential of the logarithm of a neutrix gives us back the original neutrix,
- the treatment of a product of neutrices is transformed into the treatment of a sum of external sets.

Convention: Let $\mathcal{A}$ be a neutrix. Then

$$\log \mathcal{A} \;\equiv\; \{\log x \mid x \in \mathcal{A} \;\wedge\; x > 0\}. \tag{7.18}$$

If $\mathcal{A}$ and $\mathcal{B}$ are two neutrices, we can easily verify the following identities

$$|\mathcal{A}| - \{0\} \;\equiv\; \exp(\log \mathcal{A}), \tag{7.19}$$
$$\log \mathcal{A}\mathcal{B} \;\equiv\; \log \mathcal{A} + \log \mathcal{B}. \tag{7.20}$$

From (7.19), we can conclude that

$$\mathcal{A} \equiv [-\exp(\log \mathcal{A}),\ \exp(\log \mathcal{A})]. \tag{7.21}$$

Example 7.4.1. Here are the logarithms of the neutrices we have met up to now. The number ε is assumed to be positive and infinitesimal, and the number ω is assumed to be positive and unlimited.

1. $\log \pounds \equiv]-\infty,\ \pounds]$
2. $\log \oslash \equiv]-\infty,\ \pounds[$
3. $\log \varepsilon\pounds \equiv]-\infty,\ \log(\varepsilon) + \pounds]$
4. $\log \omega\oslash \equiv]-\infty,\ \log(\omega) + \pounds[$
5. $\log \left(\pounds\varepsilon^{i-large}\right) \equiv]-\infty,\ \pounds\log(\varepsilon)[$
6. $\log \left(\pounds e^{-@\omega}\right) \equiv]-\infty,\ \oslash\omega[.$

Definition 7.4.3. *A* cut *in* $\mathbb{R}$ *is an ordered pair* (Σ, Π) *of subsets of* $\mathbb{R}$ *such that* $\Sigma \bigcup \Pi \equiv \mathbb{R}$, $\Sigma \bigcap \Pi \equiv \emptyset$ *and* Σ *is dominated by* Π.

The convex subset Σ *is called a* lower halfline *of* $\mathbb{R}$ *and the convex subset* Π *an* upper halfline *of* $\mathbb{R}$.

The next theorem is a reformulation of the classification theorem of the internal and external cuts of $\mathbb{R}$, (see [19] and Theorem 4.34 of [15]), using external numbers.

Theorem 7.4.1. *Let (Σ, Π) be a cut of $\mathbb{R}$. Then there exists a unique external number α such that*

$$\Sigma \equiv]-\infty, \alpha[\quad \vee \quad \Sigma \equiv]-\infty, \alpha].$$

Theorem 7.4.1 implies that our work is related to some other studies of the external cuts of $\mathbb{R}$ such as that in the article [119] of Zakon. The main difference is that in a sense we are concerned with exactly half of the cuts. Indeed, a cut defines a unique external number (the upper boundary of its lower halfline), while an external number defines two cuts, *i.e.* those whose lower halflines have the external number as their upper boundary, one lower halfline being open and one being closed.

Remark also the analogy with the classical construction of $\mathbb{R}$ as the collection of Dedekind cuts of the rational numbers where, similarly, we consider only half of the possible cuts , namely those whose lower halflines do not contain a greatest element. See for example [101].

Canonical form of a neutrix. The logarithms of the neutrices of Example 7.4.1 appear to be external lower halflines of $\mathbb{R}$. In fact, this is a general phenomenon because if $\mathcal{A}$ is an external neutrix then clearly $\log \mathcal{A}$ is an external set of $\mathbb{R}$ and $(\log \mathcal{A}, (\log \mathcal{A})^c)$ is an external cut of $\mathbb{R}$. Consequently, if $\mathcal{A}$ is a neutrix, Theorem 7.4.1 ensures the existence of a real p and a unique neutrix $\mathcal{P}$ such that

$$\log \mathcal{A} \equiv]-\infty, p + \mathcal{P}[\quad \vee \quad \log \mathcal{A} \equiv]-\infty, p + \mathcal{P}]. \tag{7.22}$$

Proposition 7.4.2. *Let $\mathcal{A}$ be a neutrix and let π be the external number such that*

$$\log \mathcal{A} \equiv]-\infty, \pi[\quad \vee \quad \log \mathcal{A} \equiv]-\infty, \pi].$$

Then the neutrix part of π contains $\pounds$.

Indeed,

$$@\mathcal{A} \equiv \mathcal{A} \quad \Rightarrow \quad \log \mathcal{A} \equiv \log @\mathcal{A} \equiv \log \mathcal{A} + \pounds.$$

Next, we prove that a special class of neutrices, the idempotent neutrices, generates, in some sense, all the neutrices. To be precise, we prove that any neutrix is the product of a real number and an idempotent neutrix.

Definition 7.4.4. *Let $\mathcal{A}$ be a neutrix. if*

$$\mathcal{A} \cdot \mathcal{A} \equiv \mathcal{A} \tag{7.23}$$

then $\mathcal{A}$ is called an idempotent neutrix.

Example 7.4.2. The neutrices $\pounds$, $\oslash$, $\pounds \varepsilon^{i-large}$ and $\pounds e^{-@\omega}$ are all idempotent neutrices.

Let ε be a positive infinitesimal. Then $\varepsilon \oslash$ is not an idempotent neutrix, because $\varepsilon^2 \oslash$ does not contain ε^2.

Let ω be a positive, unlimited real number. Then $\omega \pounds$ is not an idempotent neutrix, because $\omega^2 \pounds$ contains ω^2 while $\omega \pounds$ does not.

Proposition 7.4.3. *A neutrix $\mathcal{A}$ is idempotent if and only if*

$$\log \mathcal{A} \equiv]-\infty, \mathcal{P}[\quad \vee \quad \log \mathcal{A} \equiv]-\infty, \mathcal{P}] \tag{7.24}$$

where $\mathcal{P}$ is a neutrix containing £.

Proof. Let $\mathcal{A}$ be an idempotent neutrix. By Theorem 7.4.1 $\log \mathcal{A}$ has the representation given in relation (7.22), and the neutrix $\mathcal{P}$ contains £ by Proposition 7.4.2. Moreover, $2\log \mathcal{A} \equiv \log \mathcal{A}$. This means that $p \in \mathcal{P}$ in relation (7.22).

The other implication is trivial.

In Examples 7.4.1 we see that the logarithms of the neutrices £, $\oslash$, $£\varepsilon^{i-large}$ and $£e^{-@\omega}$ satisfy the relation (7.24).

We saw that the neutrices $\varepsilon\oslash$ and $\omega£$ are not idempotent. Nevertheless $\varepsilon\oslash$ is the product of the real number ε and the idempotent neutrix $\oslash$, and $\omega£$ is the product of the real number ω and the idempotent neutrix £. In fact, we have the following theorem

Theorem 7.4.2. *Let $\mathcal{A}$ be a neutrix. Then there exists an idempotent neutrix $\mathcal{I}$ and a number a such that*

$$\mathcal{A} \equiv e^a \cdot \mathcal{I}.$$

Proof. Let $\mathcal{A}$ be a neutrix. By Theorem 7.4.1 and Proposition 7.4.2, there exists a unique external number $\pi \equiv p + \mathcal{P}$ such that $\mathcal{P}$ contains £ and $\log \mathcal{A} \equiv p+]-\infty, \mathcal{P}[$ or $\log \mathcal{A} \equiv p+]-\infty, \mathcal{P}]$.

If $\log \mathcal{A} \equiv p+]-\infty, \mathcal{P}[$ then $\mathcal{A} \equiv e^p \cdot]-e^{\mathcal{P}}, e^{\mathcal{P}}[$, and if $\log \mathcal{A} \equiv p+]-\infty, \mathcal{P}]$ then $\mathcal{A} \equiv e^p \cdot [-e^{\mathcal{P}}, e^{\mathcal{P}}]$.

Moreover, $]-e^{\mathcal{P}}, e^{\mathcal{P}}[$ and $[-e^{\mathcal{P}}, e^{\mathcal{P}}]$ are idempotent neutrices by Proposition 7.4.3.

7.4.3 On the product of neutrices

It is clear that the product of two neutrices is a neutrix. In this section, we study the form of this product.

In the case where the neutrices $\mathcal{A}$ and $\mathcal{B}$ are idempotent, we have the following theorem.

Theorem 7.4.3. *The product of two idempotent neutrices is identical to one of the terms of the product.*

To prove this theorem, we need Proposition 7.4.4 below. In the statement of this proposition, the symbol $]-\infty, \mathcal{S})$ represents a lower halfline, open or closed, whose upper boundary is a given neutrix $\mathcal{S}$.

Proposition 7.4.4. *Let $\mathcal{S}$ and $\mathcal{P}$ be two neutrices such that $\mathcal{S}$ strictly contains $\mathcal{P}$. Then*

$$-\,]-\infty,\mathcal{S})+]-\infty,\mathcal{P}) \equiv]-\infty,\mathcal{S}),$$
$$-\,]-\infty,\mathcal{S}[+]-\infty,\mathcal{S}[\equiv]-\infty,\mathcal{S}[,$$
$$-\,]-\infty,\mathcal{S}[+]-\infty,\mathcal{S}] \equiv]-\infty,\mathcal{S}[,$$
$$-\,]-\infty,\mathcal{S}]+]-\infty,\mathcal{S}] \equiv]-\infty,\mathcal{S}].$$

We see that the sum of two lower halflines having the same upper boundary $\mathcal{S}$ is closed if and only if both elements of the sum are closed, and that the sum is open if and only if at least one of the terms of the sum is an open lower halfline.

Proof (of Theorem 7.4.3). Let $\mathcal{A}$ and $\mathcal{B}$ be two idempotent neutrices. By Proposition 7.4.3, $\log \mathcal{A}$ and $\log \mathcal{B}$ are two lower halflines having a neutrix as upper boundary. Relation (7.20), proposition 7.4.4 and then relation (7.21) imply that the product of $\mathcal{A}$ and $\mathcal{B}$ is equal to $\mathcal{A}$ or $\mathcal{B}$. $\qquad\square$

Below are some examples of products of concrete neutrices.

Example 7.4.3. Let ε and δ be two positive infinitesimals. Then

1. $\pounds\varepsilon^{i-large} \cdot \oslash \equiv \pounds\varepsilon^{i-large}$.

2. $\pounds\varepsilon^{i-large} \cdot \pounds\exp\left(-\frac{\oslash}{\delta}\right) \equiv \begin{cases} \pounds\exp\left(-\frac{\oslash}{\delta}\right) & if \ \ \delta\log\varepsilon = \oslash, \\ \pounds\varepsilon^{i-large} & if \ \ \delta\log\varepsilon > \oslash. \end{cases}$

Indeed, the sum of the corresponding logarithms $]-\infty, \pounds\log(\varepsilon) + \frac{\oslash}{\delta}[$ equals $]-\infty, \pounds\log(\varepsilon)[$ if $\delta\log\varepsilon$ is not infinitesimal, and $]-\infty, \frac{\oslash}{\delta}[$ if $\delta\log\varepsilon$ is infinitesimal.

3. $\varepsilon\pounds \cdot \frac{1}{\delta}\oslash \equiv \left(\frac{\varepsilon}{\delta}\oslash\right)$

The last example is not the product of two idempotent neutrices. It is also not reduced to one of the terms of the product, but to the product of one of the terms by a real number. This linearization property is a general phenomenon.

Theorem 7.4.4. *Let $\mathcal{A}$ and $\mathcal{B}$ be two neutrices. Then there exists a real number c such that*
$$\mathcal{A}\mathcal{B} \equiv c\mathcal{A} \quad \vee \quad \mathcal{A}\mathcal{B} \equiv c\mathcal{B}.$$

Proof. Let $\mathcal{A}$ and $\mathcal{B}$ be two neutrices. By Theorem 7.4.2 there exists a real number p and an idempotent neutrix $\mathcal{I}$ such that $\mathcal{A} \equiv p\mathcal{I}$, and there exists a real number q and an idempotent neutrix $\mathcal{J}$ such that $\mathcal{B} \equiv q\mathcal{J}$. By Theorem 7.4.3 $\mathcal{I}\mathcal{J} \equiv \mathcal{I}$ or $\mathcal{I}\mathcal{J} \equiv \mathcal{J}$. Hence, $\mathcal{A}\mathcal{B} \equiv q\mathcal{A}$ or $\mathcal{A}\mathcal{B} \equiv p\mathcal{B}$. $\qquad\square$

The resolution of algebraic equations involving neutrices and external numbers will be the subject of future publications. However, in Section 7.6 we will need to solve the equation $\log(u+1) - u = \frac{\pounds}{\omega}$. This equation is of the form $f(x) = \mathcal{A}$, where f is a standard function and $\mathcal{A}$ a neutrix in $\oslash$. The next proposition gives the solution of such equations.

Proposition 7.4.5. *Let n be a standard integer, and let f be a standard function of class $C^{(n)}$ in a neighbourhood of 0 and such that*

$- \forall i < n \quad f^{(i)}(0) = 0$

$- f^{(n)}(0) \neq 0.$

Let $\mathcal{A}$ be a neutrix included in $\oslash$. Then

(i) $f(\mathcal{A}) \equiv \mathcal{A}^n$

(ii) $\forall x \in \mathbb{R} \quad f(x) = \mathcal{A} \quad \Leftrightarrow \quad x = \mathcal{A}^{\frac{1}{n}}$

Proof. Let x be any real number in $\mathcal{A}$. Then $x = \oslash$. The Taylor expansion of f in x yields

$$f(x) = x^n \frac{f^n(c)}{n!},$$

where c is a real number between x and 0. The real number $\frac{f^n(c)}{n!}$ is consequently nearly standard, and thus $f(x) \in \mathcal{A}^n$. This means that $f(\mathcal{A}) \subset \mathcal{A}^n$. Let now y be a positive real number in $\mathcal{A}^n$. Put $x = (2y)^{\frac{1}{n}} \frac{n!}{f^{(n)}(0)}$. Then there exists a positive real $c \in [0, x]$ such that

$$f(x) = 2y \frac{f^{(n)}(c)}{f^{(n)}(0)}.$$

Since $f^{(n)}$ is continuous, and $f^{(n)}(0) \neq 0$, the quotient $\frac{f^{(n)}(c)}{f^{(n)}(0)}$ is infinitely close to 1, and consequently $f(x) > y$. This means that $f(\mathcal{A}) \supset \mathcal{A}^n$ by convexity, and by symmetry if n is odd. Hence $f(\mathcal{A}) \equiv \mathcal{A}^n$

Part *(ii)* follows immediately from *(i)*.

7.5 Basic analytic properties

The integral of the constant function 1 on a given interval $]a, b[$ of $\mathbb{R}$ is equal to $b - a$: the difference between the least upper bound and the greatest lower bound of the interval.

We will extent this fact to the external numbers. We define the notion of an external distance, and then the notions of least upper bound and greatest lower bound of an external interval and more generally of any collection of external numbers. Then we develop a method of integration and we will see, among others, that the integral of the constant function 1 on $]\alpha, \beta[$, where α and β are two given external numbers, is equal to $\beta - \alpha$.

Finally, we will consider external sets of functions, *external functions*, and the integration of external functions on external intervals.

7.5.1 External distance and extrema of a collection of external numbers

We denote the set of all external numbers α verifying $0 \leq \alpha$ by $\mathbb{E}^+$.

We define a mapping denoted Λ which associates to any two fixed external numbers $\alpha \equiv a + \mathcal{A}$ and $\beta \equiv b + \mathcal{B}$ an external number $\Lambda(\alpha, \beta)$ in $\mathbb{E}^+$, called the *external distance between* α *and* β. The mapping Λ is given by

$$\Lambda(\alpha, \beta) \equiv |b - a| + (\mathcal{A} + \mathcal{B}).$$

We have

$$\begin{array}{lll} \forall \alpha \in \mathbb{E} & 0 = & \Lambda(\alpha, \alpha) \\ \forall \alpha, \beta \in \mathbb{E} & 0 \leq & \Lambda(\alpha, \beta) \\ \forall \alpha, \beta, \gamma \in \mathbb{E} & \Lambda(\alpha, \gamma) \leq & \Lambda(\alpha, \beta) + \Lambda(\beta, \gamma) \end{array}$$

Remark that $\Lambda(\alpha, \alpha)$ is equal to the neutrix part of α, and that $0 = \Lambda(\alpha, \beta)$ does not necessarily imply that α and β are identical.

Definition 7.5.1. *Let Π be a collection of external numbers. We say that Π is* convex *if*

$$\forall \alpha, \beta \in \Pi, \forall \delta \in \mathbb{E} \quad (\alpha \leq \beta \ \wedge \ \delta \geq \alpha \ \wedge \ \delta \leq \beta) \ \Rightarrow \ (\delta \in \Pi).$$

Example 7.5.1. The collection of external numbers defined by $\Pi_1 \equiv \{x \mid x \leq 1 + \oslash\}$ is convex. The union (in $\mathbb{R}$) of its elements gives the external interval $] - \infty, 1 + \oslash]$.

The collection of external numbers defined by $\Pi_2 \equiv \{x + \oslash \mid x \leq 1 + \oslash\}$ is not convex in $\mathbb{E}$ because $0 \notin \Pi_2$. Nevertheless, the union (in $\mathbb{R}$) of its elements gives the external interval $] - \infty, 1 + \oslash]$.

The collection of external numbers given by $\Pi_3 \equiv \{x \in \mathbb{R} \mid x \leq 1 + \oslash\}$ is not convex in $\mathbb{E}$ because $\oslash$ is not an element of Π_3. Again, the union (in $\mathbb{R}$) of its elements gives the external interval $] - \infty, 1 + \oslash]$.

Definition 7.5.2. *Let Δ be a collection of external numbers.*
The lower convexification *of Δ is the collection*

$$\underline{conv}(\Delta) \equiv \{\xi \in \mathbb{E} \mid \exists \delta \in \Delta \ \wedge \ \xi \leq \delta\}$$

and the upper convexification *of Δ is the collection*

$$\overline{conv}(\Delta) \equiv \{\xi \in \mathbb{E} \mid \exists \delta \in \Delta \ \wedge \ \xi \geq \delta\}.$$

The convexification *of Δ, denoted by $conv(\Delta)$ is the intersection in $\mathbb{E}$ of $\underline{conv}(\Delta)$ and $\overline{conv}(\Delta)$*
The union of the elements of Δ on $\mathbb{R}$ is denoted by $\bigcup \Delta$.

Proposition 7.5.1. *Let Δ be any non–empty set of external numbers, not containing the external number $\mathbb{R}$. Then*

1. $\bigcup \underline{conv}(\Delta)$ is a lower halfline of $\mathbb{R}$.
2. $\bigcup \overline{conv}(\Delta)$ is an upper halfline of $\mathbb{R}$.
3. $\bigcup conv(\Delta)$ is an external interval whose upper boundary matches the upper boundary of $\bigcup \underline{conv}(\Delta)$ and whose lower boundary matches the lower boundary of $\bigcup \overline{conv}(\Delta)$.

Definition 7.5.3. *Let Δ be a non–empty set of external numbers.*

The least upper bound of Δ is the upper boundary of $\bigcup conv(\Delta)$. It will be denoted by $\sup(\Delta)$,

The greatest lower bound of Δ is the lower boundary of $\bigcup conv(\Delta)$. It will be denoted by $\inf(\Delta)$.

It is clear that for any set of external numbers Δ, the following relation holds

$$\sup \Delta \equiv - \inf(-\Delta). \tag{7.25}$$

Example 7.5.2. 1. If $\Pi \equiv \{x \mid x \leq 1 + \oslash\}$ and $\Gamma \equiv \{x \mid x > 1 + \oslash\}$, then $\sup \Pi \equiv \inf \Gamma \equiv 1 + \oslash$.
2. Clearly $\inf \left\{ \frac{1}{n} \mid st(n) \in \mathbb{N} \right\} \equiv \oslash$ and $\sup\{n \mid n \in \mathbb{N}, st(n)\} \equiv \pounds$.
3. We have $\sup \mathbb{R} \equiv \inf \mathbb{R} \equiv \mathbb{R}$.
4. If $\Psi \equiv \{x + \varepsilon\pounds \mid x < 2 + \oslash\}$, then $\sup \Psi \equiv 2 + \oslash$.
5. Let ε be a positive infinitesimal and let $\Sigma \equiv \{1 - \varepsilon^n + \pounds\varepsilon^{n+1}\}_{st(n)}$. Then $\sup(\Sigma) \equiv 1 + \pounds\varepsilon^{i-large}$.
6. Let ω a positive unlimited number, and let

$$\Omega \equiv \left\{ \frac{\exp(-\omega\mu)}{\omega} \mid \mu = \oslash \right\}.$$

Then

$$\sup_{\mu = \oslash}(\Omega) \equiv \pounds e^{-\oslash\omega}.$$

Remark that the notions of least upper bound and greatest lower bound of sets of external numbers are different from the common notions of least upper bound and greatest lower bound of sets of real numbers. For example, if $\Pi \equiv \{x \mid x < 1 + \oslash\}$ then $\sup(\Pi) \equiv 1 + \oslash$, yet $1 \leq 1 + \oslash$ and

$$\forall x \in \Pi \quad x < 1.$$

Nevertheless, the following theorem, extending a classical characterization of the supremum, justifies this terminology.

Theorem 7.5.1. *Let Π be a non–empty collection of external numbers and let γ be an external number. Then*

$$\gamma \equiv \sup \Pi \quad \Leftrightarrow \quad \begin{cases} (\gamma \bigcap \underline{conv}(\Pi) \equiv \emptyset) & \wedge & (\forall \alpha < \gamma, \exists \delta \in \Pi \mid \delta > \alpha) \\ & \vee & \\ (\gamma \subset \underline{conv}(\Pi)) & \wedge & (\forall \delta \in \Pi, \delta \leq \gamma). \end{cases} \tag{7.26}$$

In words, relation (7.26) means that an external number γ is the least upper bound of a collection of external numbers Π if and only if one of the following statements is true:

- γ strictly dominates all elements of Π, and any external number strictly dominated by γ is strictly dominated by an element of Π.
- γ is totally included in the convexification of Π, and γ dominates all elements of Π.

It is clear that these statements are mutually exclusive.

Similarly, an external number η is the greatest lower bound of the set Π of external numbers if and only if one of the following statements holds:

- $(\eta \cap \overline{conv}(\Pi) \equiv \emptyset) \quad \wedge \quad (\forall \alpha > \eta, \ \exists \delta \in \Pi \mid \delta < \alpha)$
- $(\eta \subset \overline{conv}(\Pi)) \quad \wedge \quad (\forall \delta \in \Pi, \ \delta \geq \eta)$

Proof (of Theorem 7.5.1). It suffices to show that
1.

$$(\gamma \cap \underline{conv}(\Pi) \equiv \emptyset) \wedge (\forall \alpha < \gamma, \ \exists \delta \in \Pi \mid \delta > \alpha)$$

$$\Updownarrow$$

$$\left(\bigcup_{\delta \in \Pi}] - \infty, \delta \right] \equiv] - \infty, \gamma[ )$$

and that 2.

$$(\gamma \subset \underline{conv}(\Pi)) \wedge (\forall \delta \in \Pi, \ \delta \leq \gamma)$$

$$\Updownarrow$$

$$\left(\bigcup_{\delta \in \Pi}] - \infty, \delta \right] \equiv] - \infty, \gamma])$$

1. ($\Rightarrow$) This implication is straightforward.

($\Leftarrow$) Trivially $(\gamma \cap \underline{conv}(\Pi) \equiv \emptyset)$.
Let $\alpha < \gamma$. Then $\alpha \subset \bigcup_{\delta \in \Pi}] - \infty, \delta]$. Put $\Delta_\alpha \equiv \{\delta \in \Pi \mid \alpha \leq \delta\}$.
Suppose, by absurdity, that for all δ in Δ_α we have $\alpha = \delta$. Put $\Delta \equiv \bigcup(\Delta_\alpha)$. Then Δ is an external number in Π, by construction. So $\Delta < \gamma$. Let $x \in]\Delta, \gamma[$. Then

$$\exists \delta_1 \in \Pi \mid x \leq \delta_1$$

and we certainly have $\delta_1 > \Delta$. Hence $\alpha < \delta_1$.
2. This equivalence is straightforward.

Concerning the notions of least upper bound and greatest lower bound of external numbers, we have the following fundamental theorem.

Theorem 7.5.2. *Every non–empty subset of* $\mathbb{E}$ *possesses a unique least upper bound and a unique greatest lower bound.*

The theorem is a direct consequence of Theorem 7.4.1

The following results deal with the least upper bound and the greatest lower bound of the image of a family of real numbers by an internal function.

Proposition 7.5.2. *Let* $f : \mathbb{R} \longrightarrow \mathbb{R}$ *be an internal, continuous and increasing function. Let* $\gamma \not\equiv \mathbb{R}$ *be an external number. Then*

$$\sup\{f(x) \mid x \leq \gamma\} \equiv \inf\{f(x) \mid x > \gamma\} \tag{7.27}$$

$$\sup\{f(x) \mid x < \gamma\} \equiv \inf\{f(x) \mid x \geq \gamma\} \tag{7.28}$$

Proof. It is clear that $] - \infty, f(\gamma)]$ is a lower halfline of $\mathbb{R}$. If this halfline is internal the proposition is trivial. Suppose now that $] - \infty, f(\gamma)]$ is an external lower halfline. Then $\{] - \infty, f(\gamma)],]f(\gamma), \infty[\}$ is an external cut of $\mathbb{R}$. By Theorem 7.4.1, the upper boundary of $] - \infty, f(\gamma)]$ is exactly the lower boundary of $]f(\gamma), \infty[$. This proves the relation (7.27).

The relation (7.28) can be proved in a similar way.

Proposition 7.5.3. *Let* $f : \mathbb{R} \longrightarrow \mathbb{R}$ *be an internal increasing function. Let* I *be an external interval of* $\mathbb{R}$. *Then*

$$\sup\{f(t) - f(s) \mid s < t, \ s, t \in I\} \equiv \sup_{t \in I}\left[\sup_{s \in I}\{f(t) - f(s)\}\right]. \tag{7.29}$$

Proof. Let $\alpha \equiv a + \mathcal{A} \equiv \sup\{f(t) - f(s) \mid s < t, \ s, t \in I\}$, and for all $t \in I$ let $\beta_t \equiv \sup_{s \in I}\{f(t) - f(s)\}$ and let $\beta \equiv b + \mathcal{B} \equiv \sup\{\beta_t \mid t \in I\}$.

It is clear that

$$\forall t \in I \quad \beta_t \leq \alpha$$

so $\beta \leq \alpha$.

Let r be a real number which is strictly larger than $\mathcal{A}$. By the relation (7.26),

$$\exists (s_r, t_r) \in I \times I \mid f(t_r) - f(s_r) > a - r.$$

Consequently,

$$a - r \leq \sup_{s \in I}\{f(t_r) - f(s)\} \leq \beta.$$

Since $\alpha \equiv \sup\{a - r \mid r > \mathcal{A}\}$ we have $\alpha \leq \beta$. Hence $\alpha \equiv \beta$.

7.5.2 External integration

Definition 7.5.4. *Let* f *be an internal function which is positive and Lebesgue integrable on an internal interval* $[c, d]$ *of* $\mathbb{R}$. *Let* I *be an external interval included in* $[c, d]$. *We define the external integral of* f *on* I *by*

$$\int_I f(x)dx \equiv \sup\left\{\int_s^t f(x)dx \mid s, t \in I \ \wedge \ s < t\right\}.$$

Proposition 7.5.4. *Let f be an internal function which is positive and Lebesgue integrable on an internal interval $[c, d]$ of $\mathbb{R}$. Let F be a primitive of f, and let I be an external interval included in $[c, d]$. Then*

$$\int_I f(x)dx \equiv \sup_{t \in I} F(t) - \inf_{s \in I} F(s).$$

Proof. By hypothesis, the primitive F is an internal function which is positive and increasing on $[c, d]$. By Proposition 7.5.3 and relation (7.25)

$$
\begin{aligned}
\int_I f(x)dx &\equiv \sup \left\{ \int_s^t f(x)dx \mid s < t,\ s, t \in I \right\} \\
&\equiv \sup \left\{ F(t) - F(s) \mid s < t,\ s, t \in I \right\} \\
&\equiv \sup_{t \in I} \left\{ \sup_{s \in I} (F(t) - F(s)) \mid s < t \right\} \\
&\equiv \sup_{t \in I} \left[F(t) - \left\{ \inf_{s \in I} (F(s)) \right\} \right] \\
&\equiv \sup_{t \in I} (F(t)) - \inf_{s \in I} (F(s)).
\end{aligned}
$$

Clearly, the integral of a positive function on an external number is not necessarily equal to zero. So we must adopt different symbols to distinguish integration over intervals which include or do not include their extrema.

Notation: Let f and $[c, d]$ be as defined in Definition 7.5.4 above. Let α and β be two external numbers such that $c < \alpha \leq \beta < d$. Then

$$
\begin{aligned}
\int_\alpha^\beta f(x)dx &\equiv \int_{[\alpha,\beta]} f(x)dx \\
\int_\alpha^{\underline{\beta}} f(x)dx &\equiv \int_{[\alpha,\beta[} f(x)dx \\
\int_{\overline{\alpha}}^\beta f(x)dx &\equiv \int_{]\alpha,\beta]} f(x)dx \\
\int_{\overline{\alpha}}^{\underline{\beta}} f(x)dx &\equiv \int_{]\alpha,\beta[} f(x)dx
\end{aligned}
$$

Example 7.5.3. 1. Let f be a positive, internal function, Lebesgue integrable on an internal interval $[c, d]$ of $\mathbb{R}$. Let F be a primitive of f, and let α and β be two external numbers such that $c < \alpha \leq \beta < d$. Then, by Proposition 7.5.4

$$\int_\alpha^\beta f(x)dx \equiv \sup_{t \in \beta} F(t) - \inf_{s \in \alpha} F(s).$$

2. Let I be an external interval. Then

$$\int_I 1 \cdot dx \equiv \Lambda(I).$$

3. Let f be a standard, strictly positive function which is integrable on $\mathbb{R}$. Then

$$\int_{\pounds} f(x)dx \equiv \int_{\mathbb{R}} f(x)dx + \oslash. \tag{7.30}$$

It is well known that the integral of a function on an internal interval is equal to the sum of the integrals of the function on an internal cut of the interval. This fact is known as the *relation of Chasles* or the *additivity of the integral*. In the following proposition we handle the case where the cut is external.

Proposition 7.5.5. *Let f be a real positive function of one real variable, and which is Lebesgue–integrable on an internal interval $[c, d]$. Let γ be an external number such that $c < \gamma < d$. Then*

$$\int_c^d f(x)dx = \int_c^\gamma f(x)dx + \int_{\overline{\gamma}}^d f(x)dx \tag{7.31}$$

and

$$\int_c^d f(x)dx = \int_c^{\overline{\gamma}} f(x)dx + \int_\gamma^d f(x)dx. \tag{7.32}$$

Since $\int_c^d f(x)dx$ is a real number and since the integrals on the right side of (7.31) and (7.32) are generally external numbers, the above relations are generally not identities.

Proof. We prove the inclusion (7.31).

Consider the real positive function h given by

$$
\begin{aligned}
h: \quad [c, d] \quad &\longrightarrow \quad [0, \int_c^d f(x)dx] \\
s \quad &\longrightarrow \quad \int_c^s f(x)dx.
\end{aligned}
$$

By relation (7.25) and Proposition 7.5.2

$$\sup\left\{ \int_c^d f(x)dx - h(s) \mid s > \gamma \right\} \equiv \int_c^d f(x)dx - \inf\{h(s) \mid s > \gamma\}$$

$$\equiv \int_c^d f(x)dx - \sup\{h(s) \mid s \leq \gamma\}.$$

Remark that the relations (7.31) and (7.32) do not imply that $\int_c^\gamma f(x)dx +$
$\int_{\overline{\gamma}}^d f(x)dx$ and $\int_c^{\overline{\gamma}} f(x)dx + \int_\gamma^d f(x)dx$ are identical external numbers. Indeed,
if ω is a positive unlimited number then $\int_{-\omega}^\omega e^x dx = e^\omega - e^{-\omega}$. Nevertheless,

$$\int_{-\omega}^{\pounds} e^x dx \equiv \oslash \quad \wedge \quad \int_{\pounds}^\omega e^x dx \equiv e^\omega + \oslash$$

$$\int_{-\omega}^{\pounds} e^x dx \equiv \pounds \quad \wedge \quad \int_{\pounds}^\omega e^x dx \equiv e^\omega + \pounds,$$

which means that $\int_{-\omega}^{\pounds} e^x dx + \int_{\pounds}^\omega e^x dx \equiv e^\omega + \oslash$ and $\int_{-\omega}^{\pounds} e^x dx + \int_{\pounds}^\omega e^x dx \equiv e^\omega + \pounds$.

7.5.3 External functions

We started this paper by considering real numbers having external properties: limited real numbers, appreciable real numbers, infinitesimals, etc... Then we studied external sets of real numbers satisfying an external property (the neutrices, the external numbers and the external intervals).

Now we make a further step: we speak about *collections of functions having the same external property*, that is to say *external sets of functions*. We already dealt informally with such sets, like the external set of functions whose values remain limited or the external set of functions whose values are infinitesimals.

An **external function** defined on an interval I of $\mathbb{R}$ is simply an external set of internal functions, all of them defined at least on I. However, we will only deal with external functions of the type given in Definition 7.5.5 below, or those external functions which are easily reduced to this type.

Definition 7.5.5. *Let p be a standard integer, and let $F : \mathbb{R}^{p+1} \longrightarrow \mathbb{R}$ be an internal function. Let I be a interval of $\mathbb{R}$ which may be internal or external, and let $\{\alpha_n\}_{1 \leq n \leq p}$ be a finite family of external numbers.*
We denote by

$$\varphi(x) \equiv F(x, \alpha_1, \ldots, \alpha_p)$$

the external set of all internal functions

$$x \longrightarrow F(x, a_1(x), \ldots, a_p(x))$$

where $a_n, \ldots, a_p$ are internal functions defined at least on the interval I, and such that $a_1(I) \subset \alpha_1, \ldots, a_p(I) \subset \alpha_p$.

Definition 7.5.6. *We say that a set Σ of internal functions is convex if, when given two graphs Γ_f and Γ_g of functions f and g of Σ, any graph Γ_h of a function which lies between Γ_f and Γ_g is the graph of a function h of Σ.*

Remark that if the function F of Definition 7.5.5 is continuous, the external function φ is convex.

Example 7.5.4. 1. The constant function $\psi(x) \equiv 1 + \oslash$ defined on the external interval $[\oslash, 1 + \oslash]$ is the external set of all the internal functions $f(x)$ defined on $[\oslash, 1 + \oslash]$ at least, with values infinitely close to 1.
An example of a non–constant internal element of ψ is the function $f(x) = 1 + \frac{\sin x}{\omega}$ restricted to $[\oslash, 1 + \oslash]$.
2. The external function

$$\lambda(s) \equiv \exp\left(-\frac{s^2}{2}(1 + \oslash)\right)$$

is the set of all internal functions $\exp\left(-\frac{s^2}{2}(1 + g(s))\right)$ where $g(s)$ is such that $g(s) = \oslash$ for any s.
3. The external function defined on $]\oslash, +\infty[$ by

$$\eta(x) \equiv \exp(-\omega x^{@}). \tag{7.33}$$

where ω is a fixed positive unlimited number, is the collection of all internal functions $\exp(-\omega x^{u(s)})$ where $u(s)$ is an internal function whose values are positive appreciable for any s in $]\oslash, \infty[$

Definition 7.5.7. *Let φ be an external function defined on an interval I and such that*

$$\forall \delta \in I \quad 0 \leq \varphi(\delta).$$

The integral of φ on I which is denoted by $\int_I \varphi(x)dx$, is the set of external numbers given by

$$\int_I \varphi(x)dx \equiv \left\{ \int_I f(x)dx \mid f \in \varphi, \ f \ Lebesgue - integrable \right\}.$$

Notation: Let α and β be two external numbers and φ be an external function. Analogously to the internal case we adopt the following notation

$$\int_\alpha^\beta \varphi(x)dx \equiv \int_{[\alpha,\beta]} \varphi(x)dx,$$

$$\int_\alpha^{\underline{\beta}} \varphi(x)dx \equiv \int_{[\alpha,\beta[} \varphi(x)dx,$$

$$\int_{\underline{\alpha}}^\beta \varphi(x)dx \equiv \int_{]\alpha,\beta]} \varphi(x)dx,$$

$$\int_{\underline{\alpha}}^{\underline{\beta}} \varphi(x)dx \equiv \int_{]\alpha,\beta[} \varphi(x)dx.$$

Remark that $\int_I \varphi(x)dx$ is in general not an external number. However, if

φ is a convex function then $\int_I \varphi(x)dx$ is an external interval of $\mathbb{R}$ bounded above and below by external numbers. Furthermore, in most practical cases we are not looking for the exact value of the integral of an external function but for an approximation of that integral, typically for an external number containing this value.

Example 7.5.5.   1. Let δ be an external number such that $0 \leq \delta$, and let α and β be two external numbers satisfying $\alpha < \beta$. Then

$$\int_\alpha^\beta \delta dx \equiv \delta \cdot (\beta - \alpha). \tag{7.34}$$

2. Let ω be a positive unlimited real number. Then

$$\int_{\exp(-\text{\textcircled{a}}\omega)}^{1+\pounds e^{-\text{\textcircled{a}}\omega}} \left(1 + \frac{\sin(\log^2 \omega x)}{\omega}\right) dx \;=\; \int_{\exp(-\text{\textcircled{a}}\omega)}^{1+\pounds e^{-\text{\textcircled{a}}\omega}} \left(1 + \frac{\pounds}{\omega}\right) dx$$

$$\equiv \left(1 + \frac{\pounds}{\omega}\right) \cdot \int_{\exp(-\text{\textcircled{a}}\omega)}^{1+\pounds e^{-\text{\textcircled{a}}\omega}} dx$$

$$\equiv \left(1 + \frac{\pounds}{\omega}\right).$$

3. For any function f which is standard, strictly positive and Lebesgue–integrable in the neighbourhood of infinity, we have

$$\int_\pounds^\infty f(x)dx \equiv \oslash.$$

This implies, among other consequences, that

$$\int_\pounds^\infty e^{-s^{\text{\textcircled{o}}}} ds \;\equiv\; \oslash. \tag{7.35}$$

4. Let ω be a positive unlimited number. Then

$$\int_\pounds^{\oslash\sqrt{\omega}} \exp\left(-\frac{s^2}{2}(1+\oslash)\right) ds = \int_\pounds^{\oslash\sqrt{\omega}} e^{-s^{\text{\textcircled{o}}}} ds = \oslash$$

The rule of Chasles (Proposition 7.5.5) is easily extended to the integration of external functions.

The following result is a majoration of the integral of $\exp\left(-\omega x^{\text{\textcircled{o}}}\right)$ on $]\oslash, +\infty[$. This result is the formulation of the *concentration lemma* (see [15]) in the context of the external calculus.

The concentration lemma is a tool used in the estimation of a class of integrals by what is classically known as the "Laplace method". The integral in the Stirling formula is a typical example.

Proposition 7.5.6 (Concentration lemma). *Let ω be a positive unlimited real number. Then*

$$\int_{\oslash}^{\infty} \pounds e^{-\omega t^{\oplus}} dt = \pounds e^{-\oplus \omega}.$$

Proof. Let r be any appreciable number. It suffices to prove that

$$\int_{\oslash}^{\infty} \exp(-\omega t^r) dt = \pounds e^{-\oplus \omega}.$$

Set $s = t^r$, then

$$
\begin{aligned}
\int_{\oslash}^{\infty} \exp\left(-\omega t^r\right) dt &= \frac{1}{r} \int_{\oslash}^{\infty} \exp\left(-\omega s + \frac{1-r}{r} \log(s)\right) ds, \\
&= \frac{1}{r} \int_{\oslash}^{\infty} \exp(-\omega(1+\oslash)s) ds, \\
&= \frac{1}{r} \left\{ \int_{\oslash}^{\infty} \exp(-\omega(1+\varepsilon)s) ds \right\}_{\varepsilon=\oslash}, \\
&= \frac{1}{r} \left\{ \sup_{d>\oslash} \left(\left[-\frac{\exp(-\omega(1+\varepsilon)s)}{\omega(1+\varepsilon)} \right]_d^{\infty} \right) \right\}_{\varepsilon=\oslash}, \\
&= \frac{1}{r} \left\{ \pounds e^{-\oplus \omega} \right\}_{\varepsilon=\oslash}, \\
&= \pounds e^{-\oplus \omega}.
\end{aligned}
$$

7.6 Stirling's formula

We now present a proof of Stirling's formula,

$$\lim_{n \to \infty} \frac{n!}{n^n e^{-n} \sqrt{n} \sqrt{2\pi}} = 1.$$

using the external symbolism developed above.

The nonstandard analogue of this formula is the fact that for any positive unlimited number ω

$$\omega! = (1+\oslash) \cdot \omega^{\omega} \cdot e^{-\omega} \cdot \sqrt{\omega} \cdot \sqrt{2\pi}$$

We know that

$$\omega! = \int_0^{\infty} e^{-t} t^{\omega} dt.$$

The integrand $I(t) = e^{-t} t^{\omega}$ has its maximum at ω and $I(\omega) = e^{-\omega} \omega^{\omega} = \oslash\!\!\!\!/$ The change of variables $t = \omega + \omega x$ **centralises** and **normalizes** the integrand, and enables us to use the concentration lemma. Indeed,

$$\int_0^\infty e^{-t} t^\omega \, dt = \omega^\omega \cdot \omega \cdot e^{-\omega} \int_{-1}^\infty \exp\left(-\omega[x - \log(1+x)]\right) dx$$

The new integrand $J(x) = \exp\left(-\omega[x - \log(1+x)]\right)$ takes its maximum in 0 and $J(0) = 1$, and for any x strictly larger than -1 we have $J(x) = \exp\left(-\omega x^{@}\right)$. Using the concentration lemma, this means that

$$\int_{-1}^\infty J(x)\,dx = \int_\oslash J(x)\,dx + \pounds e^{-@\omega}.$$

We **localize**, inside $\oslash$, the domain where the value of $J(x)$ is significant. For this purpose we solve the external equation $J(x) = @$ as follows:

$$\exp\left(-\omega[x - \log(1+x)]\right) = @ \quad \Leftrightarrow \quad x - \log(1+x) = \frac{\pounds}{\omega}$$

$$\Leftrightarrow \quad x = \frac{\pounds}{\sqrt{\omega}},$$

by Proposition 7.4.5. The change of variables $x = \frac{s}{\sqrt{\omega}}$ extends $\frac{\pounds}{\omega}$ to $\pounds$, and $\oslash$ to $\oslash\sqrt{\omega}$. As for any s in $\oslash\sqrt{\omega}$

$$\exp\left[-\sqrt{\omega}s + \log\left(1 + \frac{s}{\sqrt{\omega}}\right)\right] = \exp\left[-\frac{s^2}{2}(1 + \oslash)\right],$$

it follows that

$$\begin{aligned}
\int_\oslash J(x)\,dx &= \frac{1}{\sqrt{\omega}} \int_{\oslash\sqrt{\omega}} \exp\left(-\frac{s^2}{2}(1+\oslash)\right) ds \\
&= \frac{1}{\sqrt{\omega}}\left[\int_\pounds \left(e^{-\frac{s^2}{2}} + \oslash\right) ds + 2 \cdot \int_\pounds^{\oslash\sqrt{\omega}} \exp\left(-\frac{s^2}{2}(1+\oslash)\right) ds\right] \\
&= \frac{1}{\sqrt{\omega}}\left[\sqrt{2\pi} + \oslash\right].
\end{aligned}$$

Now, $\sqrt{2\pi} + \oslash + \pounds e^{-@\omega} = (1 + \oslash)\sqrt{2\pi}$, so we get finally

$$\omega! = \omega^\omega \cdot \sqrt{\omega} \cdot e^{-\omega} \cdot \sqrt{2\pi} \cdot (1 + \oslash).$$

7.7 Conclusion

Let us summarize the ways in which we used the external numbers and the external functions in our calculations.

First, we replaced complicated expressions by simple expressions. Indeed, by embedding a real number in an external set of real numbers, or an internal function in an external set of internal functions we got rid of burdensome parameters and variables of little importance. Depending on need, this "re-laxation" could be small, as in the case of replacing a real number by the set of all real numbers which are asymptotic to it, or quite large, as in the case of replacing a real number by the set of all real numbers with the same order of magnitude.

Secondly, we developed a strategy for the problem of the approximation of an integral: localizing the domain where the integrand yields its princi-pal contribution to the value of the integral, cutting the integration interval into sub–intervals following the behaviour of the integrand, giving values to the different integrals on the different sub–intervals, and finally reaching an approximation of the first integral.

This strategy is external at different levels. Commonly, the domain where the contribution of the integrand to the value of the integral is not negligible matches the domain where the function under consideration is appreciable, and this domain is usually an external interval, which we determine by solving an external equation. This leads to a subdivision of the interval of integration into external sub–intervals. To integrate on an external interval we embed the integrand in an external function, simpler and more handy than our integrand, and then we integrate the external function by choosing internal representatives which are integrable by quadrature. The values of the different integrals on the different external intervals are external numbers, the sum of which contains the main integral, thus being an approximation of it.

This strategy avoids the classical approaches which are less natural: ad–hoc cuts, ad–hoc majorations, and passages to limits whose existence has to be proved. As to the common practice of Nonstandard Analysis, we avoided the use of logical tools such as, for example, the permanence principles.

From a certain point of view, we have developed in this paper an interval calculus. These are fuzzy intervals which are not bounded by numbers speci-fied in every detail and which are invariant under some translations. However, the fuzziness is not without limits, since the intervals are bounded by external numbers and the translations leaving invariant the interval considered form a convex additive group whose size can be given in an explicit manner. It is exactly this measured fuzziness which is at the origin of the smoothness and the efficiency of this interval calculus.

The external calculus is a new approach to uncertainty, next to statistics, probability theory, the theory of fuzzy sets and fuzzy logic. Based on the notion of order of magnitude, it distinguishes itself by being particularly rel-evant to the imperfect perception of their size and the fuzziness of numerical quantities which are incompletely identified.

8. An external probability order theorem with applications

I.P. van den Berg

8.1 Introduction

"Almost certainly the difference between a value of a random variable and its mean is of the order of the standard deviation".

This law, known by every statistician and probabilist, is expressed only in unsatisfactory ways within classical mathematics. Below are some attempts in which we refer to the above statement as C.

First assume that "almost certainly" is understood in the usual sense of probability measure, and "order" in the Archimedian sense. In the common case in which the standard deviation is non–zero C reduces to the statement that almost certainly the outcome of a random variable is a real number; true, but almost empty. Notice however that in the case of a zero standard deviation, C is reduced to the impoverished, but still quite useful property that the random variable almost certainly takes its mean value.

Second we have the well–known statistical rule of thumb, according to which 95% of a sample lies within the confidence interval $[\mu - 2\sigma, \mu + 2\sigma]$ where μ is the mean and σ the standard deviation. Of great practical value, mathematically speaking this statement has no intrinsic meaning, the numbers 95 and 2 being chosen at random, and, in any event, is false in general.

A third (see [21]) attempt concerns sequences of random variables and the notion of order in probability. Let $(x_n)_{n \in \mathbb{N}}$ be a sequence of random variables and $(y_n)_{n \in \mathbb{N}}$ be a sequence of real numbers. Then x_n is said to be at most in probability order y_n, written $x_n = O_p(y_n)$, for $n \to \infty$ if

$$\forall \varepsilon > 0 \ \exists M \ \exists k \ \forall n \geq k \ Pr\{|x_n| \geq My_n\} \leq \varepsilon.$$

Let $(\mu_n)_{n \in \mathbb{N}}$ be the sequence of mean values and $(\sigma_n)_{n \in \mathbb{N}}$ be the sequence of standard deviations associated with $(x_n)_{n \in \mathbb{N}}$. Then it follows quite readily from Chebyshev's inequality that

$$x_n - \mu_n = O_p(\sigma_n) \quad , \text{ for } n \to \infty.$$

However, when compared with the law C this statement is certainly less simple and natural.

In this chapter we show that, if the notions "almost certainly" and "order" are given the usual nonstandard interpretation, the law C is true for every random variable. Let $a, b \in \mathbb{R}$. As in chapter 7 the number b is said to be of *the order of a* if there exists a standard integer n such that $| b | < n | a |$. A property P is said to hold *almost surely* or *almost certainly* if the Loeb-measure (see [83]) of $E \equiv \{x \mid P(x)\}$ equals 1, or equivalently

$$\forall \varepsilon \gtrless 0, \quad \exists \text{ event } I \subset E, \quad Pr(I) \geq 1 - \varepsilon$$

(more details will be given in section 8.2).

The law C will be used in this chapter to give short nonstandard proofs of some asymptotic properties of elementary probability theory. We prove a central limit theorem on infinitesimal approximation of discrete probability distributions by the continuous Gaussian curve. The theorem may not be the most inclusive, yet it has a simple form and the conditions are easily checked in practice. The theorem enables one to show in a straightforward way that concrete probability distributions like the binomial distribution and Poisson distribution for large populations are almost normal.

We give also short nonstandard versions of the "probabilistic proofs" of Stirling's formula and the Weierstrass polynomial theorem, the latter using the polynomials of Bernstein.

It is convenient to present the law C in terms of external numbers and an "external probability measure". For the meaning of external number and the related notation we refer to chapter 7. The "external probability" is an external function whose values are (sets of) external numbers. As regards the notion of "almost certainly" it corresponds to the Loeb measure, yet it yields a finer measure on the external sets. In this setting C may be written as a numerical formula (formula (8.4)). This formula neatly expresses a correspondence between external sets, which forms one of the main characteristics of the law C.

The external probability is introduced in section 8.2. Because we need only some elementary properties we do not push its development very far. Section 8.3 contains the main result of this paper, i.e. the external formulation of the law C. As a first consequence we derive a sort of Transfer (Proposition 8.3.1) property: for some type of approximative properties of random variables to be true on the whole of $\mathbb{R}$, it suffices to establish them only for those numbers whose distance to the mean are of the order of the standard deviation. In section 8.4 we present the applications: the Weierstrass theorem in 8.4.1, Stirling's formula in 8.4.2 and the central limit theorem in 8.4.3.

Though this chapter is written within the axiomatic system IST [91], up to some slight modifications the results could have been formulated as well in the far more simple axiomatic system presented in Radically Elementary Probability Theory [92].

8.2 External probabilities

Definition 8.2.1. : *Let* $< \Omega, Pr >$ *be a probability space. For every* $A \subset \Omega$ *the inner external probability* $\underline{PR}(A)$ *is defined by the external number*

$$\underline{PR}(A) \equiv \; sup \; \{Pr(I) \mid I \subset A \; internal \; and \; measurable\},$$

the *outer external probability* $\overline{PR}(A)$ is defined to be the external number

$$\overline{PR}(A) \equiv \; \inf \; \{Pr(J) \mid J \supset A \; internal \; and \; measurable\}.$$

We define the *external probability* PR by

$$PR(A) \equiv [\underline{PR}(A), \overline{PR}(A)]$$

We say that A is *externally measurable* if $\underline{PR}(A) \equiv \overline{PR}(A) (\equiv PR(A))$.

Note that the external probability of an internal measurable set I is identical to its internal probability, i.e. to $Pr(I)$. The external probability is a mapping whose values are (sets of) external numbers, so it is formally not a probability. However, the examples and remarks below suggest that the external probability measures external sets rather well.

Examples:

1) Let $\Omega = [0,1]$ and λ be the Lebesgue measure on Ω. Let $A \subset [0,1]$ be an external interval with lower extremity α and upper extremity β. Then $PR(A) \equiv \beta - \alpha$. So in this case the external probability of A corresponds to the external integral (see chapter 7) of the indicator function of A.

2) Let $\nu \in \mathbb{N}$ be unlimited and $\Omega = \{0, 1, \ldots, \nu - 1\}$ with the counting measure $pr(k) = 1/\nu$ for every $k \varepsilon \Omega$. Then $PR\{st \; k\} \equiv \pounds/\nu$ and $PR\{k = \infty\} \equiv 1 - \pounds/\nu$.

3) Let x be a standard normal distributed random variable. Then $PR\{x \simeq \infty\} \equiv \oslash$ and $PR\{x \simeq c\} \equiv \oslash$ for every limited c. If c is unlimited,

$$PR\{x \simeq c\} \;\; \equiv \frac{1}{\sqrt{2\pi}} \int\limits_{c+\oslash} e^{-t^2/2} dt \equiv \frac{e^{-c^2/2}}{\sqrt{2\pi}} \int\limits_{\oslash} e^{-uc-u^2/2} du \equiv \pounds e^{-c^2/2+\oslash c}$$

$$PR\{x > \oslash c\} \equiv \frac{1}{\sqrt{2\pi}} \int\limits_{\oslash c} e^{-t^2/2} dt \equiv \pounds e^{-|@c^2}$$

$$PR\{x = \pounds c\} \equiv 1 - \sqrt{\frac{2}{\pi}} \int\limits_{\pounds c}^{\infty} e^{-t^2/2} dt \equiv 1 + \pounds e^{-\phi c^2}.$$

The latter examples show that the external probability is finer than the Loeb–measure. For instance, $PR\{x = \pounds c\}$ and $PR\{x \in \mathbb{R}\}$ are different while their Loeb–measure both equals 1.

It is not our intention to develop here the full theory of external probability. We will point out some basic aspects only. We consider briefly the problems of possible values, of measurability, of monotony and additivity, and of external analogues to the notions of almost certainly and negligible. In the following we let $< \Omega, Pr >$ be an internal probability space; by λ is always meant the Lebesgue measure.

8.2.1 Possible values

The images of $\underline{PR}$ and $\overline{PR}$ are contained in the set of all $\alpha \equiv a + A$ with $a \in [0, 1]$ and A a neutrix such that $A \leq \oslash$. If E is externally measurable $PR(E)$ is also of the above form. In general the external probability of an external set E is equal to an external interval, whose extremities are external numbers of the above form.

8.2.2 Externally measurable sets

We have the following general condition for external measurability. Let E be an external set. Put

$$S(E) = \cup\{\,]-\infty,\, Pr(I)]\mid I \subset E,\, I \text{ internal measurable}\}$$

$$T(E) = \cup\{\,]Pr(J), \infty]\mid J \supset E,\, J \text{ internal measurable}\}$$

If $(S(E), T(E))$ is a cut of $\mathbb{R}$, then $\sup S \equiv \sup T$ and E is clearly externally measurable. Notice that this condition is not necessary. Take $E \equiv [1/4 + \oslash, 3/4 + \oslash[$ in $< [0,1], \lambda >$; if $I \subset E$ is internal and measurable, then $Pr(I) \lesssim 1/2$ and if $J \supset E$ is internal and measurable then $Pr(J) \gtrsim 1/2$; yet E is externally measurable, with $PR(E) = 1/2 + \oslash$.

Using the condition above we prove the following proposition on external measurability which suffices for many practical purposes.

Proposition 2.2:. Let $G \equiv \bigcup\limits_{st,\ n \in \mathbb{N}} A_n$ and $H \equiv \bigcap\limits_{st,\ n \in \mathbb{N}} B_n$ where $(A_n)_{n \in \mathbb{N}}$ and $(B_n)_{n \in \mathbb{N}}$ are internal sequences of measurable sets. Then G and H are externally measurable.

Proof:. It suffices to prove the galactic case. Without restriction of generality, we may assume that $(A_n)_{n \in \mathbb{N}}$ is increasing. Then $S(A) \equiv \bigcup_{st, n \in \mathbb{N}}] - \infty, Pr\, A_n]$. Now if $t > S$ for all $stn \in \mathbb{N}$ it follows that $Pr(A_n) < t$. By the Cauchy principle there exists $\nu \simeq +\infty$ such that $Pr(A_\nu) < t$. Because $A_\nu \supset A$ we have $t \in T$. Hence (S, T) is a cut and A is externally measurable. $\qquad\square$

In [8] Benoit proves a better result in a somewhat less elementary way: in a finite probability space $(S(A), T(A))$ is a cut for every galaxy or halo A, implying that in this setting every galaxy or halo is externally measurable.

8.2.3 Monotony

It is not always true, say, that if $A \subset B$, then $PR(A) \leq PR(B)$: take for instance $A \equiv]\oslash, 1 + \oslash[$ and $B \equiv [0, 1]$ in the probability space $< [0, 1], \lambda >$, where λ is the Lebesgue measure. However, monotony remains valid up to the largest neutrix involved. Indeed, let A and B be externally measurable and $N \equiv \max(N(PR(A)), N(PR(B)))$. Assume $A \subset B$. Then both $PR(A) \leq PR(B) + N$ and $PR(B) \geq PR(A) + N$.

8.2.4 Additivity

The external probability is not additive in the strict sense. If we take $A \equiv [0, 1/2 + \oslash]$ in the probability space $< [0, 1], \lambda >$ we have $1 + \oslash \equiv PR(A) + PR(A^c) \not\equiv PR(A \cup A^c) \equiv 1$. However, the following property of weak additivity holds:

Proposition 8.2.1. *Let $A, B \subset \Omega$ be externally measurable, with $A \cap B = \emptyset$. Then*

$$PR(A \cup B) = PR(A) + PR(B)$$

Proof:. Let $PR(A) \equiv a + M$ and $PR(B) \equiv b + N$, where $a, b \in \mathbb{R}$ and M and N are neutrices. Without restriction of generality we may assume that $M \subset N$. Let $\varepsilon > N$. We claim that there exist internal measurable sets K, L such that $K \subset A \cup B \subset L$ and such that $a + b - \varepsilon \leq Pr(K) \leq Pr(L) \leq a + b + \varepsilon$.

Indeed, take $K = I_1 \cup J_1$ and $L = I_2 \cup J_2$, where I_1, I_2, J_1, J_2 are internal measurable sets with $I_1 \subset A \subset I_2, J_1 \subset B \subset J_2, a - \varepsilon/2 \leq Pr(I_1) \leq Pr(I_2) \leq a + \varepsilon/2$ and $b - \varepsilon/2 \leq Pr(J_1) \leq Pr(J_2) \leq b + \varepsilon/2$.

So both

$$\overline{PR}(A \cup B) \leq \overline{PR}(A) + \overline{PR}(B) \equiv PR(A) + PR(B)$$

and

$$\underline{PR}(A \cup B) \geq \underline{PR}(A) + \underline{PR}(B) \equiv PR(A) + PR(B).$$

Hence $PR(A \cup B) = PR(A) + PR(B)$. $\qquad\square$

8.2.5 Almost certain and negligible

An internal or external set A is said to be *negligible* if $PR(A) = \oslash$ and *almost certain* if $PR(A) = 1 + \oslash$. Notice that if A is internal and measurable, then A is negligible iff $Pr(A) \simeq 0$ and A is almost certain iff $Pr(A) \simeq 1$, so our definitions are strict extensions of the classical notions.

We have the following characterizations for negligibility. An external set A is negligible if and only if

$$\forall \varepsilon \gneqq 0, \ \exists J \supset A, \text{ internal, measurable, } \quad Pr(J) \leq \varepsilon. \qquad (8.1)$$

Also, if A is a galaxy, a halo, a standardly countable intersection of galaxies or a standardly countable union of halos, then A is negligible if and only if

$$\forall I, \ I \subset A, \text{ internal, measurable } \Longrightarrow Pr(I) \simeq 0. \qquad (8.2)$$

The equivalence of (8.1) and (8.2) for the above-mentioned types of external sets has been shown by Benoit [8].

Apart from the Loeb–measure and external probability there have been other approaches to the problem of measuring external sets or properties. We mention the approaches of Nelson [91] and Benoit [8]. Though different in nature all approaches correspond with respect to the notions of "almost certain" and "negligible". Indeed, the notion of "negligible" is always shown to be equivalent to (8.1).

8.3 External probability order theorems

Theorem 8.3.1. *Let y be a positive random variable such that Ey is well-defined. Then y is almost surely of order Ey.*

Proof:. If $Ey = 0$, then $Pr\{y > 0\} = 0$. A fortiori y is almost surely of order Ey. Assume $Ey > 0$. Using the Chebyshev inequality we obtain

$$PR\{y = \oslash.Ey\} \equiv \inf s \in @ \ \ Pr\{y \geq s. \ Ey\} \leq \inf s \in @1/s \equiv \oslash$$

Hence $y = \pounds.Ey$ holds almost surely.   □

Theorem 8.3.1 may be restated to become close to the Chebyshev inequality: the external probability that a positive random variable y takes unlimited values equals Ey times a part of the set of infinitesimals, i.e. the inverses of the unlimited numbers. Written as a formula, this becomes

$$PR\{y = \oslash\} \leq \frac{1}{\oslash}.Ey.$$

The next theorem, a formalization of the law C, is the main theorem of this paper.

Theorem 8.3.2 (mass concentration theorem). . *Let x be a random variable such that Ex and , Var, x are well-defined. Then $x - Ex$ is almost surely of order $\sqrt{}$, Var, x.*

Proof:. By theorem 8.3.1 almost surely $(x - Ex)^2$ is of order , Var, x. Hence $x - Ex$ is almost surely of order $\sqrt{}$, Var, x. $\qquad\qquad\square$

Let x be a random variable. Let us write μ for the expectation of x and σ for the standard deviation of x. We will call the external set $T(\equiv T(x))$

$$T \equiv \left(\bigcup \{]-\infty, c\,[\,|\, Pr\,\{x < c\} \simeq 0\}\right) \cup \left(\bigcup \{]d, \infty\,[\,|\, Pr\{x > d\} \simeq 0\}\right)$$

the *tail* associated with the probability distribution of x
and the external set

$$M \equiv M(x) \equiv \mathbb{R} - T$$

the *mass* $M \equiv M(x)$ associated with the probability distribution of x.
Then theorem 8.3.2 states, indeed, that the mass of the random variable x is concentrated in the σ-galaxy around μ, or written as a formula

$$M \subset \mu + \pounds\, \sigma. \tag{8.3}$$

Another way of expressing theorem 8.3.2 as a formula uses the external probability. Indeed one has

$$PR\{x = \mu + \pounds\, \sigma\} = 1 + \oslash \tag{8.4}$$

The mass concentration theorem may also be restated as an inequality of Chebyshev type, and a weak, yet almost universal form of the central limit theorem. Indeed, for $\sigma > 0$ we have both

$$PR\left\{\left|\frac{x - \mu}{\sigma}\right| = \phi\right\} \leq \frac{1}{\infty} \tag{8.5}$$

and

$$M\left(\frac{x - \mu}{\sigma}\right) = \pounds. \tag{8.6}$$

The latter formula just states that the mass of a random variable, after centralization and normalization is contained in the mass of a standard normal distributed random variable. In many cases the above inclusions and inequalities are identities, i.e. we not only have $Pr\{|\ x - \mu\ |> a\sigma\} \simeq 0$ for all positive unlimited a, but also $Pr\{|\ x - \mu\ |> a\sigma\} \gtrless 0$ for all positive limited a. The standard normal distribution, just as all the probability distributions infinitely close to it, satisfy this property.

If this identity property holds, the qualification "standard" in the standard deviation gets a new significance, for σ measures the mass in the following precise sense: the width of the mass is identical to all limited multiples of σ. Also, formula (8.5) having become an identity, we obtained an "external Chebyshev equality", which, in contrast to the usual, internal Chebyshev inequality does not imply a loss of information.

Note that the above properties are universally valid on finite probability spaces, the expectation and variance being sure to exist.

The mass concentration theorem suggests that in studying approximative properties it often suffices to consider only the domain where its normalization $(x - \mu)/\sigma$ is limited. This is a sort of Transfer property. The next proposition gives a concrete example of such a Transfer property. It will be used in the proof of the central limit theorem of section 8.4.

Proposition 8.3.1. *Let x, y be two normalized random variables. Then*

1. *If $Pr\{x \leq b\} \simeq Pr\{y \leq b\}$ for all limited $b \in \mathbb{R}$, then $Pr\{x \leq b\} \simeq Pr\{y \leq b\}$ for all $b \in \mathbb{R}$.*

2. *If $Pr\{a \leq x \leq b\} \simeq Pr\{a \leq y \leq b\}$ for all limited $a, b \in \mathbb{R}$, then $Pr\{a \leq x \leq b\} \simeq Pr\{a \leq y \leq b\}$ for all $a, b \in \mathbb{R}$.*

Proof: We prove only 1). If $b = -\infty$, by the mass concentration theorem $Pr\{x \leq b\} \simeq 0 \simeq Pr\{y \leq b\}$. If $b = +\infty$, then again by the mass concentration theorem

$$Pr\{x \leq b\} = 1 - Pr\{x > b\} \simeq 1 = 1 - Pr\{y > b\} \simeq Pr\{y \leq b\}$$

Hence $Pr\{x \leq b\} \simeq Pr\{y \leq b\}$ for all $b \in \mathbb{R}$.   □

8.4 Weierstrass, Stirling, De Moivre–Laplace

We give three applications of the mass concentration theorem in the form of short nonstandard proofs of the theorem of Weierstrass on uniform approximation of a continuous function by a polynomial, Stirling's formula and a weak, yet useful form of the central limit theorem.

8.4.1 The Weierstrass theorem

As is well-known, the Weierstrass polynomial theorem has a "probabilistic" proof, using the polynomials of Bernstein, which are based on binomial coefficients. Many authors, for instance [85], observed that there exists a simplified nonstandard version of such a proof. Using the mass concentration theorem we can rewrite the nonstandard proof very neatly.

We first recall the basics of the binomial distribution. Let $\nu \in \mathbb{N}$, and let x be a random variable taking the values $\{0, \ldots, \nu\}$ and let $0 \leq p \leq 1$. Then the binomial distribution $b_{\nu,p}(n)$ is given by

$$b_{\nu,p} = pr\{x = n\} = \binom{\nu}{n} p^n (1-p)^{\nu-n} \qquad\qquad 0 \leq n \leq \nu \qquad (8.7)$$

Its mean is $Ex = \nu p$ and its standard deviation is $\sqrt{\mathrm{Var}, x} = \sqrt{\nu p(1-p)}$. We now give a nonstandard version of the theorem of Weierstrass on the uniform convergence on compact sets of a sequence of polynomials to a real continuous function. The step from the theorem below to the classical version is straightforward and will be omitted.

Theorem 8.4.1. *Let $f : [0,1] \to \mathbb{R}$ be S-continuous and limited. Then there exists a polynomial g such that $g(t) \simeq f(t)$ for all $t \in [0,1]$.*

Proof:. Let g be the polynomial of Bernstein

$$g(t) = \sum_{n=0}^{\nu} t^n (1-t)^{\nu-n} f\left(\frac{n}{\nu}\right)$$

with $\nu \simeq +\infty$. Notice that for all $t \in [0,1]$ the standard deviation of the binomial distribution is at most $\sqrt{\nu/2}$.

Applying the mass concentration theorem two times, and using the fact that f is limited and nearly constant on the halo of every $t \in [0,1]$, we have

$$
\begin{aligned}
f(t) &= \sum_{n=0}^{\nu} \binom{\nu}{n} t^n (1-t)^{\nu-n} f(t) \\
&= \sum_{n=\nu t + \pounds\sqrt{\nu}} \binom{\nu}{n} t^n (1-t)^{\nu-n} f(t) \quad + \oslash \\
&= \sum_{n=\nu t + \pounds\sqrt{\nu}} \binom{\nu}{n} t^n (1-t)^{\nu-n} f(n/\nu) \quad + \oslash \\
&= \sum_{n=0}^{\nu} \binom{\nu}{n} t^n (1-t)^{\nu-n} f(n/\nu) \quad + \oslash \\
&= g(t) + \oslash.
\end{aligned}
$$

$\square$

8.4.2 Stirling's formula revisited

Using the mass concentration theorem and some properties of the gamma–distribution with density $e^{-t}t^{\omega}/\omega!$ we can give a short proof of Stirling's formula

$$\int_0^{\infty} e^{-t}t^{\omega}dt = (1+\oslash)\omega^{\omega}e^{-\omega}\sqrt{\omega}\sqrt{2\pi} \qquad (\omega \simeq +\infty).$$

It is even shorter than the proof given in chapter 7. Part of that proof consisted in localizing the domain of the principal contribution to this integral, thus determining the appropriate change of variables. This is done almost automatically, by applying the mass concentration theorem to the gamma–distribution.

We recall that the gamma–distribution has mean ω, and standard deviation $\sqrt{\omega}$. So we have the following nonstandard probabilistic proof of the formula of Stirling:

$$
\begin{aligned}
\omega! &= \omega! \int_0^{\infty} \frac{e^{-t}t^{\omega}}{\omega!} dt \\
&= \omega! \left[\int_{\omega+\pounds\sqrt{\omega}} \frac{e^{-t}t^{\omega}}{\omega!} dt \right] \\
&= \omega^{\omega}e^{-\omega}\sqrt{\omega} \int_{\pounds} e^{-s\sqrt{\omega}} \left(1+\frac{s}{\sqrt{\omega}}\right)^{\omega} ds \\
&= \omega^{\omega}e^{-\omega}\sqrt{\omega} \int_{\pounds} e^{-s\sqrt{\omega}+\omega\log(1+s/\sqrt{\omega})} ds \\
&= \omega^{\omega}e^{-\omega}\sqrt{\omega} \int_{\pounds} e^{-(1+\oslash)s^2/2} ds \\
&= (1+\oslash)\omega^{\omega}e^{-\omega}\sqrt{\omega}\sqrt{2\pi}.
\end{aligned}
$$

8.4.3 A central limit theorem

The nonstandard version of the De Moivre–Laplace theorem states that after an appropriate rescaling, the binomial distribution is infinitely close to the standard normal distribution. Thus we have here a special case of an infinitesimal approximation of a discrete, S-continuous function by a standard continuous function.

Nonstandard analysis disposes of a general tool to prove such approximations, namely the stroboscopy lemma (see [50] and [53]). The stroboscopy lemma has in fact been used to prove the De Moivre–Laplace theorem, in [53] for instance. Below we use the stroboscopy lemma to prove a more general form of the central limit theorem. It is easy to show that the binomial distribution reduces to our case, as do other concrete probability distributions, like the Poisson distribution.

The proof of the version of the central limit theorem below is somewhat simpler than in the case of the binomial distribution, which tends to be obscured by some technicalities, proper to this distribution.

Theorem 8.4.2. *Let* $x : \Omega \to \mathbb{Z}$ *be a random variable with probability distribution* $x_n \equiv pr\{x = n\}$, *mean* $\mu \in \mathbb{R}$ *and unlimited variance* σ^2. *Assume that for every* k *of order* σ *one has that*

$$x_{[\mu]+k+1}/x_{[\mu]+k} = 1 - k/\sigma^2 + \oslash/\sigma. \tag{8.8}$$

Then for every $A \in \mathbb{R}$

$$Pr\{x \leq A\} \simeq N((A - \mu)/\sigma).$$

Notice that the difference equation (8.8) allows for some freedom and irregularities on the x_n. Not only is there no condition on the behaviour of the $x_{[\mu]+k}$ on the tail of the distribution, more precisely, for those k that are unlimited with respect to σ, but also the concrete term k/σ^2 is absorbed by the uncertain term $\oslash/\sigma$ if k is infinitely small with respect to σ, i.e. the concrete term has no influence for such k.

This gives some idea about the generality of the above form of the central limit theorem; of course the property $\sum\limits_{n=-\infty}^{\infty} x_n = 1$ provides an a fortiori bound to the irregularities of the x_n. The proof of the theorem uses the following lemma which describes some special cases of approximation by stroboscopy.

Lemma 8.4.1. *Let* $\omega \simeq +\infty$ *and* $(y_k)_{k \in \mathbb{Z}}$ *be a sequence such that* $y_o = 1$ *and such that for every* k *or order* ω *it satisfies the difference equation*

$$\frac{y_{k+1} - y_k}{1/\omega} \simeq -\frac{k}{\omega} \cdot y_k$$

Then

(i) for every limited t *we have* $y_{[t\omega]} \simeq e^{-t^2/2}$

(ii) for every limited a *and* b *we have* $\dfrac{1}{\omega} \cdot \sum\limits_{k=[a\omega]}^{[b\omega]} y_k \simeq \int\limits_a^b e^{-t^2/2}\, dt$

(iii) $\sum\limits_{k \simeq \omega \mathcal{L}} y_m = (1 + \oslash)\sqrt{2\pi} \cdot \omega$

Proof:. By the stroboscopy lemma the shadow $\overline{y}$ of the sequence $\{(k/\omega, y_k) \mid k \in \mathbb{Z}\}$ satisfies the differential equation

$$\overline{y}'(s) = -s.\overline{y}(s)$$

with initial value $\overline{y}(0) = 1$. So $\overline{y}(s) = e^{-s^2/2}$ for all $s \geq 0$. Hence $y_{[t\omega]} \simeq e^{-t^2/2}$ for every limited t.

(ii)   Using (i) and the infinitesimal approximation of a Riemann sum by the corresponding integral we obtain

$$\frac{1}{\omega} \sum_{k=[a\omega]}^{[b\omega]} y_m = \sum_{k=[a\omega]}^{[b\omega]} \exp\left(-\frac{m^2}{2\omega} + \oslash\right).\frac{1}{\omega} = \int_a^b e^{-t^2/2}\, dt + \oslash.$$

(iii)   It follows from (ii) that

$$\sum_{k=\omega\pounds} y_k = \omega.\int_\pounds e^{-t^2/2} dt = (1+\oslash)\sqrt{2\pi}.\,\omega.$$

Proof of theorem 8.4.2.. Put

$$y_k = x_{[\mu]+k}/x_{[\mu]}$$

Then for all k which are limited with respect to σ

$$\frac{y_{k+1} - y_k}{1/\sigma} = \left(-\frac{k}{\sigma} + \oslash\right) y_k$$

Notice that $y_0 = 1$ and that for all $k = \pounds\sigma$ we have at most $y_k \leq (1 + \oslash/\sigma)^{\pounds\sigma} = 1+\oslash$. So $(y_k)_{k\in\mathbb{Z}}$ satisfies for every $k = \pounds\sigma$ the difference equation

$$\frac{y_{k+1} - y_k}{1/\sigma} \simeq -\frac{k}{\sigma}.\,y_k$$

Using the mass concentration theorem and lemma 8.4.1 (iii) we obtain for $x_{[\mu]}$ the estimation

$$x_{[\mu]} = \frac{x_{[\mu]}}{\sum\limits_{k=-\infty}^{\infty} x_{[\mu]+k}} = \frac{x_{[\mu]}}{\sum\limits_{k=\pounds\sigma} x_{[\mu]+k}} = \frac{1}{\sum\limits_{k=\pounds\sigma} y_k} = \frac{1+\oslash}{\sqrt{2.\pi}\,\sigma} \tag{8.9}$$

It follows from the mass concentration theorem and lemma 8.4.1 (ii) that for every $A = \mu + \pounds\sigma$

$$Pr\{x \leq A\} = \sum_{n\leq A} x_n = x_{[\mu]}.\sum_{k=[\mu]+\pounds\sigma}^{[A-\mu]} y_k =$$

$$= \int_\pounds^{(A-\mu)/\sigma} e^{-t^2/2} dt = N\left(\frac{A-\mu}{\sigma}\right) + \oslash$$

Then by proposition 8.3.1.1 we have for every $A \in \mathbb{R}$

$$Pr\{x \le A\} = N\left(\frac{A - \mu}{\sigma}\right).\square$$

We apply theorem 8.4.2 to the Poisson distribution and binomial distribution for infinitely large index, to show that they are almost normal.

Poisson distribution:. Let $x : \Omega \to \mathbb{N}$ be a random variable and $\nu > 0$. The Poisson distribution is given by

$$x_n \equiv Pr\{x = n\} = e^{-\nu}\nu^n/n!$$

We recall that $Ex = $, Var, $x = \nu$. Assume that $\nu \simeq +\infty$. We verify that x satisfies the difference equation (8.8). Indeed,

$$\frac{x_{[\nu]+k+1}}{x_{[\nu]+k}} = \frac{[\nu]}{[\nu] + k + 1} = \frac{\nu + \pounds}{\nu + k + \pounds} = 1 - \frac{k}{\nu} + \frac{\pounds}{\nu}$$

Hence by theorem 8.4.2 we have for every $A \in \mathbb{R}$

$$Pr\{x \le A\} \simeq N\left(\frac{A - \nu}{\sqrt{\nu}}\right).$$

Binomial distribution:. Finally we prove a nonstandard version of the De Moivre–Laplace theorem. Let $\nu \simeq +\infty$, let $0 \lneqq p \lneqq 1$ and let the random variable $x : \Omega \to \mathbb{Z}$ satisfy the binomial distribution $x_n = b_{\nu,p}(n)$. We prove that

$$Pr\{x \le A\} \simeq N\left(\frac{A - p\nu}{\sqrt{\nu p(1 - p)}}\right) \tag{8.10}$$

Indeed, let $\mu \equiv Ex = \nu p$ and $\sigma \equiv \sqrt{}$, Var, $x = \sqrt{\nu p(1 - p)}$. Then $\sigma \simeq +\infty$. We verify that x satisfies the difference equation (8.8). Indeed,

$$\frac{x_{[\mu]+k+1}}{x_{[\mu]+k}} = \frac{\nu - [\nu p] - k}{[\nu p] + k + 1} \cdot \frac{p}{1 - p}$$

$$= \frac{\nu(1 - p) - k + \pounds}{\nu p + k + \pounds} \cdot \frac{p}{1 - p}$$

$$= \frac{1 - kp/\sigma^2 + \pounds/\sigma^2}{1 + k(1 - p)/\sigma^2 + \pounds/\sigma^2}$$

$$= 1 - k/\sigma^2 + \pounds/\sigma^2$$

Then (8.10) follows from theorem 8.4.2.

Notice that both in the case of the Poisson distribution and the binomial distribution the uncertainty term $\pounds/\sigma^2$ is strictly less than the uncertainty term $\oslash/\sigma$ required by theorem 8.4.2.

9. Integration over finite sets

Pierre Cartier and Yvette Perrin

9.1 Introduction

In *Radically elementary probability theory* [92] Edward Nelson shows that it is possible to obtain deep results in the theory of probabilities and stochastic processes, assuming that the probability space and the time space are both finite. Convinced by Nelson's arguments, we present in this chapter a theory of integration over finite spaces. It is known that any situation can be reduced to this one. For example [32], in order to integrate a function over an interval I of $\mathbb{R}$, we consider a finite approximation of I i.e. a finite partition $T = (t_i)_{0 \leq i \leq \nu}$ of I, infinitesimally fine (which we call a near interval) and we define the integral of f as the finite sum $\sum_{i=0}^{\nu-1} \left(t_{i+1} - t_i \right) f(t_i)$.

One of the first integration theories over finite spaces was S-integration which was developed by E. Nelson [91]. The notion of S-integrability is remarkably simple but it is too general and the class of S-integrable functions is too large. For example, one may want that integrability over an interval should not depend on the choice of a particular approximation of that interval. S-integrability does not satisfy this property. Moreover, a function can be S-integrable over two different approximations of the same interval and still have two different integrals over these approximations. That is why we have introduced a new notion of integrability, adding a new condition which is close to Lusin's measurability condition in classical integration theory. [23].

We begin this chapter with a quick survey of S-integration theory over finite spaces. Then we restrict our study to finite, precompact, metric spaces, such as near intervals and products of near intervals for example. In such a space X, we define a new class $L^1(X)$ of integrable functions which is contained in the class $SL^1(X)$ of S-integrable functions over X. This is the analogue of the space L^1 in classical measure theory. Then, we study strong convergence and convergence almost everywhere in $SL^1(X)$ and $L^1(X)$. We consider only finite (limited or not) sequences. Finally, we generalize J.H. Hartong's and C. Reder's results about averaging and give a decomposition theorem of $SL^1(X)$ into two subspaces: $L^1(X)$ and the space of quickly oscillating functions.

9.2 S-integration

9.2.1 Measure on a finite set

Let F be a finite set. A positive measure on F is a function m which is defined on F and takes positive real values. The measure of an internal set $A \subset F$ is the finite sum $m(A) = \sum_{t \in A} m(t)$ $(m(A) = 0$ iff A is empty).

We suppose $m(F)$ is a limited number, throughout this chapter.

Example: Let I be an interval of $\mathbb{R}$, with extremities a and b. We say that a sequence $T = \{a = t_0 < t_1 < \ldots < t_\nu = b\}$ is an approximation of I or a near interval if $t_{i+1} - t_i$ is infinitesimal for $0 \leq i < \nu$. The function m defined on $T \setminus \{b\}$ by $m(t_i) = t_{i+1} - t_i$ is called the Lebesgue measure defined by T. The measure of any interval $[c, d[$ of T is its length $(d - c)$. Following E. Nelson we denote by $d_T t$ the successor of t in T, so that $m(t) = d_T t$.

9.2.2 Rare sets

An internal or external subset A of F is said to be *rare* [9] if, for all appreciable numbers $a > 0$, there exists an internal set $B \subset F$ such that $A \subset B$ and $m(B) \leq a$. Thus, when A is internal, A is rare means that $m(A)$ is infinitesimal.

Example: Let T be a near interval and m be the Lebesgue measure defined by T. The monad of a point a of T is the external set of all $t \in T$ such that $|t - a|$ is infinitesimal. Such a monad is rare.

We shall say that a property holds *almost everywhere* if the set on which it is not satisfied is rare.

Proposition 9.2.1.

(i) *Any internal or external subset of a rare set is rare.*

(ii) *The union of a limited number of rare sets is rare.*

(iii) *Let $(E_n)_{n \in \underline{N}}$ be an external sequence of rare sets. Then the union $E = \bigcup_{n \in \underline{N}} E_n$ is rare.*

(iv) *Let $(A_n)_{0 \leq n \leq \nu}$ be an internal, decreasing sequence of sets in A. Suppose A_n is rare for all unlimited integers $n \leq \nu$. Let B be the union of the $A'_n s$, for all unlimited $n \leq \nu$. Then B is rare.*

Proof. Statements (i) and (ii) are obvious. Let us prove (iii) and (iv).

Let E_n be a rare set for each limited n and let $a > 0$ be an appreciable number. For each limited n, choose an internal set B_n such that $E_n \subset B_n$ and $m(B_n) \leq a/2^{n+1}$. Let $(B_n)_{0 \leq n \leq \nu}$ be an internal sequence which extends the external sequence $(B_n)_{n \in \underline{N}}$ and such that $m(B_n) \leq a/2^{n+1}$ for all $n \leq \nu$. Then $B = \bigcup_{0 \leq n \leq \nu} B_n$ is an internal set which contains E and satisfies $m(B) \leq \sum_{n=0}^{\nu} a/2^{n+1} \leq a$ and so E is rare.

Suppose the hypothesis in (iv) are satisfied, and let $a > 0$ be an appreciable number. Then we have $m(A_n) \leq a$ for all unlimited $n \leq \nu$. According to the permanence principle (Cauchy principle, see chapter 1), there exists a limited integer n_0 such that $m(A_{n_0}) \leq a$. Since the sequence $(A_n)_{0 \leq n \leq \nu}$ is decreasing we have $B \subset A_{n_0}$, so B is rare.

9.2.3 S-integrable functions

Given a function $f : F \to \mathbb{R}$; its integral is the finite sum $\sum_{t \in F} f(t)m(t)$, which we denote by $\int_F f \, dm$. The function f is said to be S-integrable if the integral $\int_F |f| dm$ is limited and if $\int_A f \, dm$ is infinitesimal for all internal rare subsets A of F. It follows immediately that $\int_A |f| dm$ is infinitesimal whenever $m(A)$ is infinitesimal.

Let $g : F \to \mathbb{R}$ be a function and c be a constant. We denote by $\{g \geq c\}$ the set of all points t of F such that $g(t) \geq c$.

Proposition 9.2.2. *Given $f : F \to \mathbb{R}$, the following statements are equivalent:*

(i) The function f is S-integrable;

(ii) For all unlimited numbers $\lambda > 0$, $\int_{\{|f|>\lambda\}} |f| dm$ is infinitesimal;

(iii) For all unlimited numbers $\lambda > 0$, $\int_F |f - f^{(\lambda)}| dm$ is infinitesimal, where
  $f^{(\lambda)} = f I_{\{|f| \leq \lambda\}}$ is the truncated function at level λ.

Proof. Equivalence between (ii) and (iii) follows immediately from the equality

$$\int_{\{|f|>\lambda\}} |f| dm = \int_F |f - f^{(\lambda)}| dm.$$

(i) $\implies$ (ii): let $\lambda > 0$ be an unlimited number. If f is S-integrable, $\int_F |f| dm$ is limited. By Čebyčev's inequality, we have

$$m\{|f| > \lambda)\} \leq \lambda^{-1} \int_F |f| dm$$

so that $m\{|f| > \lambda\}$ is infinitesimal and then so is $\int_{\{|f|>\lambda\}} |f| dm$, since f is S-integrable.

(ii) $\implies$ (iii): let $\lambda > 0$ be an unlimited number, and $F(\lambda)$ denote the set $\{t : |f(t)| > \lambda\}$. For all internal sets A in F, we have

$$(1) \qquad \int_A |f| dm = \int_{A \cap F(\lambda)} |f| dm + \int_{A \setminus F(\lambda)} |f| dm.$$

By hypothesis $\int_{A \cap F(\lambda)} |f| dm$ is infinitesimal. Furthermore, $\int_{A \setminus F(\lambda)} |f| dm$ is smaller that $\lambda m(A)$. If $m(A)$ is infinitesimal we can choose an unlimited number λ, such that $\lambda m(A)$ is infinitesimal. Then it follows from (1) that $\int_A |f| dm$ is infinitesimal.

Now suppose $A = F$ in (1). Inequality (2) holds

$$(2) \qquad \int_F |f| dm \leq \int_{F(\lambda)} |f| dm + \lambda m(F).$$

Since $m(F)$ is limited and $\int_{F(\lambda)} |f| dm$ is infinitesimal, the integral $\int_F |f| dm$ is smaller than λ^2 for all unlimited $\lambda > 0$; therefore it is limited.

Remarks: By proposition 9.2.2, it follows that all numerical functions which take limited values are S-integrable.

Equivalently, one may say that f is S-integrable or $|f|$ is S-integrable or f^+ and f^- are S-integrable.

Let us give now some classical results.

Proposition 9.2.3.

(i) If $\int_F |f| dm$ is limited , f takes limited values almost everywhere.
(ii) If $\int_F |f| dm$ is infinitesimal, f takes infinitesimal values almost everywhere.
(iii) If f is S-integrable and takes infinitesimal values almost everywhere, then $\int_F |f| dm$ is infinitesimal.

Proof. (i) Assume that $\int_F |f| dm$ is limited. For all integers $n \geq 0$, let F_n denote the internal set $\{|f| > n\}$. The set where f takes unlimited values is the monad $H = \bigcap_{n \in \underline{N}} F_n$. The internal sequence (F_n) is decreasing. By Čebyčev's inequality $m(F_n) \leq \frac{1}{n} \int_F |f| dm$. It follows that F_n is rare for all unlimited integers n. By the permanence principle, for all appreciable numbers $\varepsilon > 0$ there exists a limited integer n such that $m(F_n) \leq \varepsilon$. Since $H \subset F_n$ we conclude that H is rare.

(ii) Suppose $\int_F |f| dm$ is infinitesimal. Let $G_n = \{|f| \geq \frac{1}{n}\}$ for $n \in \mathbb{N} \setminus \{0\}$. The external set on which f does not take infinitesimal values is the galaxy $G = \bigcup_{n \in \underline{N}} G_n$. By Čebičev's inequality, we have $m(G_n) \leq n \int_F |f| dm$. Therefore, if n is limited G_n is rare and it follows from 9.2.1 that G is rare.

(iii) Suppose f is S-integrable and infinitesimal almost everywhere. Then we have $m(G_n) \leq \frac{1}{n}$ for all limited integers n. By the permanence principle, there exists an unlimited integer μ such that $m(G_\mu) \leq \frac{1}{\mu}$ is infinitesimal. Furthermore

$$\int_F |f| dm = \int_{F \setminus G_\mu} |f| dm + \int_{G_\mu} |f| dm.$$

Since f is S-integrable and $m(G_\mu)$ is infinitesimal, the integral $\int_{G_\mu} |f| dm$ is infinitesimal. Besides $\int_{F \setminus G_\mu} |f| dm \leq \frac{1}{\mu} m(F)$ is infinitesimal and so is $\int_F |f| dm$.

Theorem 9.2.1 (Radon-Nykodym decomposition). *Let $f : F \to \mathbb{R}$. Suppose $\int_F |f| dm$ is limited. There exists a decomposition $f = g + h$, where g is S-integrable and f vanishes almost everywhere.*

Proof. We need the following lemma about convergence of monotone, bounded sequences.

Lemma 9.2.1. *Let μ be an unlimited integer and $(a_n)_{0 \leq n \leq \mu}$ be a monotone internal sequence of limited numbers. There exists an unlimited integer $\lambda \leq \mu$ such that $a_n \simeq a_\lambda$ for all unlimited $n \leq \lambda$.*

Let us prove this Lemma. We suppose that the sequence is increasing and we fix a limited integer $n > 0$. There exists a limited integer $p > 0$ such that, for all limited integers $i \geq p$, we have $a_i - a_p < \frac{1}{n}$. Otherwise there would exist an external sequence of limited integers $p(0) < p(1) < ... < p(i) < ...$ such that $a_{p(i+1)} - a_{p(i)} \geq \frac{1}{n}$ from which $a_{p(i)} - a_{p(0)} \geq \frac{1}{n}$. By the permanence principle, there would exist an unlimited integer i such that $a_{p(i)} - a_{p(0)} \geq \frac{1}{n}$ and so $a_{p(i)}$ would be unlimited.

By the permanence principle (Cauchy principle), there exists an unlimited integer $\lambda(n)$ such that $p \leq i \leq \lambda(n) \implies a_i - a_p < 1/n$. In particular, for all unlimited i and j smaller than $\lambda(n)$, we have $|a_i - a_j| < \frac{1}{n}$.

Robinson's lemma implies that the sequence $(\lambda(n))_{n \in \mathbb{N}}$ of unlimited integers is bounded below by an unlimited integer λ. And for unlimited integers $i \leq \lambda$ and $j \leq \lambda$ we have $|a_i - a_j| \leq \frac{1}{n}$. This last inequality holds for any limited integer n. So, $a_i - a_j$ is infinitesimal.

We now turn to theorem 9.2.1 Let a_n denote the integral $\int_{\{|f| < n\}} |f| dm$. For all integers n, a_n is smaller than $\int_F |f| dm$. By 9.2.1, there exists an unlimited integer λ such that $a_m \simeq a_\lambda$ for all unlimited integers $m \leq 1$. Denoting by B the set $\{|f(| < \lambda\}$ and by C its complementary set in F, we can write $f = g + h$ with $g = fI_B$ and $h = fI_C$. By 9.2.3, C is rare, and h vanishes outside of C. Moreover, for all t, we have $|g(t)| \leq \lambda$ and for all unlimited integers $n \leq \lambda$

$$\int_{\{|g| \geq n\}} |g| dm = \int_{\{n \leq |f| < \lambda\}} |f| dm = a_\lambda - a_n$$

is infinitesimal and so g is S-integrable.

9.3 Convergence in $SL^1(F)$

9.3.1 Strong convergence

Let $SL^1(F)$ denote the linear space of S-integrable functions over the finite set F. We consider $SL^1(F)$ with the L^1 norm

$$\|f\|_1 = \int_F |f| dm.$$

We adopt Nelson's definition for the convergence of numerical sequences: the numerical sequence $(a_n)_{0 \le n \le \nu}$ converges to a if, for all unlimited integers $n \le \nu$, we have $a_n \simeq a$.

We shall say that the sequence $(f_n)_{0 \le n \le \nu}$ of elements of $\mathbb{R}^F$, L^1-converges if $\|f_n - f_\nu\|_1$ is infinitesimal for all unlimited integers $n \le \nu$ and that it converges almost everywhere if the subset of F, on which it does not converge, is rare.

Theorem 9.3.1. *The space $SL^1(F)$ is closed in $\mathbb{R}^F$. That is to say: if the sequence $(f_n)_{0 \le n \le \nu}$ in $\mathbb{R}^F$, L^1-converges and if f_n is S-integrable for all limited numbers n, then f_n is S-integrable for all integers $n \le \nu$.*

Proof. Let $\mu \le \nu$ be an unlimited integer and a be an appreciable positive number. For all unlimited $n \le \nu$ we have $\|f_n - f_\mu\|_1 \simeq 0$. Then by the permanence principle (Cauchy principle) there exists a limited number m such that $\|f_m - f_\mu\| \le a$. Since f_m is S-integrable $\|f_m\|_1$ is limited and so is $\|f_\mu\|_1$.

Let A be a rare internal subset of F. We have:

$$\int_A |f_\mu| dm \le \int_A |f_m - f_\mu| dm + \int_A |f_m| dm \le \|f_m - f_\mu\|_1 + \int_A |f_m| dm \le 2a.$$

Hence $\int_A |f_\mu| dm$ is infinitesimal, for it is smaller than every appreciable number $a > 0$. We have proved that f_μ is S-integrable.

9.3.2 Convergence almost everywhere

Proposition 9.3.1.

Let $(f_n)_{0 \le n \le \nu}$ be a sequence of numerical functions on F. It converges almost everywhere iff, for all appreciable numbers $a > 0$ and all unlimited integers $n \le \nu$ we have $m\{t : \max_{n \le i \le \nu} |f_i(t) - f_\nu(t)| > a\} \simeq 0$.

Proof. For all integers $n \le \nu$, and all real numbers $a > 0$, let $D(n, a)$ denote the internal set of points t such that $\max_{n \le i \le \nu} |f_i(t) - f_\nu(t)| > a$.

The external set D of points t such that $(f_n(t))_{0 \le n \le \nu}$ does not converge is the union of the sets $D(n, \frac{1}{p})$ where n is unlimited and $p \ge 1$ is limited.

The sequence $(f_n)_{0 \le n \le \nu}$ converges almost everywhere iff D is rare. If this is the case, each of the sets $D(n, a)$ which is contained in D is rare, thus $m[D(n, a)] \simeq 0$.

Suppose now $m[D(n, \frac{1}{p})] \simeq 0$ for $n \le \nu$ illimited and $p \ge 1$ limited. If p is fixed the sequence of sets $D(n, \frac{1}{p})$ decreases with n. Proposition 9.2.1(iv) shows that D is rare.

Proposition 9.3.2. *Suppose that the internal sequence $(f_n)_{0 \leq n \leq \nu}$ is L^1-convergent. Then, there exists a subsequence which converges almost everywhere in F.*

Proof. Let I_n denote the internal set of integers r such that $n \leq r \leq \nu$ and $\max_{r \leq i \leq \nu} \|f_i - f_r\|_1 \leq 2^{-n-1}$.

For any limited n , I_n contains all unlimited integers $r \leq \nu$, by the definition of L^1-convergence. Thus, it is not empty and the permanence principle implies that there exists an unlimited integer $\mu \leq \nu$ such that I_n is not empty for all $n \leq \mu$. Let $\varphi(n)$ denote the smallest element of the internal set I_n (for $n \leq \mu$). The sequence $(\varphi(n))_{0 \leq n \leq \mu}$ is increasing and we have $\varphi(n) \geq n$, for every $n \leq \mu$. Put $g_n = f_{\varphi(n)}$. By construction, we have $\|g_n - g_{n+1}\|_1 \leq 2^{-n-1}$ *for* $n \leq \mu$. Let

$$u_n = \max_{n \leq i \leq \mu} |g_i - g_\mu|$$

$$v_n = \sum_{n \leq i \leq \mu} |g_i - g_{i+1}|.$$

We have

$$\|v_n\|_1 = \sum_{n \leq i \leq \mu} \|g_i - g_{i+1}\|_1 \leq \sum_{n \leq i \leq \mu} 2^{-i-1} \leq 2^{-n},$$

and since $0 \leq u_n \leq v_n$, it follows that $\|u_n\|_1 \leq 2^{-n}$. From Čebyčev's inequality we conclude that $m\{u_n \geq \frac{1}{p}\} \leq pr^{-n}$ is infinitesimal for all limited integers p and all unlimited $n \leq \mu$. Hence, the sequence $(g_n)_{0 \leq n \leq \mu}$ converges almost everywhere in X.

9.3.3 Averaging

Let $\mathcal{P}$ be an internal partition of F and f be a numerical function on F. We denote by $E^{\mathcal{P}}[f]$ the function which is constant on each atom A of $\mathcal{P}$ and takes the value $\frac{1}{m(A)} \int_A f \, dm$. We shall call $E^{\mathcal{P}}[f]$ the average of f relative to $\mathcal{P}$. If $\mathcal{P} = \{X\}$, we denote by $E[f]$, instead of $E^{\mathcal{P}}[f]$, the average of f. If m is a probability measure ($m(F) = 1$), $E^{\mathcal{P}}[f]$ is the conditional expectation of f relative to the partition $\mathcal{P}$.

We say that a partition $\mathcal{Q}$ is finer than a partition $\mathcal{P}$, if any atom of $\mathcal{Q}$ is contained in an atom of $\mathcal{P}$.

Proposition 9.3.3. *If f is S-integrable then $E^{\mathcal{P}}[f]$ is S-integrable for all internal partitions $\mathcal{P}$ of F.*

Proof. Let $\mathcal{P}$ be a partition of F and $g = E^{\mathcal{P}}[f]$ be the average of f relative to $\mathcal{P}$. If an internal set A is the union of atoms of $\mathcal{P}$, then $\int_A (u - E^{\mathcal{P}}[u]) dm = 0$ and $|E[u]| \leq E^{\mathcal{P}}[|u|]$, for all functions $u \in \mathbb{R}^F$. So we have:

(1) $$\int_A |g|dm \le \int_A E^P[\|f\|]dm = \int_A |f|dm.$$

Applying this inequality with $A = F$, we obtain that $\int_F |g|dm$ is limited since so is $\int_F |f|dm$. Now, let λ be an unlimited positive number and apply (1) with $A = \{|g| > \lambda\}$. We deduce that $\int_A |g|dm$ is limited, and by Čebyčev's inequality that $m(A) \le \lambda^{-1}\int_A |g|dm$ is infinitesimal. Since f is S-integrable, it follows that $\int_A |f|dm$ is infinitesimal and from (1) it follows that $\int_A |g|dm \simeq 0$. By proposition 9.2.2 we conclude that g is S-integrable.

Proposition 9.3.4. *Let P and Q be two internal partitions of F and f be a numerical function on F. Let $f_P = E^P[f]$ and $f_Q = E^Q[f]$. Suppose Q is finer than P, then*

    i) $f_P = E^P[f_Q]$
    ii) $E[f_Q^2] - E[f_P^2] = E[(f_Q - f_P)^2];$ *so* $E[f_Q^2] \ge E[f_P^2].$

Proof. Any atom A of P is the union of atoms of Q, so we have

$$\int_A f\,dm = \int_A f_Q\,dm$$

then $f_P = E^P[f_Q]$.
    Now

$$E[(f_Q - f_P)^2] = E[f_Q^2] - 2E[f_Q f_P] + E[f_P^2]$$

and, for all functions $g \in \mathbb{R}^F$,

$$E[f_P g] = E[f_P E^P[g]].$$

In particular for $g = f_Q$, we have $E^P[g] = f_P$, so $E[f_P g] = E[f_P^2]$ and $E[(f_Q - f_P)^2] = E[f_Q^2] - E[f_P^2]$.

9.3.4 Martingales relative to a function

Let f be a numerical function on F and $(f_n)_{0 \le n \le \nu}$ be a sequence in $\mathbb{R}^F$. We say that $(f_n)_{0 \le n \le \nu}$ is a martingale relative to f, if there exists an internal sequence of partitions P_n of F such that $f_n = E^{P_n}[f]$ for every $n \le \nu$ and P_n is finer than P_m whenever $n \ge m$.

Proposition 9.3.5. *Let f be a S-integrable function and $(f_n)_{0 \le n \le \nu}$ be a martingale elative to f. Then, every function f_n is S-integrable and there exists an unlimited integer $\mu \le \nu$ such that the sequence $(f_n)_{0 \le n \le \mu}$ L^1-converges.*

Proof. By 9.3.3, every function f_n is S-integrable.

The proof of the second statement conveniently splits into two cases:

Case 1: The function f takes only limited values then so does every f_n, and thus $E[f_n^2]$ is limited. By proposition 9.3.4 the sequence of limited numbers $E[f_n^2]$ is increasing. By lemma 9.2.1 there exists an unlimited number $\mu \leq \nu$ such that the sequence $(E[f_n^2]^2)_{0 \leq n \leq \mu}$ is convergent. Applying proposition 9.3.2 and Hölder's inequality, we obtain:

$$\|f_\mu - f_n\|_1 \leq m(F)^{1/2}(E[(f_\mu - f_n)^2])^{1/2} \leq m(F)^{1/2}(E[f_\mu^2] - E[f_n^2])^{1/2}.$$

Hence, if $n \leq \mu$ is unlimited $\|f_\mu - f_n\|_1$ is infinitesimal, and then $(f_n)_{0 \leq n \leq \mu}$ is L^1-convergent.

General case: For every integer k, let $f^{(k)}$ denote the truncated function $fI_{\{|f| \leq k\}}$. Also, let E_n denote the operator $f \to E_n f = E^{\mathcal{P}_n}[f]$. For all limited integers k, $f^{(k)}$ takes limited values.

In that case we have just shown that there exists an unlimited integer $\mu(k)$ such that $(E_n[f^{(k)}])_{0 \leq n \leq \mu(k)}$ is L^1-convergent. There exists an unlimited integer μ which is smaller than $\mu(k)$ for each limited k. Hence, for all limited k, the sequence $(E_n[f^{(k)}])_{0 \leq n \leq \mu}$ is L^1-convergent.

By definition of a martingale, $f_n = E_n[f]$. Put $f_{n,k} = E_n[f^{(k)}]$. For any numerical function g on F and any partition $\mathcal{P}$, we have $\|E^{\mathcal{P}}[g]\|_1 \leq \|g\|_1$. We conclude that:

$$\|f_n - f_{n,k}\|_1 \leq \|f - f^{(k)}\|_1$$

and by the triangular inequality, that the following inequality (1) holds

$$\|f_n - f_\mu\|_1 \leq \|f_n - f_{n,k}\|_1 + \|f_{n,k} - f_{\mu,k}\|_1 + \|f_{\mu,k} - f_\mu\|_1 \leq \|f_{n,k} - f_{\mu,k}\|_1 + 2\|f - f^{(k)}\|_1$$

Let n be an unlimited integer such that $0 \leq n \leq \mu$. For all limited integers $k > 0$, we have

(2) $$\|f_{n,k} - f_{\mu,k}\|_1 \leq \frac{1}{k}$$

since the martingale $(f_{n,k})_{0 \leq n \leq \mu}$ is L^1-convergent. By the permanence principle (Cauchy principle) there exists an unlimited integer k for which inequality (2) holds. For such a number k, $\|f - f^{(k)}\|_1$ is infinitesimal since f is S-integrable. From inequality (1) we deduce that $\|f_n - f_\mu\|_1 \simeq 0$ for every unlimited $n \leq \mu$ and conclude that $(f_n)_{0 \leq n \leq \mu}$ is L^1-convergent.

Proposition 9.3.6. *Let $(f_n)_{0 \leq n \leq \mu}$ be a martingale relative to f. For all positive numbers λ, we have:*

$$m\left[x : \text{Max}, _{n \leq i \leq \nu}|f_i(x) - f_n(x)| \geq \lambda\right] \leq \frac{2\|f_\nu - f_n\|_1}{\lambda}$$

Proof. Let, for $n \leq \nu$,

$$A = \{x : \text{Max}, \ _{n \leq i \leq \nu}(f_i(x) - f_n(x)) \geq \lambda\}$$

and $g_n = \sum_{n \leq i \leq \nu} \eta_i(f_{i+1} - f_i)$, with $\eta_i(x) = 1$ if for all j such that $n \leq j < i$, $f_{j+1}(x) - f_j(x) < \lambda$, and $\eta_i(x) = 0$ otherwise.

For $0 \leq j \leq i$, f_j is constant on each atom of the partition $\mathcal{P}_i$, then so is the function η_i. Hence, we have:

$$E^{\mathcal{P}_i}[\eta_i(f_{i+1} - f_i)] = \eta_i E^{\mathcal{P}_i}[f_{i+1} - f_i] = 0$$

and

$$E[g_n] = \sum_{n \leq i < \nu} E[\eta_i(f_{i+1} - f_i)] = \sum_{n \leq i < \nu} EE^{\mathcal{P}_i}[\eta_i(f_{i+1} - f_i)] = 0$$

Moreover, we have

$$g(x) \geq \lambda \quad \text{if} \quad x \in A$$
$$g(x) = f_\nu(x) - f_n(x) \quad \text{if} \quad x \in X \setminus A.$$

Then

$g \geq \lambda I_A + (f_\nu - f_n)I_{F \setminus A}$. Integrating, we obtain the inequality:

$$\lambda m(A) \leq \int_{F \setminus A} (f_n - f_\nu) dm \leq \|f_n - f_\nu\|_1.$$

Similarly, we prove that

$$m\left[\{x : \text{Max}, \ _{n \leq i \leq \nu}(f_n(x) - f_i(x)) \geq \lambda\}\right] \leq \frac{\|f_n - f_\nu\|_1}{\lambda}$$

and we deduce immediately the proposition 9.3.6.

Theorem 9.3.2. *Let f be an S-integrable function and $(f_n)_{0 \leq n \leq \mu}$ be a martingale relative to f, which is L^1-convergent. Then, this martingale converges almost everywhere.*

Proof. It is an immediate consequence of propositions 9.3.1 and 9.3.6.

9.3.5 Commentary

It is a common practice to use averages of functions when we are dealing with different finite approximations of sets. For example, let $[a, b]$ be an interval of $\mathbb{R}$, T and U be two near intervals approximating $[a, b]$ such that T is contained in U. For any $t \in T^- = T \setminus \{b\}$ let I_t denote the set: $\{u \in U : t \leq u < t + d_T t\}$. The intervals I_t ($t \in T^-$) form a partition of $U^- = U \setminus \{b\}$. With any function $f : T \to \mathbb{R}$ we associate its extrapolated function defined by $\gamma f(u) = f(t)$ for any $u \in I_t$ and any $t \in T^-$. Conversely, with any function $g : U \to \mathbb{R}$, we associate its average $\pi g : T^- \to \mathbb{R}$ defined by $\pi g(t) = \frac{1}{d_T t} \int_{I_t} g(u) \, d_U u$.

Our purpose is to define a notion of integrability such that for any integrable functions f, g, both $\pi \circ \gamma(f)$ and $\gamma \circ \pi g$ are integrable and infinitely close to f and g respectively. S-integrability does not satisfy this property. Let us give an example: let ν be an unlimited integer, the two near intervals $T = \left\{ \dfrac{2k}{2\nu}, \ 0 \le k \le \nu \right\}$ and $U = \left\{ \dfrac{k}{2\nu}, \ 0 \le k \le 2\nu \right\}$ are approximations of the interval [0,1] in $\mathbb{R}$. The function f, defined on U by

$$\begin{cases} f\left(\frac{2k}{2\nu}\right) & = & 1 \\ f\left(\frac{2k+1}{2\nu}\right) & = & -1 \end{cases},$$

is S-integrable over U, with L^1-norm $\|f\|_1 = 1$ while the function $\gamma \circ \pi(f)$ is null.

This is why we are going to give a more restricted definition of integrability.

9.3.6 Definitions

Let X be a finite set. A function $d : X \times X \to \mathbb{R}$ is said to be a distance if it satisfies the usual axioms:

$d(x, x) = 0$
$d(x, y) = d(y, x) > 0 \quad$ for $\quad x \ne y$
$d(x, z) \le d(x, y) + d(y, z)$

We consider now a finite set X provided with a distance d. The diameter of an internal subset A of X is $\max \{d(x, y) : x \in A$ and $y \in A\}$. The monad of a point $x \in X$ is the external set of all the points $y \in X$ such that $d(x, y)$ is infinitesimal.

We say that a finite metric space X is precompact if its diameter is limited and if the following statement holds:

For all appreciable numbers $r > 0$, there exists a limited cover of X : $\{A_1, \ldots, A_n\}$ such that for $1 \le i \le n$ the diameter of A_i is less than r.

Proposition 9.3.7.

(i) If X is a finite metric precompact space, then every internal subset of X is a metric precompact space.

(ii) Every near interval which has a limited diameter is a metric precompact space, with respect to the distance $d(x, y) = |x - y|$.

(iii) Let $(X_i)_{1 \le i \le \nu}$ be an internal sequence of finite metric precompact spaces (we denote by d_i the distance on X_i), then the product space $X = \prod_{i=1}^{\nu} X_i$ with distance d defined by $d(x, y) = \sum_{i=1}^{\nu} \frac{1}{2^i} d_i(x_i, y_i)$ is a finite metric precompact space.

Proof. (i) is an immediate consequence of the definitions. To prove (ii) we notice that an interval in $\mathbb{R}$, with lenght L can be covered by n intervals with length L/n.

We now prove (iii) , assuming that ν is unlimited. Let M be the maximum of the diameters of the $X_i's$. Choose an appreciable number $r > 0$ and an integer $N \leq \nu$ such that $2^{-N}M \leq r/2$. Since $X_1, \ldots, X_n$ are precompact, for each i, there exists a cover $\mathcal{U}_i$ of X_i by a limited number of subsets whose diameters are less than $r/2$. Let $A_1, \ldots, A_N$ be in $\mathcal{U}_1, \ldots, \mathcal{U}_N$ respectively and let $A = \{x = (x_1, \ldots, x_\nu) \ x_1 \in A_1, \ldots, x_N \in A_N\}$. For x and y in A, we have:

$$d(x,y) = \sum_{i=I}^{N} 2^{-i} d_i(x_i, y_i) + \sum_{i=N+1}^{\nu} 2^{-i} d_i(x_i, y_i)$$

$$\leq \frac{r}{2} \sum_{i=I}^{N} 2^{-i} + \sum_{i=N+1}^{\nu} 2^{-i} M$$

$$\leq \frac{r}{2} + 2^{-N} M \leq r.$$

When $A_1, \ldots, A_N$ vary, we obtain a cover of X by a limited number of subsets which have a diameter less than r.

From here to the end of this chapter, we consider a finite, metric precompact space X with a positive measure m such that $m(X)$ is limited.

9.3.7 Quadrable sets

For $x \in X$, let $\mu(x)$ denote the monad of x. The boundary of a subset $A \subset X$ is the external set $Fr_X(A)$ of all points $x \in X$, such that $\mu(x) \cap A \neq \emptyset$ and $\mu(x) \cap (X \setminus A) \neq \emptyset$.

Lemma 9.3.1. *Let A and B be two internals subsets of X. We then have:*

(1) $$Fr_X(A) = Fr_X(X \setminus A),$$

(2) $$Fr_X(A \cup B) \subset Fr_X(A) \cup Fr_X(B).$$

If A contains B, then

(3) $$Fr_X(B) \subset Fr_X(A) \cup Fr_A(B).$$

Proof. Formula (1) is obvious. Let us prove (2). Let x be a point of the boundary of $A \cup B$. Its monad $\mu(x)$ intersects $X \setminus (A \cup B)$, so it intersects both $X \setminus A$ *and* $X \setminus B$. Furthermore $\mu(x)$ intersects $A \cup B$, so it intersects A or B. Hence x belongs to $Fr_X(A)$ or to $Fr_X(B)$.

Suppose now that A contains B. Let x be a point of $Fr_X(B)$ which does not belongs to $Fr_X(A)$ and let us show that x belongs to $Fr_A(B)$. We have $\mu(x) \cap B \neq \emptyset$ so $\mu(x) \cap A \neq \emptyset$ and since $x \notin Fr_X(A)$ it follows that $\mu(x) \subset A$. Furthermore $\mu(x) \cap (X \setminus B) \neq \emptyset$, so $\mu(x) \cap (A \setminus B) \neq \emptyset$. Hence, $x \in Fr_A(B)$.

A subset of X is said to be *quadrable* if it is internal and its boundary is rare.

Proposition 9.3.8.

(i) *If A and B are quadrable subsets of X, then $X \backslash A$ and $A \cup B$ are quadrable; this means that the collection of all quadrable subsets of X is a Boolean ring.*

(ii) *If A is quadrable in X and B is quadrable in A, then B is quadrable in X.*

Proof. Statement (i) follows from formulas (1) and (2) and from the fact that the union of two rare sets is rare. Statement (ii) is an immediate consequence of formula (3).

9.3.8 Partitions in quadrable subsets

Our purpose is to prove 9.3.3. We need the following two lemmas.

Lemma 9.3.2. *Let T be a near interval and $f : T \to \mathbb{R}$ be an increasing function which takes only limited values. Then, there exists a rare subset E of T such that f is S-continuous at each point of $T \setminus E$.*

Proof. Let a be the origin and b be the end of the near interval T. For all limited integers $n \geq 1$, let E_n be the set of points $x \in T$ for which there exist s and t such that $t > s, s \simeq x, t \simeq x$ and $f(t) - f(s) > \frac{1}{n}$. Put $E = \bigcup_{n \in \mathbb{R}} E_n$. It is obvious that f is S-continuous at every point of $T \setminus E$. Let us show that every subset E_n is rare, from which it will follow that E is rare. Let $s_1 < t_1 < \cdots < s_m < t_m$ be points of T such that $f(t_1) - f(s_1) > 1/n, \ldots, f(t_m) - f(s_m) > 1/n$. Necessarily $m \leq n(f(b) - f(a))$, then E_n is the union of a limited number of monads. Since every monad is rare, so is E_n.

Lemma 9.3.3. *Every ball with an appreciable radius ρ contains a quadrable ball with the same centre and a radius $\rho' > \rho/2$.*

Proof. Let a be a point of X and T be a near interval with limited extremities and which contains all the distances between the points of X. For all $r \in T$, let $f(r)$ be the measure of the closed ball $B(a, r)$ with center a and radius r. We can apply lemma 9.3.1 to the function f; there exist a number ρ' such that $\rho/2 < \rho' < \rho$ and a ball $B(a, \rho')$ on which f is S-continuous. Let

$$B_n = \{x \in X : \rho' - 1/n < d(a, x) \leq \rho' + 1/n\}.$$

If n is limited B_n contains the boundary of the ball $B(a, \rho')$, and if n is unlimited, the measure $f(\rho' + 1/n) - f(\rho' - 1/n)$ of B_n is infinitesimal. Then the boundary of $B(a, \rho')$ is rare.

Theorem 9.3.3. *For all appreciable numbers $r > o$ there exists an internal partition of X by a limited number of quadrable subsets, each one having a diameter smaller than r.*

Proof. Fix an appreciable number $r > 0$. Since X is precompact, there exists a limited integer $n \geq 1$, and points $x_1, \ldots, x_n$ of X such that $X = B_1 \cup \ldots \cup B_n$ where B_i is the ball $B(x_i, r/8)$). Suppose we have obtained quadrable sets $C_1, \ldots, C_k$ with diameter less than r, any two of them being disjoint and $B_1 \cup \ldots \cup B_k$ being contained in $C_1 \cup \ldots \cup C_k$. For $k = 0$, the hypothesis is empty. Let us show how to continue the construction for $0 < k < n$. Let Y be the metric space $X \setminus (C_1 \cup \ldots \cup C_k)$. If $B_{k+1} \cap Y = \emptyset$ put $C_{k+1} = \emptyset$, and if $B_{k+1} \cap Y \neq \emptyset$ choose a point y_{k+1} in this set and apply lemma 9.3.3, to the space Y and the ball $B\left(Y_{k+1}, \frac{r}{4}\right) \cap Y$ of this space which contains $B_{k+1} \cap Y$.

In the space Y there exists a ball C_{k+1} with centre y_{k+1} and radius $r_{k+1}, (r/4 < r_{k+1} < r/2)$ which is a quadrable subset of Y. Its diameter is smaller than r. By proposition 9.3.8(i) Y is a quadrable subset of X. By proposition 9.3.8(ii) C_{k+1} is a quadrable subset of X. By construction C_{k+1} and $C_1 \cup \ldots \cup C_k$ are disjoint and $B_1 \cup \ldots \cup B_k \cup B_{k+1}$ is contained in $C_1 \cup \ldots \cup C_k \cup C_{k+1}$.

Finally, for $k = n$, we obtain quadrable sets $C_1, \ldots, C_n$, with diameter less than r, any two of which are disjoint, and $X = B_1 \cup \ldots \cup B_n$ is contained in $C_1 \cup \ldots \cup C_n$. This proves theorem 9.3.3.

Proposition 9.3.9. *There exists an internal sequence $(\mathcal{P}_n)_{0 \leq n \leq \nu}$ of partitions of X which satisfies the following properties:*

(i) The partition $\mathcal{P}_m$ is finer than $\mathcal{P}_n$, for $0 \leq n \leq m \leq \nu$.

(ii) If n is limited, the partitions $\mathcal{P}_n$ is composed of a limited number of quadrable subsets.

(iii) If n is unlimited the atoms of $\mathcal{P}_n$ have infinitesimal diameters.

Proof. Keep $\mathcal{P}_0 = \{X\}$ (X is quadrable because its boundary is empty). Let n be a limited integer and suppose we have partitions $\mathcal{P}_0, \ldots, \mathcal{P}_n$ such that each of them is composed of a limited number of quadrable sets and $\mathcal{P}_k$ is finer than $\mathcal{P}_j$ when $0 \leq j \leq k \leq n$. Each atom A of $\mathcal{P}_n$ is a finite, metric, precompact space. We can apply theorem 9.3.3 to the space A : we obtain a partition $\mathcal{Q}_A$ of A into a limited number of sets which are quadrable in the space A and therefore quadrable in the space X, by proposition 9.3.8. We can also suppose that the diameter of these subsets are smaller than $1/(n + 1)$. The union $\mathcal{P}_{n+1} = \cup \{\mathcal{Q}_A, A \in \mathcal{P}_n\}$ is a partition of X composed of a limited number of quadrable subsets whose diameters are less than $1/n+1$. Using the external recurrence principle , and the saturation principle, we construct an internal sequence $(\mathcal{P}_n)_{0 \leq n \leq \mu}$ of partitions of X, which satisfies the following properties:

(i) When n is limited, the partition $\mathcal{P}_n$ is composed of a limited number of quadrable sets whose diameters are less than $1/n$.

(ii) When n and m are limited and n is less than m, the partition $\mathcal{P}_m$ is finer than $\mathcal{P}_n$.

Let I be the set of all integers $n \leq \mu$ such that each atom of $\mathcal{P}_n$ has a diameter smaller than $1/n$ and $\mathcal{P}_j$ is finer than $\mathcal{P}_i$ for $i \leq j \leq n$. It is internal and contains all the limited numbers. By the permanence principle, there exists an unlimited integer $\nu \leq \mu$ such that I contains all the integers less than ν. We observe that, if n is unlimited, the atoms of $\mathcal{P}_n$ have infinitesimal diameters. So the sequence $(\mathcal{P}_n)_{0 \leq n \leq \nu}$ satisfies the required properties.

9.3.9 Lebesgue integrable functions

A function $f : X \to \mathbb{R}$ is said to be S-continuous on a subset A of X if $f(x) \simeq f(y)$ for all x and y in A such that $d(x, y)$ is infinitesimal.

We say that f is almost continuous on X if there exists a rare subset E of X such that f is S-continuous in $X \setminus E$. We say that f is Lebesgue integrable or L-integrable if it is S-integrable and almost continuous.

Let $L^1(X)$ denote the space of L-integrable functions on X, normed by the L^1-norm.

Proposition 9.3.10. *Let $(f_n)_{0 \leq n \leq \nu}$ be a sequence in $\mathbb{R}^X$, which converges almost everywhere in X. Suppose f_n is almost continuous for all limited integers n. Then f_n is almost continuous for all integers $n \leq \nu$.*

Proof. For every limited integer n, let E_n be a rare subset of X such that f is S-continuous on $X \setminus E_n$ and let F be a rare subset such that the sequence $(f_n(x))_{0 \leq n \leq \nu}$ converges for all $x \in X \setminus F$. By proposition 9.2.1, there exists a rare set G which contains F and the sets E_n.

Let x and y be two points of $X \setminus G$ such that $x \simeq y$. The two sequences $(f_n(x))_{0 \leq n \leq \nu}$ and $(f_n(y))_{0 \leq n \leq \nu}$ are convergent and we have $f_n(x) \simeq f_n(y)$ for all limited integers n. By Robinson's lemma, there exists an unlimited integer $\mu \leq \nu$ such that $f_\mu(x) \simeq f_\mu(y)$. If $\lambda \leq \nu$ is an unlimited integer, then we have $f_\lambda(x) \simeq f_\mu(x)$ and $f_\lambda(y) \simeq f_\mu(y)$, so $f_\lambda(x) \simeq f_\lambda(y)$. This proves that the function f_λ is continuous on $X \setminus G$ for any unlimited integer $\lambda \leq \nu$.

Theorem 9.3.4. *Let the space $\mathbb{R}^X$ be normed with the L^1-norm, then $L^1(X)$ is closed in $\mathbb{R}^X$.*

Proof. Let $(f_n(t))_{0 \leq n \leq \nu}$ be a sequence which converges in $\mathbb{R}^X$ with the L^1-norm. We suppose f_n is L-integrable if n is limited. We have proved in theorem 9.3.1 that $SL^1(X)$ is a closed subspace of $\mathbb{R}^X$, thus f_μ is S-integrable for all unlimited $\mu \leq \nu$. By proposition 9.3.2, there exists an internal subsequence $(f_{\varphi(n)})_{0 \leq n \leq \mu}$ which converges almost every where in X. We have $\varphi(n) \geq n$ for all $n \leq \mu$, so $\varphi(\mu) = \mu$. Moreover if n is limited so is $\varphi(n)$, then $f_{\varphi(n)}$ is almost continuous, and by proposition 9.3.10 f_μ is S-continuous. Therefore f_μ belongs to $L^1(X)$ for all unlimited $\mu \leq \nu$.

9.3.10 Average of L-integrable functions

Definition 9.3.1. *We say that an internal partition $\mathcal{P}$ of X is infinitely fine if each of its atom has an infinitesimal diameter. We shall call a "dissection" of X, any sequence $(\mathcal{P}_n)_{0 \leq n \leq \mu}$ of partitions of X which satisfies statements (i), (ii) and (iii) of proposition 9.3.9.*

Theorem 9.3.5. *Let $f : X \to \mathbb{R}$ be an S-integrable function. Then the three following properties are equivalent:*

(i) f is Lebesgue integrable.
(ii) For all infinitely fine partitions, the function $f - E^{\mathcal{P}}[f]$ is infinitesimal almost everywhere.
(iii) For all infinitely fine partitions $\mathcal{P}$, $\|f - E^{\mathcal{P}}[f]\|_1$ is infinitesimal.

Proof. Since every L-integrable function is the difference of two positive L-integrable functions, we can suppose that f is positive.

(i) $\Rightarrow$ (ii) Let $n \geq 1$ be an integer, and let D_n denote the set of all points x of X for which the following property holds: if J is the atom of $\mathcal{P}$ which contains x, then we have:

$$| f(x) - \frac{1}{m(J)} \int_J f \, dm \, | > \frac{1}{n}.$$

Let us prove that D_n is rare for all limited integers $n \geq 1$. By proposition 9.2.1(iii), statement (ii) will follow.

By hypothesis, f is almost continuous so there exists a sequence $(E_p)_{0 \leq p \leq \pi}$ of subsets of X, with the following two properties:

for all limited p, f is S-continuous on $X \setminus E_p$ and
for all unlimited $p \leq \pi$, $m(E_p)$ is infinitesimal.

We can construct a sequence of positive reals α_p such that for all unlimited $p \leq \pi$, α_p and $m(E_p)/\alpha_p$ are infinitesimal.

Let x be a point of X and J be the atom of $\mathcal{P}$ which contains x. Then we have

$$(1) \qquad f(x) = \frac{1}{m(J)} \int_J f \, dm - \sum_{i=1}^{3} \varepsilon_i(p, x),$$

with the following definitions:

$$\varepsilon_1(p, x) = \frac{1}{m(J)} \int_{J \cap (X \setminus E_p)} (f(t) - f(x)) \, dm,$$

$$\varepsilon_2(p, x) = -\frac{1}{m(J)} m(J \cap E_p) f(x),$$

$$\varepsilon_3(p, x) = \frac{1}{m(J)} \int_{J \cap E_p} f \, dm.$$

Moreover let A_p, B_p, C_p, and T_p denote the following subsets respectively:

$$A_p = \cap\left\{I \in \mathcal{P} : m(I \cap E_p) \geq \alpha_p m(I)\right\},$$

$$B_p = \cap\left\{I \in \mathcal{P} : \int_{I \cap E_p} f \, dm \geq \frac{m(I)}{3n}\right\},$$

$$C_p = \left\{t \in X : f(t) > \frac{1}{3n\alpha_p}\right\},$$

$$T_p = E_p \cup A_p \cup B_p \cup C_p.$$

Assume that x belongs to $X \setminus T_p$, and first, that p is limited. The function f is S-continuous on $X \setminus E_p$, and the diameter of the atom J containing x is infinitesimal, then for all $t \in J \cap (X \setminus E_p)$, $f(t) - f(x)$ is infinitesimal, thus, all the more so we have $|f(t) - f(x)| \leq \frac{3}{n}$. Integrating this inequality we obtain:

$$(2) \qquad\qquad |\varepsilon_1(p,x)| \leq 1/3n.$$

Since x does not belong to A_p, we have $m(J \cap E_p) < \alpha_p$ and since x does not belong to C_p, $f(x)$ is less than $1/3n\alpha_p$. Thus

$$(3) \qquad\qquad |\varepsilon_2(x,p)| \leq 1/3n.$$

Finally, since x does not belong to B_p, we have

$$(4) \qquad\qquad |\varepsilon_3(x,p)| \leq 1/3n.$$

From (1), (2), (3) and (4), we conclude that:

$$|f(x) - \frac{1}{m(J)} \int_J f \, dm| \leq 1/n.$$

In other words, x does not belong to D_n.

So D_n is contained in the intersection of all the sets T_p for $p \geq 1$ limited. Let us show now that $m(T_p)$ is infinitesimal for any unlimited integer p. By the definition of the sets E_k, $m(E_p)$ is infinitesimal. Furthermore,

$$m(A_p) = \sum_{I \in \mathcal{P}, I \subset A_p} m(I) \leq \frac{1}{\alpha_p} m(A_p \cup E_p) \leq \frac{1}{\alpha_p} m(E_p) \simeq 0,$$

$$m(B_p) = \sum_{I \in \mathcal{P}, I \subset B_p} m(I) \leq 3n \int_{B_p} f \, dm \leq 3n \int_{E_p} f \, dm.$$

The number n is limited, E_p is rare and f is S-continuous, so $3n \int_{E_p} f \, dm$ is infinite-simal and so is $m(B_p)$.

Finally, by Čebičev's inequality we have:

$$m(C_p) \leq 3n\alpha_p \int_X f\,dm \simeq 0.$$

We have proved that each of the four internal subsets E_p, A_p, B_p and C_p is rare, thus, so is their union T_p.

(ii) $\Rightarrow$ (i)   Let $(\mathcal{P}_n)_{0 \leq n \leq \mu}$ be a dissection of X, and $(E_n f)_{0 \leq n \leq \mu}$ be the martingale relative to f. By proposition 9.3.7, for all limited integers n, the function $E_n f$ is S-integrable. This function is a combination of a limited number of characteristic functions of quadrable disjoints sets A_i, therefore it is S-continuous on the complement of the rare set $\cup_i Fr A_i$. The sequence $(E_n f)$ L^1-converges to f, and since $L^1(X)$ is closed it follows that f belongs to $L^1(X)$.

By proposition 9.2.3 we have equivalence between (ii) and (iii).

Corollary 9.3.1. *Let f be a Lebesgue integrable function and $(\mathcal{P}_n)_{0 \leq n \leq \nu}$ a dissection of X. Then, the martingale $(E_n f)_{0 \leq n \leq \nu}$ relative to f L^1-converges to f.*

Proof. Indeed, for any unlimited $n \leq \nu$, the partition $\mathcal{P}_n$ is infinitely fine, $f - E_n f$ is L-integrable, so $\|f - E_n f\|_1$ is infinitesimal.

9.3.11 Decomposition of S-integrable functions

Definition 9.3.2. *A function $h : X \to \mathbb{R}$ is said to be quickly oscillating if it is S-integrable and if $\int_A h\,dm$ is infinitesimal for all quadrable subsets A of X.*

Proposition 9.3.11. *Let h be a quickly oscillating function, $(\mathcal{P}_n)_{n \leq \mu}$ be a dissection and $(E_n h)_{n \leq \mu}$ be the martingale relative to h. Then there exists an unlimited integer $\nu \leq \mu$ such that for any $n \leq \nu, E_n h$ is infinitesimal almost everywhere.*

Proof. Suppose n is a limited integer. The partition $\mathcal{P}_n$ has a limited number of atoms. Let us call B the union of those ones which have an infinitesimal measure. Then B is rare. Let $t \in X \setminus B$ and A_t be the atom of $\mathcal{P}_n$ which contains t , then A_t is quadrable, $m(A_t)$ is appreciable and $\int_{A_t} h\,dm$ is infinitesimal, so $E_n h(t) = \frac{1}{m(A_t)} \int_{A_t} h\,dm$ is infinitesimal. Therefore $E_n h$ is infinitesimal almost everywhere, and thus its norm $\|E_n h\|_1$ is infinitesimal since $E_n h$ is S-integrable. By Robinson's lemma, there exists an unlimited integer $\mu \leq \nu$ such that $\|E_n h\|_1$ is infinitesimal for all integers $n \leq \mu$. So, for all $n \leq \mu$, $E_n h$ is infinitesimal almost everywhere.

Proposition 9.3.12. *A Lebesgue integrable function which is quickly oscillating is infinitesimal almost everywhere.*

Proof. With the previous notation, let ν be unlimited such that $E_\nu[h]$ is infinitesimal almost everywhere. The partition $\mathcal{P}_\nu$ is infinitely fine, thus, by theorem 9.3.5, $h - E_\nu[h]$ is infinitesimal almost everywhere and so is h.

Theorem 9.3.6. *For an S-integrable function f, there exist a Lebesgue integrable function g and a quickly oscillating function h such that $f = g + h$.*

If $f = g' + h'$ is another such decomposition of f then we have $g(x) \simeq g'(x)$ and $h(x) \simeq h'(x)$ almost everywhere.

Proof. Let $(\mathcal{P}_n)_{n \leq \mu}$ be a dissection of X and $(E_n f)_{n \leq \mu}$ be the martingale relative to f. By proposition 9.3.5, there exists an unlimited integer $\nu \leq \mu$ such that the sequence $(E_n f)_{n \leq \nu}$ L^1-converges. By theorem 9.3.4, $E_\nu f$ belongs to $L^1(X)$. Put $g = E_\nu f$ and $h = f - g$. The function h belongs to $SL^1(X)$ because f and g so do. It remains to be shown that h is quickly oscillating.

Let A be a quadrable set,
$$B = \cup\{E \in \mathcal{P}_\nu : E \subset A\}, \text{ and}$$
$$C = \cup\{E \in \mathcal{P}_\nu : E \cap A \neq \emptyset \text{ and } E \cap (X \setminus A) \neq \emptyset\}.$$

Then, we have $A = B \cup C$, $\int_E h \, dm = 0$ for all atoms E of $\mathcal{P}_\nu$, so $\int_B h \, dm = 0$. Since each atom of $\mathcal{P}_n$ has an infinitesimal diameter , C is contained in the boundary of A and hence, C is rare. Consequently $\int_C h \, dm$ is infinitesimal. Therefore $\int_A h \, dm = \int_B h \, dm + \int_C h \, dm$ is infinitesimal.

The second statement of the proposition follows from proposition 9.3.12.

9.3.12 Commentary

We will illustrate the theorem 9.3.6 by explaining how *grey results from the mixture of black and white*. Suppose we are given a partition of X into two internal subsets W (for "white") and B (for "black"). *Suppose that the sets B and W are quadrable.* The eye does not distinguish the details in the interior of a set whose diameter is infinitesimal, nor in a monad. The monad of a point in the interior of W will be seen as totally white; that of a point in the interior of B will be seen as totally black. The boundary is neither white nor black since the monad of any point contains white and black, but this boundary is rare and the limit between black and white will be sharp.

Let us now consider the case where B and W are not quadrable so that white and black are intermingled. Let us apply theorem 9.3.6 to the characteristic function I_B of B which assures us that there exists an infinitely fine partition $\mathcal{P}$ such that $g = E^{\mathcal{P}}[I_B]$ is L-integrable. If x is a point of X, and A is the atom of $\mathcal{P}$ containing x, then we have $g(x) = m(A \cap B)/m(A)$ so that $g(x)$ *measures the proportion of black in the neighbourhood of x*, or the *level of grey* at x. Let us notice two particular cases only:

a) g is a simple function, (i.e.) there exists a partition $(A_1, \ldots, A_n)$ of X such that n is limited , the A_i are quadrable and g takes a constant value on each A_i. The numbers $c_1, \ldots, c_n$ represent levels of grey and since the A_i are quadrable, the limit between two greys is sharp.

b) g is a continuous function. Now black and white shade off gradually.

If g is merely almost continuous, it will be possible to see clear transitions in the level of grey.

9.4 Conclusion

Given a finite metric space X we have defined three notions of integrability, to which there correspond three external sets:

  – the set $S(X)$ of all functions f such that $\int_X |f| dm$ is limited,
  – the set $SL^1(X)$ of S-integrable functions on X,
  – the set $L^1(X)$ of L-integrable functions on X.

We have the following inclusion:

$$S(X) \supset SL^1(X) \supset L^1(X)$$

and the two decomposition theorems:

1. For any function $f \in S(X)$, there exist a function g belonging to $SL^1(X)$ and a function h which is infinitesimal almost everywhere, such that:

$$f = g + h$$

2. For any function $f \in SL_1(X)$, there exist a function $h \in L^1(X)$ and a function $\ell \in SL^1(X)$ which is quickly oscillating, such that

$$f = k + \ell$$

and this decomposition is unique to within infinitesimals.

Let us substitute for X an infinitely fine partition Y of X and define a distance γ and a measure μ on Y, in the following way: for all A and B in Y, let $\gamma(A, B) = \max\{d(x, B), d(x', A) : x \in A$ and $x' \in B\}$, and $\mu(A) = \sum_{a \in A} m(a)$.

Let $N(X)$ denote the set of all the L-integrable functions on X whose L^1-norm is infinitesimal, and let $N(Y)$ denote the analogous subspace in $L^1(Y)$. Then $L^1(Y)/N(Y)$ is isomorphic to $L^1(X)/N(X)$. This is a stability theorem for L-integrable function spaces.

10. Ducks and rivers: three existence results

Francine Diener and Marc Diener

The beginnings of nonstandard analysis in France are strongly related to the study of singular perturbations of the Van der Pol equation [110] and its ducks [29, 12]. This equation is an interesting and simple example of a *slow-fast equation*, in other words, an equation of the form:

$$\frac{dy}{dx} = \frac{1}{\varepsilon} f(x, y) \tag{10.1}$$

where $\varepsilon > 0$ is a fixed i-small number, f a near-standard function, and $f_0 := {}^0\!f$. We will study here a typical equation of such a kind and this will provide us with the opportunity to look at some of the tools that have been developed for their study.

10.1 The ducks of the Van der Pol equation

10.1.1 Definition and existence

The slow curve For $(x, y) \in \mathcal{D} := \mathbb{R} \times] - \infty, 0[$, consider the slow-fast equation

$$\frac{dy}{dx} = \frac{1}{\varepsilon} \left((1 - x^2) + \frac{a - x}{y} \right) \tag{10.2}$$

where a is a real parameter, the interesting values of which are here close to 1, say $a \in [0, 2]$. The *slow curve* of (10.2) is the standard curve $\mathcal{L}$ (figure 10.1) given by $(1 - x^2) + \frac{{}^0a - x}{y} = 0$, in other words the curve $y = \tilde{y}(x)$, with

$$\tilde{y}(x) = ({}^0a - x)/(x^2 - 1).$$

The name of slow-curve is related to the fact that, for limited (x_0, y_0) not belonging to the halo of $\mathcal{L}$, the derivative $\frac{dy}{dx}$ of the solution starting at this point is i-large, because the factor $(1 - x^2) + \frac{a - x}{y}$ is not i-small and $\frac{1}{\varepsilon}$ is i-large: the solution is then "fast", and stays fast as long as the halo of $\mathcal{L}$ has not been reached. In section 10.2 we will clarify the sense in which solutions staying in the halo of the slow curve are really slow.

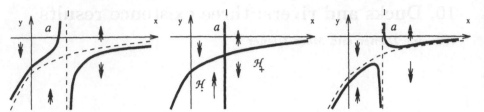

Fig. 10.1. The slow curve of equation (10.2) for various values of $a \in [0, 2]$, and the corresponding behaviour for the solution starting from the point M_0.

Behaviour of a maximal solution Consider in particular a *maximal* solution $\hat{y}$ of (10.2) starting at a point $(x_0, y_0) \in \mathcal{D}$, near-standard in $\mathcal{D}$ and not i-close to $\mathcal{L}$. Notice first that the classical theorem describing the prolongation of a solution for a locally lipschitz equation implies that the graph of any maximal half-solution, $\hat{y}$, leaves any compact subset of $\mathcal{D}$; as usual, a maximal half-solution means, , a solution defined on an interval I of type $[x_0, x_+[$ or $]x_-, x_0]$ (with $x_\pm$ possibly equal to $\pm\infty$) such that there exists no solution equal to $\hat{y}$ at x_0 and defined on an interval strictly larger then I. As $\mathcal{D}$ contains a compact set which includes all points which are near-standard in $\mathcal{D}$, we see that $(x, \hat{y}(x))$ must leave such a compact set, both for $x < x_0$ and for $x > x_0$, and thus reach some point which is i-large or i-close to the x-axis.

Let $\mathcal{G}$ be the external set of all near-standard points in $\mathcal{D} - \mathcal{L}$, that is, the set of points $(x, y) \in \mathcal{D}$ which are neither i-large, nor such that $y \simeq 0$, nor i-close to $\mathcal{L}$. As long as $(x, \hat{y}(x))$ stays in $\mathcal{G}$, the derivative of the solution is i-large and thus $x \simeq x_0$: if not, $\hat{y}(x) - \hat{y}(x_0)$ would be i-large (we have $\hat{y}(x) - \hat{y}(x_0) = (x - x_0)\hat{y}'(\xi)$ for a ξ between x and x_0). In other words, as long as $(x, \hat{y}(x))$ stays in $\mathcal{G}$, the graph of $\hat{y}$ is i-close to the vertical line $x = x_0$.

Attracting and repelling branches It is easy to determine the behaviour of $\mathcal{L}$ as a function of ${}^{\circ}a$; one simply finds the sign of $((1 - x^2) + (a - x)/y)/\varepsilon$ in $\mathcal{G}$, and deduces from it the direction of the solutions outside the halo of $\mathcal{L}$. In figure 10.1, one may see that, for ${}^{\circ}a \neq 1$ and $x > -1$, $\mathcal{L}$ consists of two branches; one, denoted by $\mathcal{L}_-$, belongs to the halfplane $x < {}^{\circ}a$, and the other, denoted by $\mathcal{L}_+$, belongs to the halfplane $x > {}^{\circ}a$.

From the sign of the derivative of the solutions in $\mathcal{G}$, one deduces that $\mathcal{L}_-$ is *attracting*: for $-1 \lessapprox x \lessapprox {}^{\circ}a$, while x increases, and as long as it stays in $\mathcal{G}$, the point $(x, \hat{y}(x))$ moves towards $\mathcal{L}_-$. Thus, as soon as the point is in the halo to $\mathcal{L}_-$, say at $(x_1, \hat{y}(x_1))$, it stays in the halo of $\mathcal{L}_-$ as long as x is not i-close to ${}^{\circ}a$; in other words, $\hat{y}(x) \simeq \tilde{y}(x)$ for $x_1 \leq x \lessapprox {}^{\circ}a$. One says that the solution $\hat{y}$ is slow and attracting for such values of x.

Similarly, $\mathcal{L}_+$ is *repelling*, or "attracting for decreasing x ", if, whenever $\hat{y}(x_1) \simeq \tilde{y}(x_1)$, $\hat{y}(x) \simeq \tilde{y}(x)$ for all $x \leq x_1$ (such that ${}^{\circ}a \lessapprox x$); the solution $\hat{y}$ is said to be repelling for these values of x.

When $°a = 1$, the two branches "join" at the "critical" point with abscissa $x_c = 1$. $\mathcal{L}$ then consists of the straight line $\{x = 1\}$ and a branch of the hyperbola $y = \tilde{y}(x)$, with

$$\tilde{y}(x) = \frac{-1}{1+x},$$

the component $\mathcal{H}_-$ in $x < 1$ being attracting, and the component $\mathcal{H}_+$ in $x > 1$ repelling.

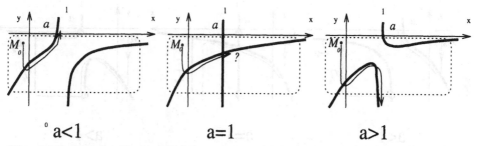

Fig. 10.2. Idea for the proof of the existence of ducks for some values of a. The part of the solution outside of M_0 that is contained in the (external) region of non-near-standard points (hatched) has to travel continuously from points (x, y) such that y is i-large to points such that y is i-small, and thus have points such that y is appreciable, in which case x has to be i-large, and the solution has to stay infinitely close to the repelling part of the slow curve. The solution is a duck.

The problem We know that $(x, \hat{y}(x))$ will leave the region of those points in $\mathcal{D}$ which are near-standard. Now comes the question: how will this occur when $a \simeq 1$? Will we have $\hat{y}(x) \simeq 0$ for some $x \simeq °a(= 1)$ as is the case when $°a < 1$; or will we have $\hat{y}(x)$ i-large (negative) for some $x \simeq °a(= 1)$ as is the case when $°a > 1$. Or will there exist $x_1 \gtrsim 1$ such that $\hat{y}(x) \simeq \tilde{y}(x)$ for $x \leq x_1$, although $\mathcal{H}_+$ is repelling? Such a solution is what we call a *duck* (or *canard*), more precisely, a solution which is *attracting for some* $x \lesssim x_c$ and *repelling for some* $x \gtrsim x_c$.

From the study of the behaviour of the solutions just given, it is not difficult to see that ducks must exist; more exactly, there must exist a value $a_* \simeq 1$ such that, for $a = a_*$, equation (10.2) has a duck, the value a_* being called a *duck value* (of the parameter). To *understand* that such duck values must exist, consider a fixed initial point $M_0 = (x_0, y_0) \in \mathcal{D}$, with $-1 \lesssim x_0 \lesssim 0(\leq °a)$, and the maximal solution $\hat{y}_a$ through this point M_0 (figure 10.2). We have seen that for $°a \neq 1$ and $x \geq x_0$, when $(x, \hat{y}_a(x))$ is no longer near-standard in $\mathcal{D}$, then $x \simeq °a$, $\hat{y}_a(x)$ is i-small if $°a < 1$, and i-large if $°a > 1$. By continuity, such a $(x, \hat{y}(x))$ that is not near-standard in $\mathcal{D}$, will have, for the values of a such that $a \simeq 1$, to travel along a continuous path connecting these two types of points that stays inside the region of points that are not near-standard in $\mathcal{D}$ (this region has been sketched on figure 10.2

as the points laying outside of the doted rectangular), so the solution has to
be a duck for some values of a.

As the property "$(x, \hat{y}_a(x))$ is not near-standard in $\mathcal{D}$" does not charac-
terise a unique x for each value of a, we still have to make a little effort to
formalise this reasoning. Let us associate with each value of a the intersection
of the graph of the solution $\hat{y}_a$ with a segment $\mathcal{T}_a$ that depends continuously
on a. The choice of such a segment is described in figure 10.3. The inter-

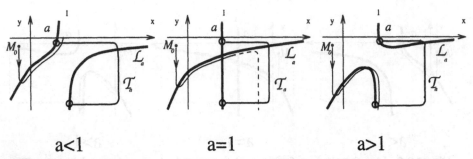

$$a<1 \qquad\qquad a=1 \qquad\qquad a>1$$

Fig. 10.3. A segment transversal to the graph of solutions and meeting the graph
of the solution $\hat{y}_a$ outside of M_0. Let $\mathcal{L}_a$ be the zero-isocline of (10.2) (by definition,
$\mathcal{L} = \mathcal{L}_{\circ_a}$). The endpoints of $\mathcal{T}_a$ are chosen on $\mathcal{L}_a$ and the segment is completed as
indicated, with a vertical segment at $x = x_2$ ($^\circ x_2 > 1$ limited) and two horizontal
segments $y = y_-$ and $y = y_+$ such that for any $a \in [0, 2]$, one has $^\circ y_+ < \tilde{y}(m_a)$ and
$\tilde{y}(x_2) < ^\circ y_- < 0$.

section $N_a = (x_a, \hat{y}_a(x_a))$ of the graph $\hat{y}_a$ with $\mathcal{T}_a$ depends continuously on
the parameter a. When a changes from $a \lesssim 1$ to $a \gtrsim 1$, N_a travels from the
halo of one end to the halo of the other end of the transversal segment $\mathcal{T}_a$.
By continuity, there exists a value a_* of the parameter for which N_{a_*} is no
longer in the halo of either end and, more precisely, a value of a for which
the intersection with the vertical segment with abscissa $x = x_2 \gtrsim 1$ of $\mathcal{T}_{a_*}$
is nonempty. For this value of a at least, the solution $\hat{y}_a$ is a duck such that
$\hat{y}_a(x) \simeq \tilde{y}(x)$ for all x such that $1 \lesssim x \lesssim x_2$.

10.1.2 Application to the Van der Pol equation

The slow-fast Van der Pol equation with parameter $a \in \mathbb{R}$ is the equation

$$\varepsilon\ddot{x} + (x^2 - 1)\dot{x} + x = a \tag{10.3}$$

where $\varepsilon > 0$ is supposed to be i-small. One is indebted to Lienard [79] for the
idea of introducing the following auxiliary variable

$$u = \varepsilon\dot{x} + F(x),$$

with $F(x) = x^3/3 - x$, which reduces equation (10.3) to $\dot{u}$ $(= \varepsilon\ddot{x} + F'(x)\dot{x}) =$
$a - x$. In other words, one considers the system

$$\begin{cases} \varepsilon \dot{x} & = \ u - F(x) \\ \dot{u} & = \ a - x \end{cases} \qquad (10.4)$$

which is a slow-fast system: outside of the halo of the slow curve $\mathcal{C}$ (figure 10.4) with equation $u = F(x)$, the speed $\dot{x}$ is i-large. Observe that, on the contrary, for (x, u, a) limited, the component $\dot{u} = a - x$ is limited, and thus, outside of the halo of $\mathcal{C}$, the slope $\dot{u}/\dot{x}$ of the vector field associated with the system (10.4) is i-small. By adapting the reasoning of the beginning of the previous paragraph on the behaviour of the solutions inside and outside of the halo of the slow curve, one sees that the solutions of (10.4) have alternately rapid parts for which u stays quasi-constant, and slow parts, when the solution is trapped in the halo of the slow curve $\mathcal{C}$, which is attracting for $|x| \gtrsim 1$ and repelling for $|x| \lesssim 1$. Let us analyse more carefully the parts of the solutions

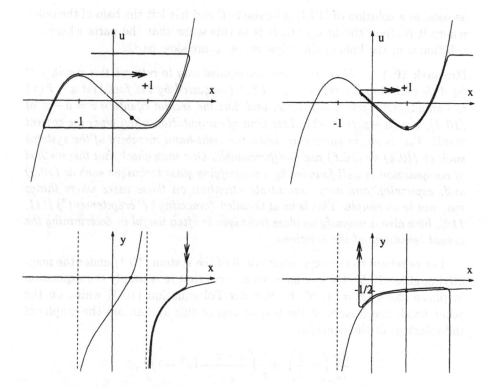

Fig. 10.4. Behaviour in the phase space $(y = \dot{x})$ and in the Lienard space $u = \varepsilon \dot{x} + F(x)$, with $F(x) = x^3/3 - x$.

that we have called slow and explain this terminology. To do so, let us examine the behaviour of a solution with a "magnifying glass" for (x, u) i-close close to $\mathcal{C}$, i.e. let

$$y := (u - F(x))/\varepsilon. \qquad (10.5)$$

We get $\varepsilon \dot{y} = \dot{u} - F'(x)\dot{x}$, or

$$\begin{cases} \dot{x} & = & y \\ \varepsilon \dot{y} & = & a - x - (x^2 - 1)y \end{cases} \tag{10.6}$$

that is a new slow-fast system. Now, it is the second component that is i-large outside of the halo of the slow curve $^{\circ}a - x - (x^2 - 1)y$. The trajectories of the second system are, for $y \neq 0$, the solutions of $\varepsilon \frac{dy}{dx} (= \varepsilon \frac{\dot{y}}{\dot{x}}) = (a-x)/y - (x^2 - 1)$, that is, precisely, the solutions of equation (10.2) that we studied in the previous paragraph. In particular, we have shown that outside the rapid parts during which x changes only by an infinitesimal amount, the solutions $\hat{y}$ are such that $\hat{y}(x) \simeq \tilde{y}(x) := (^{\circ}a - x)/(x^2 - 1)$. In other words

$$\dot{x} \simeq \frac{^{\circ}a - x}{x^2 - 1}$$

as soon as a solution of (10.4) is i-close to $\mathcal{C}$ and has left the halo of the point where it reached the halo of $\mathcal{C}$. It is in this sense that the parts where the solution is in the halo of the "slow curve" $\mathcal{C}$ are slow parts.

Remark 10.1.1. *There is a mnemotechnical way to retrieve this result, just by differentiating the relation $u - F(x)$ (suggested by the fact that $u - F(x)$ is i-small). One gets $\dot{u} = F'(x)\dot{x}$, and thus the second equation $\dot{u} = a - x$ of (10.4), $\dot{x} = (a - x)/(x^2 - 1)$. This kind of computation often gives the correct result. This is so, in particular, when the right-hand members of the systems such as (10.4) or (10.6) are S-differentiable. One may check that this method of computation is well-founded by a magnifying glass technique such as (10.5) and, especially, one may concentrate attention on those cases where things may not be so simple. This is what is called "crackling" ("crépitement") [111, 112]; here also a magnifying glass technique is often useful in determining the actual behaviour of the solutions.*

Let us return to the expression (10.6) of the system (10.4) under the magnifying glass (10.5). On one hand we observe that it is simply the expression in phase space ($\dot{x} = y$) of the Van der Pol equation (10.3) while, on the other hand, the images of the trajectories of this system are the graphs of the solutions of the equation

$$\frac{dy}{dx} \left(= \frac{\dot{y}}{\dot{x}} \right) = \frac{1}{\varepsilon} \left(\frac{a - x}{y} - (x^2 - 1) \right),$$

which is the equation (10.2) for which we have shown the existence of ducks. Let us interpret this existence result for ducks in relation to equation (10.2). Let a_* be a duck value, and let $y = \hat{y}(x)$ be a duck, that is a solution which is i-close to the slow curve $\mathcal{H}$ for $x \in [x_-, x_+]$, with $x_- \lesssim 1 \lesssim x_+$. In other words $\hat{y}(x) \simeq -1/(x + 1)$ for these values of x. Let $\hat{\gamma}$ be a solution of (10.4) corresponding to the solution $\hat{y}$ under the change of variable (10.5), that is, a solution such that $\hat{u} = F(\hat{x}) + \varepsilon \hat{y}(\hat{x})$ where $\hat{\gamma}(t) = (\hat{x}(t), \hat{u}(t))$. As

$\hat{y}(x) \simeq -1/(x+1)$ is limited, we see, in particular, that $\hat{u}(t) \simeq F(\hat{x}(t))$, for all $\hat{x}(t)$ in $[x_-, x_+]$. Thus, we have demonstrated the existence of a solution that is infinitely close to the slow curve $\mathcal{C}$ with equation $y = F(x)$ and that is first attracting (for $x \gtrsim 1$), and then repelling (for $x \lesssim 1$): it is a *duck* of the system (10.4).

10.1.3 Duck cycles: the missing link

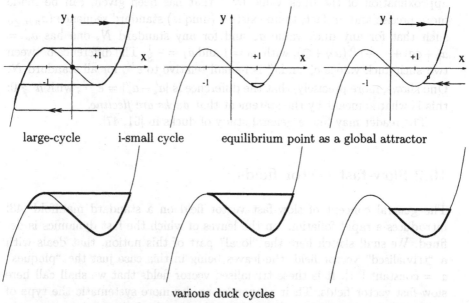

large-cycle   i-small cycle   equilibrium point as a global attractor

various duck cycles

Fig. 10.5. The ducks are the missing link between the large-cycles and the i-small cycle that occur just before the Hopf bifurcation.

The main property of the Van der Pol equation, for $a = 0$, is that it exhibits a limit-cycle. The existence of this cycle is easy to understand in the system (10.4). Indeed, we have pointed out that outside of the halo of the slow curve, the solution moves almost horizontally; from left to right above $\mathcal{C}$ ($\dot{x} > 0$), and from right to left below ($\dot{x} < 0$). On the other hand, as $\dot{u} = a - x$, we see that u decreases when $x > a$, and increases when $x < a$. It is easy to deduce from this that as long as $|^\circ a| < 1$, the system exhibits a cycle i-close to the "rounded parallelogram" $\mathcal{P}$ shown in figure 10.5. On the contrary, for $|^\circ a| > 1$, the equilibrium point $(a, F(a))$ is a global attractor, and there is no longer any cycle. As ε is not zero, the solutions, and, in particular, the limit-cycles vary continuously with the parameter a. The amplitude of the limit cycle must thus decrease continuously (not, of course, S-continuously) to let the cycle vanish. This disappearance, which actually occurs for $a = 1$, is a Hopf bifurcation, i.e. the amplitude of the cycle that circles around the equilibrium point decreases and the stability of the equilibrium point when the cycle vanishes into it, is reversed. The transition from "large-cycle" (i-- close to $\mathcal{P}$) to i-small cycle can only occur through cycles that run along the attracting part of $\mathcal{C}$ then along its repelling part, the reversal of the vertical motion of the solution for $x \simeq 1$ being possible due to the fact that $a \simeq 1$. *The cycles of medium size are necessarily ducks.*

It has to be noted that if a_* is a duck value, then $a_* = 1 - \frac{\varepsilon}{8} + \varepsilon(1 + \oslash)$: the transition from large cycles to i-small cycles through ducks takes place quite apart from the Hopf bifurcation, that itself takes place at $a = 1$. The approximation of the duck value by 1 that has been given, can be made more precise and, in fact, there exists a (unique) standard sequence $(a_n)_{n \geq 0}$ such that for any duck value a_* and for any standard N, one has $a_* = a_0 + \varepsilon a_1 + \cdots + \varepsilon^N(a_N + \oslash)$, with $a_0 = 1$ and $a_1 = -\frac{1}{8}$. The difference between two such duck values a'_* and a''_* is i-small relative to ε^N, for all standard N. One shows, more precisely, that the difference is $|a'_* - a''_*| = e^{-\frac{h}{\varepsilon}}$, with $h \gtrless 0$: this is what is meant by the statement that *ducks are fleeting.*

The reader may find a general study of ducks in [51, 47].

10.2 Slow-fast vector fields

The general concept of slow-fast vector field on a standard manifold [48] introduces a rapid foliation, on the leaves of which the fast dynamics is defined. We shall sketch here the "local" part of this notion, that deals with a "trivialised" vector field, the leaves being in this case just the "plaques" $x =$ constant [71]. It is these trivialised vector fields that we shall call here slow-fast vector fields. Their properties render more systematic the type of reasoning that we used for the ducks of the Van der Pol equation.

Definition 10.2.1. *A slow-fast vector field on an open standard subset U of $\mathbb{R}^p_x \times \mathbb{R}^q_y$ is a vector field $f\frac{\partial}{\partial x} + \frac{1}{\varepsilon}g\frac{\partial}{\partial y}$ associated with a slow-fast system*

$$\begin{cases} \dot{x} &= f(x,y) \\ \dot{y} &= \frac{1}{\varepsilon}g(x,y) \end{cases} \tag{10.7}$$

where f and g are S-differentiable functions defined on U, and $\varepsilon > 0$ is i-small.

10.2.1 The fast dynamic

As f and g are assumed to be S-differentiable, they are locally lipschitz, and they have C^1 shadows f_0 and g_0. So one has local existence and uniqueness of the solutions of the vector field $f\frac{\partial}{\partial x} + g\frac{\partial}{\partial y}$, so as for the vector fields that define the fast and the slow dynamics that we shall consider below. Each connected component of the intersection of U with the affine subspace

$x = x_0 =$ constant is called a *fast plaque* of abscissa x_0. Each fast plaque is parametrized by $y \in \mathbb{R}^q$. The *fast dynamic* on any plaque of abscissa x_0 is the vector field $g_0(x_0, y)\frac{\partial}{\partial y}$ associated with the equation

$$y' = g_0(x_0, y). \tag{10.8}$$

Example: On $U = \mathbb{R}_x \times \mathbb{R}^2_{(y_{-1}, y_{-2})}$, consider the slow-fast vector field associated with the system

$$\begin{cases} \dot{x} &= 1 \\ \varepsilon\dot{y}_1 &= xy_1 - y_2 + \varepsilon \\ \varepsilon\dot{y}_2 &= y_1 + xy_2 \end{cases} \tag{10.9}$$

The plaques are the planes $x = x_0$, for any $x_0 \in \mathbb{R}$. The trajectories of the fast dynamics are circles if $x_0 = 0$, and spirals otherwise, that move towards the equilibrium point $(x_0, 0, 0)$ if $x_0 < 0$ and towards infinity if $x_0 > 0$ (figure 10.7).

10.2.2 Application of the fast dynamic

Denote by γ a solution of the slow-fast system (10.7) and by ρ a maximal trajectory of the fast dynamic. Assume that $^{\circ}\gamma(0)$ exists in U and that $^{\circ}\gamma(0) = \rho(0)$. Comparing the second equation of system (10.7) and equation (10.8) suggests the relation $\tau = \varepsilon t$ between the variable t of γ and the variable τ of ρ, and the study of those τ for which $\gamma(\frac{\tau}{\varepsilon}) \simeq \rho(\tau)$, a relation that by assumption holds for $\tau = 0$. It can be shown that as long as τ is near-standard in the domain of ρ, and that $\rho(\tau)$ stays thus near-standard in U, $\gamma(\frac{\tau}{\varepsilon})$ is defined and i-close to $\rho(\tau)$. By permanence (Fehrele's principle), there thus exists two instants τ_- and τ_+ in the domain $\mathcal{D}(\rho)$ of ρ and *not* near-standard in $\mathcal{D}(\rho)$, with $\tau_- < 0 < \tau_+$, and such that these properties stays true for all $\tau \in [\tau_-, \tau_+]$. As ρ is a standard trajectory of a standard dynamics, $\rho(\tau_\pm)$ is a point that is infinitely close to the limit set $\omega_\pm$ of ρ or a point that is not near-standard in U (if $\omega_\pm$ is empty or non compact); this is the general situation corresponding to the behaviour observed for the Van der Pol equation, where all half trajectories become either infinitely close to the slow curve or become i-large.

The simplest case is when the limit set considered is just a point. This leads to the idea of a slow manifold.

Definition 10.2.2. *Let (x_0, y_0) be an equilibrium point of the fast dynamics. This point is called* attracting *(resp.* repelling, *resp.* hyperbolic*) iff all the eigenvalues of $A := (g_0)'_y(x_0, y_0)$ have negative (resp. positive, resp. zero) real part. The point (x_0, y_0) is called* critical *iff the linear mapping A is not invertible.*

In example (10.9), the point $(x_0, 0, 0)$ is attracting if $x_0 < 0$ and repelling if $x_0 > 0$. Observe that for $x_0 = 0$, the equilibrium $x_0 = 0$ is not critical, nor is it hyperbolic: only the eigenvalues $\pm i$, of $A = (g_0)'_y$ have zero real part.

Definition 10.2.3. *One calls* slow manifold *of the slow-fast vector field associated with (10.7), and denotes it by* $\mathcal{L}$, *the set of all equilibrium points of the slow dynamics; one denotes by* $\mathcal{L}_*$ *the subset of* $\mathcal{L}$ *of all non-critical equilibrium points,* $\mathcal{L}_*^- \subseteq \mathcal{L}_*$ *the subset of all attracting points, and finally by* $\mathcal{L}_*^+ \subseteq \mathcal{L}_*$ *the subset of all repelling points.*

The slow manifold $\mathcal{L}$, with equation $g_0(x, y) = 0$, may be singular at the critical points, but $\mathcal{L}_*$ is smooth. More precisely, the restriction to $\mathcal{L}_*$ of the projection $(x, y) \mapsto x$ is a local chart of $\mathcal{L}_*$ in the neighbourhood of each of its points. In other words, in the neighbourhood of each $(x_0, y_0) \in \mathcal{L}_*$, there exists a differentiable function $\tilde{y}(x)$ such that $y_0 = \tilde{y}(x_0)$ and $(x, \tilde{y}(x)) \in \mathcal{L}_*$ for $x \in \mathcal{D}(\tilde{y})$. We shall use this property to define a new dynamic on $\mathcal{L}_*$, called the slow dynamic of the slow-fast dynamic associated with (10.7). Observe that the slow manifold $\mathcal{L}$ is defined from the fast dynamics (it is the set of all equilibrium point of it) and not from the slow dynamics that is simply defined on it.

10.2.3 The slow dynamic

The *slow dynamic* associated with (10.7) is the vector field tangent to $\mathcal{L}_*$, the image of which by the projection $(x, y) \mapsto y$ leads to the differential equation

$$\dot{x} = f_0(x, \tilde{y}(x)) \tag{10.10}$$

for all $x \in \mathcal{D}(\tilde{y})$, where $\tilde{y}$ is a smooth function having its graph contained in $\mathcal{L}_*$. As the equation of $\mathcal{L}_*$ is $g_0(x, y) = 0$, and as

$$\tilde{y}'(x) = \left[(g_0)'_y(x, \tilde{y}(x)) \right]^{-1} \circ (g_0)'_x(x, \tilde{y}(x)),$$

the projection $(x, y) \mapsto x$ leads to the following differential equation

$$\dot{y} = - \left[(\partial g_0)'_y(x, y) \right]^{-1} (g_0)'_x(x, y) f_0(x, y) \tag{10.11}$$

for $(x, y) \in \mathcal{L}_*$. In other words, the slow dynamic is

$$\frac{\partial}{\partial x} f_0 - \frac{\partial}{\partial y} \left[(g_0)'_y \right]^{-1} (g_0)'_x f_0. \tag{10.12}$$

Example: In [7, 10], E. Benoit studied, for various "generic" values of the real parameters a and b, the slow-fast vector field $\mathbb{R}^2_{x_1, x_2} \times \mathbb{R}_y$ associated with

$$\begin{cases} \dot{x}_1 &= ax_2 + by \\ \dot{x}_2 &= 1 \\ \varepsilon \dot{y} &= -(y^2 + x_1) \end{cases} \tag{10.13}$$

In that example the slow manifold $\mathcal{L}$ is the parabolic cylinder with equation $y^2 + x_1 = 0$. The points of that manifold such that $y > 0$ are attracting, and those such that $y < 0$ are repelling; the critical points are those of the straight line $y = 0$; one has $\mathcal{L}_* = \mathcal{L} \cap \{(x, y) \mid y \neq 0\}$.

The projection $(x, y) \mapsto x$ of the slow dynamic leads to

$$\begin{cases} \dot{x}_1 &= ax_2 \pm \sqrt{-x_1} \\ \dot{x}_2 &= 1 \end{cases} \tag{10.14}$$

that is defined for $x_1 < 0$, with $\pm$ chosen according to the sign of y on the chosen branch of $\mathcal{L}_*$. For $(x, y) \in \mathcal{L}_*$, the slow dynamic is

$$(ax_2 + by)\frac{\partial}{\partial x_1} + \frac{\partial}{\partial x_2} - \frac{1}{2y}(-ax_2 - (b+2)y)\frac{\partial}{\partial y}$$

that is proportional (with reversed orientation for $y < 0$) to the vector field

$$2y(ax_2 + by)\frac{\partial}{\partial x_1} + 2y\frac{\partial}{\partial x_2} + (ax_2 + (b+2)y)\frac{\partial}{\partial y}.$$

Here, the projection $(x_1, x_2, y) \mapsto (x_2, y)$ turns out to be a more appropriate local chart of $\mathcal{L}_*$ (it extends to all $\mathcal{L}$). The image of the slow dynamic by this projection is associated with the system

$$\begin{cases} x_2' &= 2y \\ y' &= ax_2 + (b+2)y \end{cases} \tag{10.15}$$

which is a linear. The case where this system exhibits a saddle is studied in [7], and the case where it is a node is considered in [10].

10.2.4 Application of the slow dynamic

Once more, let γ be a maximal solution of the slow-fast system (10.7), and assume that $^\circ\gamma(0) \in \mathcal{L}_*$. Let λ be the maximal solution of the slow dynamic such that $\lambda(0) = ^\circ\gamma(0)$. It is easy to see that, as long as $\gamma(t)$ stays near-standard in $\mathcal{L}_*$ and t stays limited, $\lambda(t)$ is defined and $\lambda(t) \simeq \gamma(t)$. Conversely, it can be shown that, for $t \geq 0$, as long as $\lambda(t)$ stays near-standard in the *attracting part* $\mathcal{L}_*^-$ of $\mathcal{L}_*$, and as long as t is limited, the solution $\gamma(t)$ is defined and $\gamma(t) \sim \lambda(t)$. By permanence (Fehrele's principle), this results extends up to some $t \geq 0$ i-large or such that $\lambda(t)$ (and $\gamma(t)$) are no longer near-standard in $\mathcal{L}_*^-$: the halo of any positive half trajectory λ_+ of the attracting slow dynamic is thus an external trap that "captures", for increasing t, any solution γ that enters it, and this up to a point where $\lambda_+(t)$ is no longer near-standard in $\mathcal{L}_*^-$ or until an i-large time interval has run out. Of course, one has a similar result for the negative half trajectories of the slow dynamic on the repelling part $\mathcal{L}_*^+ \subseteq \mathcal{L}_*$ and for the behaviour of the solutions from an initial point i-close to $\mathcal{L}_*^+$ and for decreasing t.

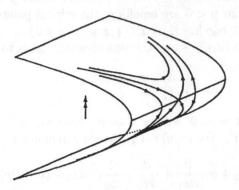

Fig. 10.6. A slow dynamic on a slow surface that suggests the existence of ducks.

Example:

Let us consider the slow-fast system of Benoit (10.13) with $a = 0$ and $b = -2$: the linear system (10.15) exhibits a saddle; the trajectories of the slow dynamic have been sketched in figure 10.6. On the attracting part of $\mathcal{L}_*$ ($y > 0$), they draw the solutions that have become close to them, either up to the halo of the "fold-line" of the surface $(x_1, y) = (0, 0)$, or up to points such that (x_1, y) is i-large ($x_1 < 0$ and $y > 0$). Benoit has shown that some solutions that have reached the halo of the critical point $(0, 0, 0)$ stay close to the slow surface on its repelling part, and are thus ducks: they form the continuous transition between the behaviour of the two end-points of the line-segment $\mathcal{T}$ sketched in figure 10.6.

10.3 Robust ducks

Let us consider first the slow-fast system of $R_x \times \mathbb{R}_y^2$

$$\begin{cases} \dot{x} &= 1 \\ \varepsilon \dot{y}_1 &= x y_1 - y_2 \\ \varepsilon \dot{y}_2 &= y_1 + x y_2 \end{cases} \tag{10.16}$$

the slow curve of which, with equation $y = 0$, is also a trajectory γ_0. This slow curve is attracting for $x < 0$ and repelling for $x > 0$; the trajectory γ_0 is thus a duck of this system. The results for the ducks of the Van der Pol equation and more generally for the ducks of a slow-fast vector field on $\mathbb{R}^2$ (the ducks are fleeting), suggest that an i-small perturbation that is not exponentially small would destroy the ducks, that is, that a the system perturbed in such a way would have no ducks. Now G. Wallet and C. Lobry observed [82] that for

$$\begin{cases} \dot{x} &= 1 \\ \varepsilon \dot{y}_1 &= xy_1 - y_2 + \varphi_1 \\ \varepsilon \dot{y}_2 &= y_1 + xy_2 + \varphi_2 \end{cases} \qquad (10.17)$$

with φ_1 and φ_2 assumed to be i-small, for example, $\varphi_1 = \varepsilon$ and $\varphi_2 = 0$, there still exists solutions that are i-small for $x \lesssim 0$ and that stay i-small for some $x \gtrsim 0$. Such a robustness of some ducks had been already noticed by M.A. Shishkova [105], but one had to wait for M.A. Neishtadt [89, 90] to have a general existence result for ducks in such a situation. The main difference compared with the case of $\mathbb{R}^2$ is that the point $(0,0,0)$ where the attractivity of the slow curve is reversed is no longer a critical point. Only the real part of the eigenvalue of $f'_y(0,0,0)$ vanishes, and thus the differential remains an invertible linear map. An hypothesis is however essential: the analyticity of the differential system.

It has to be pointed out that in the example of system (10.17) with $\varphi_1 = \varepsilon$ and $\varphi_2 = 0$, one finds numerical solutions that are i-small for all $x \lesssim 1$, and also numerical solutions that are i-small for all $x \gtrsim -1$. But no trajectory which is slow (=i-small) for $x \lesssim 0$ seems to want to stay i-small beyond the "buffer point" $x = 1$: it is at this point that there exists in this case a question of "fleeting" solutions. By embedding this system in a one-parameter family, and adjusting this parameter carefully, one produces fleeting ducks that overstep the buffer point. The buffer point phenomenon has been analysed in [30, 49] using resurgent methods.

In the next paragraph, we shall first show how Callot's hills-and-dales method (see section 2.4) makes it easy to understand why, for φ_1 and φ_2 analytic, the system (10.17) exhibits ducks, and moreover why the values $x = \pm 1$ have a special role. It will thus be an application of complex analytic methods applied to a real number problem. After this we shall consider the method itself in terms of real number slow-fast systems.

10.3.1 Robust ducks, buffer points, and hills-and-dales

It is easy to check that if one lets $y = y_1 + iy_2 \in \mathbb{C}$, then for $(\varphi_1, \varphi_2) = (\varepsilon, 0)$, the trajectories of (10.17) are solutions of equation

$$\varepsilon \frac{dy}{dx} = (x+i)y + \varepsilon \qquad (10.18)$$

defined, a priori, for x in $\mathbb{R}$. But, this equation is obviously also defined for all x in $\mathbb{C}$, and the restriction of any (complex) solution to a real interval I that is defined on a complex neighbourhood of I is a solution of the real (y_1, y_2) system. So we shall consider the question of the existence of slow complex solutions of (10.18). As the equation of the slow curve is $y = 0$, we shall consider the associated real-valued function (also called the landscape function), $R : \mathbb{C} \to \mathbb{R}$, given by

$$R(t+is) = \Re \int_0^{t+is} ((x+i)y)'_y(x,0)dx$$

$$= \frac{1}{2}(t^2 - (s+1)^2 + 1)$$

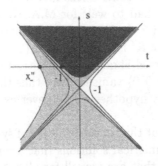

Fig. 10.7. The near-standard points in the two hatched domains, can be reached from the real points x'_- and x''_-, respectively, along paths upon which the function R "S-decreases".

So the function $R(x)$ has a saddle point at $x = -i$ that has the same "height" ($R(-i) = 1$) as the real points $x = -1$ and $x = 1$. The other level curves of R have been sketched in figure 10.7 in which have been hatched two (standard open) domains for which all near-standard points can be reached by paths that "S-decrease" the values of R from the values at two real points denoted by x'_- and x''_- respectively: here we have chosen $x'_- > -1$ and $x''_- < -1$. The hills-and-dales method ensures that any (complex) solution, slow at x'_- (respectively at x''_-) exists and stays slow on any near-standard point in that domain. As the domain is simply connected, the solution is uniform (which is also a consequence of the fact that the equation is defined for all x and linear in y, but we do not want to use this hypothesis explicitly). The intersection of the domain with the real axis is a segment $[x'_-, x'_+]$ (or $[x''_-, x''_+]$): it is on this segment that the method ensures the existence of a duck. Notice that as a consequence of the saddle point at $x = i$ one has that $x'_- = x'_+$ (as $-1 \le x'_- \le 0$), when $x''_+ = 1$ independently of the value of x''_- (provided $x''_- < -1$). This does not show that the vector field (10.17) has no ducks on an larger interval $[x_-, x_+]$, with $x_- \lessgtr -1$ and $x_+ \gtrless 1$; but does suggest that the buffer point at $x = \pm 1$ is related to the saddle point at $x = -i$. Observe that the saddle point $x = -i$ is a critical point of the slow (complex) curve $\{y = 0\}$.

The reader is referred to the study originally done by J.L. Callot [28] for the application of this method to the general case of an analytic equation and more precisely to the case of a system with *several complex* unknowns.

10.3.2 An other approach to the hills-and-dales method

We now restrict the function R to the real axis, $s = 0$, so that we are dealing with a real function of a real variable. With increasing t, the slow curve of the corresponding system (1.17) is attracting as long as we are descending in the sense of the level curves,(10.17), (10.18). On the other hand, whenever we move up through the level curves with increasing t, the slow curve is repelling. Let us see how this simple observation can be generalised and can make it possible to understand the existence of the complex slow solutions for all points that can be reached from the initial point $x_- \in \mathbb{R}$ "S-descending" the function R, using just the trick of the external trap that is the halo of an attracting slow curve.

Let us recall briefly how to explore the complex solutions using real paths. Let $g(x, y)$ be an analytic function of x and y, and consider the differential equation

$$\frac{dy}{dx} = g(x, y). \tag{10.19}$$

Let $t \mapsto \gamma(t)$ be a smooth path in $\mathbb{R}^2$, defined for real t, $\gamma(t) = (\gamma_1(t), \gamma_2(t))$, and denote by $\gamma_c(t) = \gamma_1(t) + i\gamma_2(t) \in \mathbb{C}$. If $x \mapsto \hat{y}(x)$ is an analytic solution of (10.19), then it is easy to see that $t \mapsto \hat{Y}(t) := (\Re\hat{y}(\gamma_c(t)), \Im\hat{y}(\gamma_c(t)))$ is a solution of the equation

$$\frac{dY}{dt} = G(t, Y) \tag{10.20}$$

where the components G_1 and G_2 of G are the real and imaginary parts of $g(\gamma_c(t), Y_1 + iY_2) \cdot \gamma_c'(t)$ with $\gamma_c'(t) = \gamma_1'(t) + i\gamma_2'(t)$. We shall call equation (10.20) the (real) restriction of equation (10.19) to the path γ. It is the converse of this remark that is of interest here: using Cauchy's existence and uniqueness theorem for local solutions of any complex analytic differential equation and the principle of analytic continuation of solutions, it is not difficult to show that, given any initial condition (x_0, y_0), it is possible to prolong the local solution of (10.19) from this point along the path γ originating at x_- for $t = 0$ if and only if it is possible to prolong the solution of the restriction (10.20) from $(0, Y_0)$, with $Y_0 = (\Re(y_0), \Im(y_0))$. This brief review may help one to see how to use the following proposition to understand the hills-and-dales method.

Proposition 10.3.1. *Let $\varepsilon dy/dx = f(x, y)$ be a complex analytic differential equation, where f has a shadow f_0 defined on the standard open set U. Let $y = \tilde{y}(x)$ be a slow curve of this equation, i.e. a complex standard curve defined on an open subset $\mathcal{D}(\tilde{y})$ of $\mathbb{C}$, and such that $f_0(x, \tilde{y}(x)) \equiv 0$. Assume that $\mathcal{D}(\tilde{y})$ is simply connected. Let $R(t + is) = \Re \int_{x_0}^{t+is} f_0'{}_y(x, \tilde{y}(x))dx$ be the hills-and-dales function associated with $\tilde{y}$ that vanishes at x_0 in $\mathcal{D}(\tilde{y})$. Let γ be an S-differentiable path with shadow γ_0. The path γ S-descends the levels of the function R if and only if the slow curve $y = \tilde{y} \circ \gamma_0(x)$ of the restriction to γ of the equation under consideration is attracting.*

Proof. For all (t, y) such that $t \in \mathcal{D}(\gamma) \subset \mathbb{R}$ and $h > 0$,

$$
\begin{aligned}
R(\gamma(t+h)) - R(\gamma(t)) &= \Re \int_{\gamma_c(t)}^{\gamma_c(t+h)} (f_0)'_y dx \\
&= \Re \int_t^{t+h} (f_0)'_y(\gamma_c(t), \tilde{y}(\gamma_c(t)) \cdot \gamma'_c(t) dt \\
&= \int_t^{t+h} \Re \left((f_0)'_y(\gamma_c(t), \tilde{y}(\gamma_c(t)) \cdot \gamma'_c(t) \right) dt.
\end{aligned}
$$

One arrives at the desired conclusion simply by observing that the function to be integrated is equal to the common real part of the complex conjugate eigenvalues of the linear mapping $G'_y(t, \tilde{Y}(t))$, with $\tilde{Y}(t) := (\Re\tilde{y}(\gamma_c(t)), \Im\tilde{y}(\gamma_c(t)))$, that is, the function the graph of which is the slow curve of the restriction to γ of the equation being considered. $\square$

10.4 Rivers

One of the first methods taught in an elementary lecture on ordinary differential equations is that any solution of a linear equation $dy/dx = ax + b(x)$ is of form $y = Ce^{ax} + y_0(x)$ where $y_0(x)$ is a *particular* solution of the equation; moreover, if $a \neq 0$ and if $b(x)$ is a polynomial, one can choose a polynomial for y_0 with the same degree as $b(x)$. In other words, in that case, the equation has a unique polynomial solution, and all other solutions move toward it or depart from it exponentially when x goes to $\pm\infty$. The notion of rivers permits one to find analogous results in a more general context, among others, for equations of the type $Y' = R(X, Y)$, where R is the ratio of two polynomials, a case where it is no longer possible to solve the equation by integrals. In this section, we shall study the Liouville equation which will serve as a model, and we shall be especially concerned to give the proof of the existence of rivers: we shall have to consider two types of strategies for the two types of rivers of that equation, strategies that enable one also to deal with the general case [54, 17].

10.4.1 The rivers of the Liouville equation

Let us consider the differential equation

$$\frac{dY}{dX} = Y^2 - X \tag{10.21}$$

named after Liouville who gave it as an example of an equation with no solution that can be expressed as an integral of a classical function [80]. Figure 10.8 (a), in which are drawn the plots of some solutions, summarizes nicely

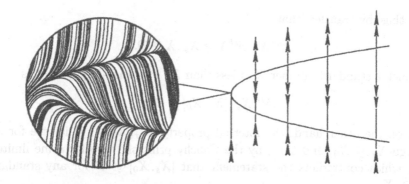

Fig. 10.8. The Liouville equation has a unique repelling river (top) and an infinity of attracting ones (bottom). Under the macroscope, their shadow is respectively the upper slow curve and the lower slow curve.

what we want to stress about the behaviour of the solutions at $X = +\infty$. One can guess from it what is indeed true for the equation: *there exists a (unique) solution $\hat{Y}_+$ defined for all X large enough that is asymptotic to $+\sqrt{X}$ at $X = +\infty$, the other solutions departing for it exponentially when X increases; and there exist infinitely many solutions $\hat{Y}_-$ defined for X large enough and asymptotic to $-\sqrt{X}$, and that gather together exponentially when X increases.* By transfer, if a river exists, then a standard one must also exist (and a unique one has thus to be standard), so we shall restrict ourselves to the study of the existence of standard rivers.

Recall that two functions $F(X)$ and $G(X)$ are said to be *asymptotic* to each other at $X = +\infty$ if and only if they are both defined for all large enough X and are such that the limit of their ratio when X tends to $+\infty$ is equal to 1; when this is so, one writes $F(X) \sim G(X)$ ($X = +\infty$). If F and G are standard, it is an easy exercise to show that this is equivalent to the fact that F and G are defined for all i-large X and that $F(X) = G(X)(1 + \oslash)$ for all i-large X. Actually, it suffices to prove the existence of an i-large X_0 such that this is true for all i-large X *less* than X_0. Indeed, it suffices to apply the following lemma to the function $H(X) = F(X)/G(X) - 1$.

Lemma 10.4.1. *Let H be a standard function. If there exists some i-large real number $X_0 > 0$ such that $H(X)$ exists and is i-small for all i-large $X \le X_0$, then there exists some X_s such that $H(X)$ is defined for all $X \ge X_s$ and $\lim_{X \to +\infty} H(X) = 0$.*

Proof.

   Existence: Let $\mathcal{D}$ be the domain of H. As H is standard, so is $\mathcal{D}$. Let us argue by contradiction; assume that

$$\forall X_s \ \exists X \geq X_s \ X \notin \mathcal{D}$$

and thus, by transfer, that

$$\forall^{st} X_s \ \exists^{st} X \geq X_s \ X \notin \mathcal{D}.$$

As such a standard number X is less than X_0 which is i-large, one has

$$\forall^{st} X_s \ [X_s, X_0] \not\subset \mathcal{D}.$$

But on the other hand, the internal property $[X_\sigma, X_0] \subset \mathcal{D}$ is true for all i-large $X_\sigma \leq X_0$, and thus, by the Cauchy principle, also for some limited X_σ, which contradicts the statement that $[X_s, X_0] \not\subset \mathcal{D}$ for any standard $X_s \geq X_\sigma$.

Limit: As H is standard, by transfer it suffices to check that for all standard $\varepsilon > 0$, there exists an X_σ such that, for all $X \geq X_\sigma$, $H(X) \in [-\varepsilon, +\varepsilon]$, and thus $X \in H^{-1}([-\varepsilon, +\varepsilon])$. It is enough to apply the same reasoning as previously to $\mathcal{D} = H^{-1}([-\varepsilon, +\varepsilon])$.   □

Remark 10.4.1. *The analogy between the two parts of the proof suggests that this method could be extended to a large class of* syntactically *analogous results: this is precisely the purpose of the* generalized transfer *of I.P. van den Berg [18].*

The fact that the solutions $\hat{Y}_+$ and $\hat{Y}_-$ that we are looking for are such that $\hat{Y}_\pm(X) = \pm\sqrt{X}(1 + \oslash)$ for all i-large X suggests that we examine the Liouville equation (10.21) under a $(1, \frac{1}{2})$-macroscope, meaning that we should make change of variables as follows,

$$\left[\begin{array}{ccc} x & = & \varepsilon X \\ y & = & \varepsilon^{\frac{1}{2}} Y \end{array} \right. \tag{10.22}$$

with any i-small $\varepsilon > 0$. The Liouville equation becomes:

$$\varepsilon \frac{dy}{dx} = y^2 - x, \tag{10.23}$$

which is slow-fast, with a slow curve with equation $y = \pm\sqrt{x}$ (figure 10.8 b)). The appreciable values of (x, y) correspond to i-large values of (X, Y). Denote by $\hat{y}_+$ and $\hat{y}_-$ the solutions of (10.23) that correspond to the solutions $\hat{Y}_+$ and $\hat{Y}_-$ that we are looking for. As $\hat{Y}_\pm(X) = \pm\sqrt{X}(1+\oslash)$ for i-large X we deduce that $\hat{y}_\pm(x) \simeq \pm\sqrt{x}$ for all appreciable x; this implication, incidentally, is an equivalence if this property holds for all i-small $\varepsilon > 0$. In other words, we are looking for standard solutions of the Liouville equation that, under the macroscope, are i-close to one or other of the branches of the slow curve . We shall have to use different strategies for the "attracting river" $\hat{Y}_-(X) \sim -\sqrt{X}$ and for the "repelling river" $\hat{Y}_+(X) \sim +\sqrt{X}$. Passing to the general case [16] will then be just a matter of formulating the general result !

10.4.2 Existence of an attracting river

Looking at figure 10.8 suggests that any solution through any point (X_0, Y_0) with large enough X_0 and close enough to the graph of $Y = \sqrt{X}$ will do.

Choice of X_0: Let k_- and k_+ be any standard numbers such that $k_- < -1 < k_+ < 0$. The curves $Y = k_\pm X$ are invariant by the macroscope (10.22), and examining them with it shows that they are the boundary of a domain $\mathcal{D}$ such that $\mathcal{D}(X_0) := \mathcal{D} \cap \{X \geq X_0\}$ is positively invariant for the Liouville equation (for increasing X) for any i-large X_0 . By the Cauchy principle, there thus exists a limited X_0 such that the domain $\mathcal{D}(X_0)$ is also positively invariant. After having possibly replaced X_0 by some larger standard number, we may assume that X_0 is standard: it is such an X_0 that we shall choose for the theorem, and we let $Y_0 = -\sqrt{X_0}$. Now let $\hat{Y}_-(X)$ be the maximal solution of the Liouville equation such that $\hat{Y}_-(X_0) = Y_0$. It is standard as X_0 and Y_0 are standard.

Properties of $\hat{Y}_-$: As $\mathcal{D}(X_0)$ is positively invariant, the graph of $\hat{Y}_-$ stays in $\mathcal{D}(X_0)$, for all $X \geq X_0$, and as $\mathcal{D}(X_0) \cap \{X \leq X_1\}$ is compact for any $X_1 \geq X_0$, $\hat{Y}_-$ is necessarily defined on $[X_0, X_1]$ for all $X_1 \geq X_0$, and thus on $[X_0, +\infty[$. Let us finally show that $\hat{Y}_-(X) \sim -\sqrt{X}$, or, equivalently, that $\hat{Y}_-(X) = -\sqrt{X}(1 + \oslash)$ for all i-large X. Assume, by contradiction, that this is not the case for some i-large X_1. Let $\varepsilon := 1/X_1$ and consider the image y_- of Y_- by the macroscope 10.22 for this value of ε. We have $x_1 := \varepsilon X_1 = 1$ which is appreciable, and by assumption, $\hat{y}_-(x) \not\simeq -\sqrt{x_1}$; nevertheless $k_-\sqrt{x_1} < \hat{y}(x_1) < k_+\sqrt{x_1}$. Since, for appreciable $x > 0$, the slow curve $-\sqrt{x}$ is attracting (see figure 10.9), $\hat{y}(x)$ is, for $0 \lessgtr x_2 \lessgtr x_1$, no longer between $k_- x_2$ and $k_+ x_2$; coming back to the original scale, this implies for $X_2 := x_2/\varepsilon$ (, which is i-large), that $(X_2, \hat{Y}_-(X_2)) \notin \mathcal{D}(X_0)$, which contradicts the fact that $\mathcal{D}(X_0)$ is positively invariant.

10.4.3 Existence of a repelling river

Looking once more at figure 10.8 suggests that this time we should choose the solution $\hat{Y}$ through a point $(X_0, +\sqrt{X_0})$ for an i-large X_0 (this is precisely what has been done to draw this picture: an initial condition has been chosen outside of the picture, with large abscissa) Nevertheless, nothing ensures that this solution will be asymptotic to $+\sqrt{X}$ for $X > X_0$ (actually, it is *not*). As we expect the solution sought to be unique, this solution will have to be standard (which is not the case for the solution $\hat{Y}$). This suggests that we choose for $\hat{Y}_+$ *the shadow of the solution* $\hat{Y}$: this we shall do, and in doing so, we must check that the solution $\hat{Y}$ is defined, limited, and S-continuous for some X.

Domain of $\hat{Y}$: Now let k_- and k_+ be two standard numbers such that $0 < k_- < 1 < k_+$. The curves $Y = k_\pm X$ are still invariant by the macroscope 10.22 and examining them with it shows that any solution through an i-large

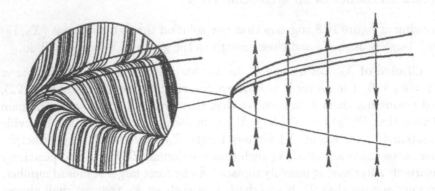

Fig. 10.9. The curves $k_- X^r$ and $k_+ X^r$ at the two scales considered. For X i-large *decreasing*, the solution cannot leave the region between the two curves.

point (X_1, Y_1) between these two curves is defined and has its graph between them for all i-large $X > 0$ such that $X \leq X_1$. By the Cauchy principle, there thus exists a limited X_2 such that this internal property stays true, not only for i-large $X \in [0, X_1]$, but also for $X \in [X_2, X_1]$. In particular, the solution $\hat{Y}$ such that $\hat{Y}(X_0) = \sqrt{X_0}$ (with i-large X_0) is defined on $[X_2, X_0]$ and its graph on the interval is between the two curves $Y = k_\pm X$. Notice that, as the slow curve $y = +\sqrt{x}$ is repelling, $\hat{Y}(X) \simeq +\sqrt{X}$ for all i-large $X \lesssim X_0$, and in particular $\hat{Y}(X)/\sqrt{X} \simeq 1$ for all i-large $X \lesssim X_0$.

The solution $\hat{Y}_+$: The solution $\hat{Y}$ is thus S-differentiable on $[X_2, X_0]$; so it has a shadow $\hat{Y}_+$ that is defined on ${}^i[X_2, X_0] =]{}^\circ X_2, +\infty[$, and this shadow is a standard solution of the Liouville equation. Let $X \geq 1$ be in the common part of the domains of $\hat{Y}$ and $\hat{Y}_+$. If X is limited, we have $\hat{Y}(X) \simeq \hat{Y}_+(X)$, and, as $\hat{Y}(X)$ is appreciable (as it is larger than $k_- X \geq k_-$), $\hat{Y}(X)/\hat{Y}_+(X) \simeq 1$. By Fehrele's principle, there thus exists an i-large X_3 such that $\hat{Y}(X)/\hat{Y}_+(X) \simeq 1$ for all $X \leq X_3$. So it follows that

$$\sqrt{X}/\hat{Y}_+(X) = \sqrt{X}/\hat{Y}(X) \cdot \hat{Y}(X)/\hat{Y}_+(X) \simeq 1$$

for all i-large $X \leq X_3$: by virtue of lemma 10.4.1 the solution $\hat{Y}$ is thus asymptotic to $\sqrt{X}$ at $X = +\infty$.

11. Teaching with infinitesimals

André Deledicq

Why has it always been "impossible" to teach Analysis, ever since the infinitesimal calculus appeared, three centuries ago ? For the first two centuries, it was because the tools of the infinitesimal calculus, though effective and splendid, were impossible to justify by consistent discursive reasoning. Thereafter, once we knew how to talk rigourously about limits and how to define the topological structure of $\mathbb{R}$, it was because of the great technical and conceptual difficulties which proved so disturbing for beginners. Our own experience leads us to think that the emergence of nonstandard analysis makes the task of teaching analysis so much more possible. In practice, we suggest that the beginnings of analysis be approached in two stages:

- The first level (here called "level 0") is based on the numerical manipulation of orders of magnitude so leading to familiarity with small and large numbers and with their behaviour with respect to elementary operations and functions. This allows one to grasp the fact that some numbers are negligible in comparison with others and to encounter notions of closeness and of asymptotic or local behaviour which leads, in turn, to the algebra of limits, the continuity of functions,the study of sequences and Taylor (asymptotic) expansions.
- The second level (here called "level 1") is geared more towards the introduction of elementary topological notions and, particularly, towards definitions and results related to the completeness of $\mathbb{R}$. Thus we get down to problems whose difficulty we ought not to hide and whose ontological nature the title of Dedekind's fundamental article aptly conveys: *Was sind und was sollen die Zahlen ?* We believe that the *shadow* concept now becomes very useful didactically and should be introduced geometrically.

It might be claimed that this treatment in two stages is already the practice in the curricula of a fair number of countries; a great difference here, however, is that the first level takes place *inside mathematics*, and not outside as occurs nowadays with intuitively treated calculator or graphical illustrations having no visible link with the formulations and techniques of Analysis which will be needed later on.

We shall develop here some arguments which will give the reader an idea of what the teaching of analysis in a nonstandard context might involve; our work will sometimes be rather detailed and at others deliberately sketchy, the reader having been provided with several references below. In this way we shall try to give an idea of the progress we have made, at the same time highlighting the central ideas we have encountered as suggested by the titles of the sections and subsections below.

11.1 Meaning rediscovered

Is not the fact that the teaching of classical analysis has always presented severe didactical problems due to what might be called the divorce of convenience between infinitesimal intuition and the formal expression of its concepts and results ?

Is the definition of a limit really "sensible" ? And what of that of a Cauchy sequence or of the compactness of an set ? And what about uniform convergence ?

11.1.1 Continuity having continuity troubles

Let us take the continuity of a function as an example. The generality of students are understandably repelled when confronted with the classical definition.

Whenever, for all x_0 in an interval I,

$$\forall \varepsilon > 0 \; \exists \alpha > 0 \; (|x - x_0| < \alpha \Rightarrow |f(x) - f(x_0)| < \varepsilon)$$

the function f is said to be continuous on I.

Commonly this concept is reduced to a manageable level by appeal to more pragmatic explanations based either, globally, on the intermediate value theorem, or, locally, on a "definition" similar to the one given by Cauchy in 1823:

> Whenever i is an infinitely small quantity, when the difference $f(x + i) - f(x)$ is always, inside an interval, an infinitely small quantity, f is said to be a continuous function of the variable x on this interval.

The idea that *a continuous function is a function that varies very little when the variable varies very little* is, of course, a very good idea and even *the* good one.... But, in the classical statement of analysis, the problem is that this idea is not close enough to the mathematical formulation: the translation of the "closeness" in terms of bounds seems to be, at first, rather natural but, to be significant, it has to come with a play on quantifiers that makes it a logically complex object. For example, if it can be said that we stay close to π in the interval $(\pi - 0.001, \pi + 0.001)$, the statement that this is still true in the interval $(\pi - 1, \pi + 1)$ is questionable; the fact that there is a positive ε such that $f(x)$ stays inside the interval $(\pi - \varepsilon, \pi + \varepsilon)$ only translates that f stays limited (?) despite naming the dummy variable with the fifth greek letter;

and the fact that for any ε positive, $f(x)$ is inside the interval $(\pi - \varepsilon, \pi + \varepsilon)$, is a good way of showing that $f(x) = \pi$, but we know that this is far from being obvious to students even at university level(see [98, 106, 43]).

11.1.2 The "wooden language" of limits

The situation is thus inexorably muddled by the *interlocking* of quantifications. We have to struggle with a sentence such as

$$\forall \varepsilon \; [\exists \alpha \; [\text{something on } \alpha] \Rightarrow [\text{something on } \varepsilon]],$$

and not one such as

$$[\forall \varepsilon \text{ something on } \varepsilon \;] \Rightarrow [\exists \alpha \text{ something on } \alpha],$$

which would be easier to explain. It is undeniable, in fact, that the possibility of analysing a proposal apart from each of its components helps to understand it; juxtaposition or succession of difficulties is less complex than their interlocking. Unfortunately then, the classical language of limits is a "wooden language" (see [39]) in which it is not possible to break the proposal

$$f(x) \text{ converges to 3 when } x \text{ approaches infinity}$$

into two meaningful parts

$$f(x) \text{ converges to 3 does not mean anything,}$$
$$\text{nor does } f(x) \text{ approaches infinity !}$$

However everyone is quite aware , teachers as well as students, that these bits of sentences could mean something; they even should, if mathematics would agree to be less "standard" than usual. So we have some American didacticians (Dubinsky for instance) daring to brave danger by proposing to illustrate, using computer manipulations, the fact that "$f(x)$ converges to 3" or that "$f(x)$ approaches infinity".

11.1.3 A second marriage between intuition and formalism

It so happens that the introduction of nonstandard vocabulary gives (mathematical) sense to expressions naturally used in day-to-day life. Words like *small, large, negligible, close, appreciable,* may thus be part of our mathematical vocabulary. In order to avoid mistakes, we propose to precede them by the prefix *i-* to indicate that they belong to a theory in which they will be used to express, for example, the elementary notion of limit. Then

$$f(x) \text{ converge to 3 when } x \text{ approaches infinity}$$

is equivalent to

$$f(x) \text{ is i-close to 3 when } x \text{ is i-large,}$$

this last sentence can be split into two proposals having each a sense and linked together by a simple logical connection. Experience in teaching students in first year in University since 1989 (see [45]) and in high school classes (16-17 years old) at the IREM of Paris 7 (see [2]) show that it is both possible and interesting to introduce nonstandard notions in teaching analysis to beginners (see [42]). The most immediate consequence has been developed in this paragraph: the mathematical context makes it possible to avoid the schizophrenia resulting from the too great divorce between intuition derived from practical situations and the mathematical formulation of the elementary concepts of analysis which these situations illustrate.

11.2 Analysis, level 0: the evidence of orders of magnitude

Blaise Pascal said it with emphasis and religiosity

" For, after all, what is man in nature ? A nothingness with regards to infinity, a whole with regards to nothingness, a middle between nothing and the whole, infinitely far from understanding the extremes; the end of things and their principles are, for him, invincibly hidden in an unfathomable secret, as well unable to see the nothingness he was pulled from as the infinity he is swallowed up in.
What is he going to do then, except noticing some appearances of the middle of things, in an eternal despair of knowing neither their principle nor their end."

Actually, the sensible position is not to become anxious about this very true remark, but to notice the universality of this phenomenon and to formulate it in a simple way, in translating the possibility of distinguishing three kinds of things:

− those that are within our scale,
− those that are much larger than us,
− those that are much smaller than us.

A mathematical theory modelling this situation is now known: it is the one explained in this book. At an elementary level, as has been done in chapter 1, and as we have it done with students , we can content ourselves with stating one axiom, some definitions and inferring from these some rules; we have gathered these together in the next paragraph (see also [38, 40]).

11.2.1 Minimal rules and the vocabulary of calculus

* *Principle of differentiation of orders of magnitude (in* N*)*

In N, there are two classes of integers that are not void: *limited* integers including 0 and 1, and integers called *i-large*. These two classes, or "orders of magnitude", are such that...,

..., *any i-large integer is larger than any limited integer,*

..., *the sum and the product of two limited integers is limited,*

..., *if n is a limited integer, 2^n is a limited integer.*

From these three basic properties, we can easily prove the next few consequences:

* *If n is i-large, each integer greater than it is i-large.*

* *If n is limited, each integer smaller than it is limited.*

* *An integer is i-large if and only if it is larger than any limited integer.*

* *The difference between two limited integers is limited.*

The existence of orders of magnitude in N and the field properties of R mean that orders of magnitude exist in R. We shall introduce *i-large, i-small, limited* and *appreciable* for real numbers as in chapter 1.

From the principle of differentiation between orders of magnitude in N, and with the set of properties of operations in R, we can deduce (the proof is easy) the properties that mathematicians of the 18th century used without any problem and that every mathematician (and physicist) would like to use in their calculations without mental reservations: the so-called Leibniz rules (see chapter 1) Notice that being unable to say anything *a priori* about the product xy, when x is i-small and y i-large, is no more nor less embarrassing than being unable to say anything a priori about the sum $x + y$ when x is positive and y negative. Two real numbers can be more or less close one to another. In order to translate the *closeness relation*, we have at our disposal two notions that naïve analysis does not always distinguish and that are nevertheless of a very different nature: one, linked more to addition, translates "absolute" closeness and the other one, linked more to multiplication, translates "relative" closeness:

* When the difference $x - y$ is i-small x and y are called i-close; we then write $x \simeq y$.

* When the quotient u/v is i-close to 1, u and v are said to be *equivalent;* we write then $u \sim v$.

* When the quotient a/b is i-close, a is called *negligible* in comparison with b; this is often written $a \ll b$.

* When a quotient x/y is limited, x and y are said to be *comparable*

11.2.2 Colour numbers

So presented, nonstandard analysis is easy to approach; all classical theorems stay valid and, in particular,

R and N remain what they have always been !

Therefore nothing has changed and what was classically true is still true in infinitesimal calculus. What is new is our ability to see differences between objects which classical axioms see as being the same: exactly as if numbers were created in colour, but our "classical" senses (i.e. our classical terms and techniques) can only perceive them in black and white. The only difference with classical concepts is that $\mathbb{R}$ is no longer seen as being homogeneous; nonstandard glasses distinguish within it three large classes structured by their behaviour with respect to order and operations. Within this first level of familiarisation, we are manipulating therefore very algebraically oriented rules of calculation; we are playing with the large, the limited and the infinitesimal, as if we were playing with the positive and the negative numbers. Corresponding mathematical activities focus on notions of limit, asymptotic behaviour, continuity and local behaviour. At this very simple level, three difficulties, serious ones in classical analysis, are well attacked and positively assimilated by a nonstandard presentation:

- the metaphysics of infinity, mathematically underlaid by the notations $+\infty$ and $-\infty$,
- the distinction between i-closeness and equivalence,
- the work on sequences of functions.

To enjoy these advantages, those who already have formed a very strong mental image of the set of real numbers have to pay a conceptual price; the lowest price, at least so it seems to us in the light of our work, is to reformulate the principle of induction. Let us look at these four points successively (see also [41, 42]).

11.2.3 The algebraic game of *huge*

The metaphysics of infinity. Is there a teacher who never had to face this question (the "first" problem within analysis): How much is $0 \times \infty$?

In a classical context such a question may have a meaning, namely: a function that converges to zero multiplied by a function that approaches infinity, gives what as the limit of their product? And one is naturally lead to the formulation (a dangerous one) by the expression "approaches infinity". But *infinity has nothing to do with the question actually asked or with the algebra of limits*. This appears clearly in the nonstandard context where large numbers and small numbers can exist. Precisely, there is no problem to making quite clear, successively, that

- $0 \times A$ makes 0 for any A (even if A is large),
- the product $\varepsilon \times A$ for ε small and A large takes values depending (of course) on the relative values of ε and of A.

For example, if ε is $1/A$, $\varepsilon \times A$ is 1, but, if ε is $1/A^2$, $\varepsilon \times A$ is small, and if ε is $1/\sqrt{A}$, $\varepsilon \times A$ is large. Thus infinity does not appear at the first stage of infinitesimal analysis as we have defined it. Infinity must therefore appear at the next stage, and it is of no little interest to be able to keep its appearance to the place where it becomes really essential.

Let us be clear: the vocabulary of infinity in elementary analysis is an artifact due to the "daltonian" nature of classical mathematics. Distinguishing orders of magnitude allows us to stay at a level of conceptualisation of an algebraic nature, so, in particular, not requiring us to call upon the notion of completeness.

Arithmetical closeness and geometrical closeness. It is known that relationships between infinitely small or infinitely large numbers and the operations ($+$, $-$, $\times$, $/$) have always been difficult to learn because of their partial incompatibility. At a time when the explicit existence of infinitely small and infinitely large numbers was claimed to be untenable, that of infinitely small and infinitely large functions was nevertheless rescued for didactical reasons. So a statement on two "principles of substitution" of infinitely small numbers is to be found in Duhamel's *éléments de calcul infinitésimal* (1860):

> Theorem 1: The limit of the sum of infinitely small positive quantities
> is not changed when these quantities are replaced by others whose
> ratios with the former have respectively the unity as a limit.
> Theorem 2: The limit of the ratio between two infinitely small quan-
> tities is not changed when these quantities are replaced by others
> which are not equal to the former but whose ratios with them have
> respectively the unity as a limit.

These two theorems do express well the properties of *equivalent* functions; and those who have had to teach Taylor (asymptotic) expansions know quite well the problems students have in constructing the fields of validity of them. But these problems are not surprising as long as one denies that infinitesimal quantities exist as numbers (and not only as functions), as Leibniz himself denied it:

> "I do not believe real infinite magnitudes exist nor real infinitesimal
> ones; they are fictions only, but useful fictions in order to abbreviate
> and to speak generally, just as imaginary roots in algebra ... "

This standpoint proved to be very prejudicial to intuition, as we can see in the first Leibnitzian papers, for instance, in the first "request" put by Jean Bernouilli in *Analyse des infiniment petits pour l'intelligence des lignes courbes*, signed by the Marquis de l'Hospital:

> "It is required that two quantities which differ one from the other by
> an infinitely small quantity may be taken one for the other; or, what
> is the same, that a quantity increased or decreased only by another
> quantity infinitely smaller than it , may be considered as staying the
> same."

The request is actually double. Its first part translates *arithmetical* i-closeness: x and y are i-neighbours (and could therefore be taken equally one for the other) when $x - y$ is i-small. But this request would not be sufficient to reach a conclusion, for instance, in the proof about the "difference" of a product; indeed, on the one hand, Bernouilli does not take $x + dx$ for x, or for $x + 2dx$, although that would be allowed by the first part of his request, but on the other hand, he simply removes $dxdy$, not because the difference $xdy + ydx + dxdy$ minus $xdy + ydx$ is infinitely small, but because $dxdy$ divided by dx is infinitely small whereas ydx divided by dx is not so. It is actually the second part of his request which is useful here (and which is not, therefore, the same thing as the first part); the quotient now plays the main role, the one played previously by the difference, and we could talk about *geometrical* i-closeness; however we have said that we prefer to talk about *equivalence*.

It should not be thought, however, that Bernouilli was really confusing arithmetical closeness and equivalence; actually, in his own context, since infinitely small numbers did not really exist, he is quite right in claiming that these two requests are the same since they concern only *quantities* which means, in modern terms, *appreciable numbers*.

And this explains also the origin of students' mistakes who, in standard analysis, do not have available to them examples of couples of near numbers which are not equivalent (like ε, i-small, and 2ε) or are equivalent without being near (like n, i-large, and $n + 1$)

Numbers or functions ?. Let us prove the following property:

The function $x \mapsto \sin \frac{1}{x}$ cannot be extended continuously about 0.

In the nonstandard context you simply have to exhibit two i-small numbers x_0 and x_1 for which $\sin \frac{1}{x_0}$ and $\sin \frac{1}{x_1}$ are not i-close; if we choose an i-large integer, then $x_0 = \frac{1}{2n\pi}$ and $x_1 = \frac{1}{\frac{\pi}{2} + 2n\pi}$ serve this purpose. Notice the characteristic difference with what would be the corresponding classical proof: instead of the *numbers* $\frac{1}{2n\pi}$ and $\frac{1}{\frac{\pi}{2} + 2n\pi}$, we would have to introduce two sequences $u_n = \left(\frac{1}{2n\pi}\right)$ and $v_n = \left(\frac{1}{\frac{\pi}{2} + 2n\pi}\right)$ and to refer to a theorem about limits of sequences resulting from values of a function...

The point, already mentioned in the last section where we saw *equivalence of numbers* replacing *equivalence of functions* rises here again. Thus, it is no slight advantage to be able to reason systematically about numbers instead of reasoning about functions.

"Pragmatic induction" and "Bound obstacle". Let us speak now about an obstacle which never fails to appear at the beginning of a study of nonstandard inspiration: "Any number in N is limited, since according to the induction principle, if n is limited, $n + 1$ is limited as well: therefore no i-large number can exist !" In order to overcome this obstacle (in Bachelard's sense in *La formation de l'esprit scientifique*, 1938), I appeal to pragmatism by proposing three assertions in which I believe personally and simultaneously very strongly.

1. *A wall of 3 bricks high can exist.*
2. *If a wall of n bricks high can exist, then a wall of n+1 bricks high can exist.*
3. *Integers measuring the heights of walls that can exist, do not fill up* N *(in other words : there are some heights that walls cannot reach).*

The induction principle formulated in standard mathematics seems to be in contradiction with this belief. Yet the practical limitation of this principle stands out for any sensible human being having an optimistic but empirical vision of progress. And the "right" formulation is the following:

Pragmatic (nonstandard) induction principle:
If a property is true for some integer N_0, and if we demonstrate that, being true for the integer n, it is true for $n + 1$, then this property is true for all *limited* integers larger than N_0.

So we have at our disposal a pattern of argument which is both realistic and in perfect consistency with classical notions. Understanding this new statement also helps to remove another obstacle to learning nonstandard analysis: the lack of an upper bound for i-small numbers !

Let us go back for a while to the main argument that prevented classical mathematicians from believing in the possibility of the coexistence of different orders of magnitude: If a number ε is "very small", then infinitesimal calculus has practical interest only if $\varepsilon + \varepsilon$ is "very small" as well. But then 3ε is "very small" also, and so is 4ε, and 5ε, and so on... "by induction", each multiple of ε should thus be "very small".

But this is in contradiction with a statement that we would not like to do without and with which the name of Archimedes is associated: given two numbers (ε and 7 for instance), there is necessarily a multiple of the smaller one which exceeds the larger one so that there is an integer n such that $n\varepsilon$ is bigger than 7. Thus, this multiple ($n\varepsilon$) of ε, exceeding 7, will not be "very small" ! To get out of this apparent paradox seems so simple today that one must wonder why nobody thought of it earlier (aren't things always so ?). Of course if ε is "very small", 4ε, 10ε, 52ε are "very small" as well, ... but not ALL multiples of ε are "very small". For if we suppose there are some "very small" numbers (such as ε), we have also to suppose that there are some "very large" numbers (for example, its inverse $\frac{1}{\varepsilon}$). And the factor n in $n\varepsilon$ can itself be "very large". More precisely, we might say: when n is not "very large" and ε is "very small", $n\varepsilon$ is in fact "very small" ! On the other hand, when n is "very large" and ε is "very small", the product $n\varepsilon$ can be "very small" or not... And it is this very interesting situation which is the subject of many problems in analysis. The transition from the i-small order of magnitude to the appreciable one can be explained by properties similar to the following one:

- *When you pass from one order of magnitude to another using a certain number of very small stages, then the number of stages is "very large"*, in other words, after a "very small" or "limited" number of small stages, one is still in the same order of magnitude.
- When objects possibly have, a priori, different orders of magnitude, then *if the difference between the two objects is very small, it means that these two objects are of the same order of magnitude*. Ultimately, if one wishes to *formalize*, in a satisfactory way, the orders-of-magnitude concept, then one needs to assert *the impossibility of giving a "numerical" definition of their bounds*, i.e. a definition using only the relation of order and ordinary arithmetical operations. On the other hand, it is possible to state properties concerning the results of certain operations; for example: *If n is "limited" and if ε is "i-small", then $n + \varepsilon$ is also "limited"*.

In fact, the *Leibniz rules* formalize correctly what we have in mind concerning the manipulation of orders of magnitude..

11.3 Analysis, Level 1: Completeness and the shadows concept

Students or pupils should not be uneasy as long as one is discussing limits defined at infinity and at zero, in the nonstandard context. On the other hand, problems begin to arise when one takes an interest in series approaching 3 for example; if one actually means that 3 is the limit because all terms of i-large suffix are i-close to 3, it is impossible to avoid any longer the thought that (for an i-small ε) $3 + \varepsilon$ should also be a limit of the series since all terms of i-large suffix are equally i-neighbours of $3 + \varepsilon$. What is then so special about the number 3 over against $3 + \varepsilon$? Answering this question allows one to enter the field of *analysis* truly. For it is now time to put forward the first non-algebraic property of $\mathbb{R}$; previously it was an ordered field in which one distinguished three classes of numbers but it now becomes something where all Cauchy sequences converge, or, equivalently , where any bounded subset has a least upper bound, or speaking nonstandardly, where there are numbers, called *standard*, which have the property of being both very dense but relatively isolated from each other because *any limited real is i-close to one and only one standard real.*

11.3.1 The geometrical game of *almost*

After the *game of HUGE*, level 0, we can now begin the *game of ALMOST*, level1. We believe we managed to show (see [43, 2]) that we are close, here, to one of the strongest arguments for introducing analysis in a nonstandard way: the possibility of introducing an intermediate stage when putting in place the new concept of limit. According to this scheme, a first level allows one to work out the concept of *orders of magnitude*, and a second tackles the concept which classical analysis recognizes as *completeness* and which in nonstandard analysis is related to the *theorem of the standard part*, or, more geometrically, to the existence of a *shadow*.

Here, geometrical intuition brings with it powerful conceptual help as in modern textbooks where more and more examples of series of geometrical objects "approaching" a "limit" appear; interlocking squares, for example, whose sides decrease with a constant ratio, or approximations of fractals... and indeed these images seem to help in a positive way the visualization of the fundamental concepts of analysis (see [70, 106]). But I think we have not paid attention to one particular difficulty: the concept illustrated by such situations is not the one whose mathematical expression looks like the classical definition of a limit. Indeed, to be able to formalize the fact that a sequences of subsets of $\mathbb{R}^2$ approaches some subset, one should, at least, make explicit a topology on the set of subsets of $\mathbb{R}^2$. Such topologies exist (see chapter 6) but they are far from being easily accessible so that some of them have for long been considered as different until their nonstandard translation showed their identity. But, once again, the nonstandard context is able to smooth difficult ground. Indeed, the notion of the standard part of a number seems as delicate to handle as the notion of the shadow of a subset of $\mathbb{R}^2$ seems to be obvious. We are convinced that the notion of *shadow* is a real "concept" (in G. Vergnaud's sense, see [114]) and that its introduction allows one to reach more easily what we have called here "analysis, level 1". Here are three examples which, from our point of view, show quite well how the fact of having (good) mental images associated with the shadow concept leads to greater lucidity.

11.3.2 Examples

Situations enlightened by shadows Example 1:   The shadow of the graph of $x \mapsto x^n$, for n i-large, is a kind of square-angle (Fig. 11.1, left). One may recognize here, in particular, the shadows of the points in the graph of the abscissa $1 + \frac{1}{n}$ (abscissa nearly 1, ordinate nearly e), $1 - \frac{1}{n}$ (ordinate nearly $1/e$), $1 - \frac{1}{\sqrt{n}}$ (abscissa nearly 1, ordinate nearly 0), ..., . This simple image allows one to overcome easily the surprise that grips some mathematicians, even, at times, professional ones as we know by experience, when asked the question *where do the graphs of $x \mapsto x^n$ and $x \mapsto e^x$ intersect* ? Not so rarely one is told that these graphs never cross "since the exponential, already bigger than 1 for x positive and rather small, increases much faster than any power". But "you just have" to draw the previous graph and the graph of the exponential on the same picture to see that they cross at least in one point whose abscissa is quite close to 1; a little reasoning shows that there will

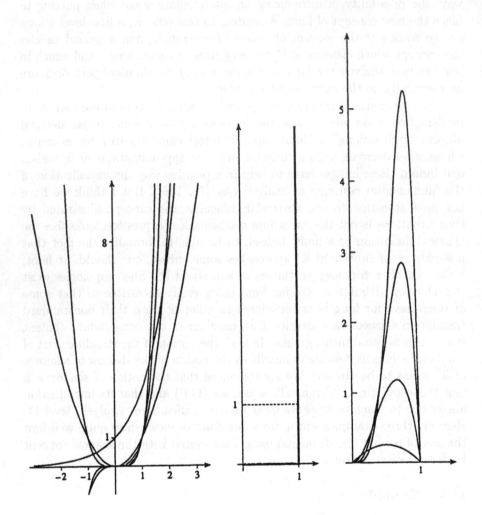

Fig. 11.1. Graphs of the functions $x \mapsto x^n$ and $x \mapsto \exp x$, for various values of n (left), graphs of the functions $f_n : x \mapsto n^3 x^n (1-x)$ (right), and their (common) shadow for n i-large and $x \geq 0$ (middle).

be another point in common whose abscissa is i-large (and therefore is not represented in the drawing). We have then understood why graphs of $x \mapsto x^n$ and of $x \mapsto e^x$ have two points of intersection for large n; and unchallengeable experience shows that this intersection is already there for $n = 3$!

Example 2: The above example also demonstrates an obvious fact about which classical analysis cannot say anything: *the shadow of the graph of a function is not always the graph of a function.* And this fact is fundamental to the understanding of phenomena that classical mathematics treats under the heading of *uniform convergence.* Let us take the sequences of functions $f_n : x \mapsto n^3 x^n (1 - x)$ (Fig. 11.1 right) as an example. It is true that this sequence approaches the zero function when n approaches infinity. However, the integral of f_n from 0 to 1 approaches infinity which is seen as a paradox in the classical context. But, as far as measuring an area is concerned, it is not the function that should be focused on but its graph; so, without a definition of what might be the limit of the graph of f_n, the mind rejects it. Within the nonstandard context, this phenomenon is not paradoxical any more, it is "classical": actually, the shadow of the graph of f_n for n i-large, is the same kind of square-angle as the one in the first example, so that it is not surprising that the area of a very thin but very long surface may be very large.

Shape depends on scale. Example 3: *What is the shape of the graph of $x \mapsto x^n$ for x close to 0 and n close to 1 ?*

Here we have a practical question which classical analysis is unable to answer ! For classical analysis translates neighbourhood problems into limit problems; so, as far as simultaneous closeness is concerned, it ought to wonder about simultaneous convergence with which, however, it does not know how to deal! What classical analysis is able to say is what happens when one convergence precedes the other. So, when x is close to 0, the graph of $x \mapsto x^n$ is, for each n positive, close to its tangent at 0, the latter being horizontal; the graph should then look like an horizontal line.

But, when n is close to 1, the graph of $x \mapsto x^n$ is, for each x, close to the one of $x \mapsto x$. The graph should then look like the first bisector.

For given values of x and n, actual calculation allows one of course to draw the graph. But some uneasiness remains as long as we leave it at that. What does the nonstandard view add ?

First of all, a clear answer concerning the "theoretical shape" of the graph:

- for n i-close to 1, the graph of $x \mapsto x^n$ is i-close to the first bisector
- seen under a microscope, in the neighbourhood of zero, the graph of $x \mapsto x^n$ is i-close to an horizontal segment. For more details , see [44]. Contrary to what a naive reading might lead one to believe, these two assertions are not at all contradictory in a nonstandard context; the phenomenon of changing shape according to the order of magnitude is natural. For example,the graph of $x \mapsto x^2$ looks like a parabola from the standard point of view,

like an horizontal line when seen from very near in the neighbourhood of 0, and like a vertical line when seen from very far away.

And these changes of shape are shown by very simple calculations interpreting real changes of scale (i.e. "true" i-small or i-large numbers). Once again, practical experience is at hand to reinforce the validity of the nonstandard model because, actually, engravers can draw a vertical line by making little strokes which are all horizontal.

More: it is easy to demonstrate that the shadow of the graph of $x \mapsto x^n$ on $[0, \alpha]$, α i-small, is a segment joining the point $(0,0)$ to the point $[\alpha, °(\alpha^n)]$, which is horizontal or oblique. Is it really possible to express simply this fact in classical analysis?

11.3.3 Brave new numbers

Suppose that the geometrical concept of shadow has been assimilated. This means that various situations helped to give a meaning to this concept and that this meaning was made concrete by mathematical definitions and properties, without creating epistemological or didactical obstacles (see [43]). Let us recall that the shadow of a set E is the standardization of the points almost intermingled with the points of E. With such a definition, we can go back to $\mathbb{R}$ and the standard numbers: each limited number is almost equal to a standard number. But how should one visualize the standard numbers which seem, as we saw previously, to have more right than others of being the limit of certain sequences ? A first approach would consist in supposing them to be *all the possible results one gets from zero and one, using elementary operations and applying elementary functions* ($+, -, \times, /,$ exp, sin,...,). This is an interesting idea, all the more in so far as it allows one to illustrate a fundamental property which ensures the consistency of nonstandard mathematics: *starting with the integers, and using all the elementary classical functions and operations, it is impossible to construct two numbers which are i-close.* Fortified by this impossibility, one can indeed assume the existence of halos of numbers around each definable number without running a risk of contradiction. It is this idea which we tried to develop a little by imagining that standard mathematical objects were exactly those which were manufactured in a "Big Factory of the Universe" where they would all be manufactured according to certain "standards". But, to paraphrase Aldous Huxley, those would only be the "brave new numbers"; fortunately, nonstandard numbers would be there to testify that the world is heterogeneous, even though those who only have standard eyes only see their shadows.

References

1. R. N. Anderson. Starfinite representations of measure spaces. *Trans. Amer. Math. Soc.*, 271:667–687, 1982.
2. T. Antoine, A. Beaumont, A. Deledicq, J.-L. Forgues, and M. Diener. L'analyse au lycée avec le vocabulaire infinitésimal. Rapport pédagogique, I.R.E.M. de Paris 7, Université Paris 7, 75251 Paris Cedex 05, 1992.
3. J. Aubin and H. Frankowska. *Set-Valued Analysis*. Birkhäuser, Boston, 1990.
4. H. Barreau and J. Harthong, editors. *La mathématique non standard.* éditions du CNRS, 1989.
5. I. Benelmamoun. Contribution à la théorie des equations différentielles dans un espace de dimension infiniment grande. Thèse de doctorat, Université de Nice, Laboratoire CNRS J. A. Dieudonné, URA 168, F06108 NICE Cedex 2, 1991.
6. I. Benelmamoun, E. Benoit, and C. Lobry. Une version non standard du théorème de F. Riesz. *C. R. Acad. Sci.*, 316:653–656, 1993.
7. E. Benoit. Systèmes lents-rapides de $\mathbb{R}^3$ et leurs canards. In *IIIe rencontre de géométrie du Schnepfenried*, pages 159–192. Astérisque 109–110, Société Mathématique de France, 1983.
8. E. Benoit. Probabilité d'évènements externes. Lemmes de permanence. Prépublication 214, Université de Nice, Mathématiques, Laboratoire C.N.R.S J-A Dieudonné, 06108 Nice Cedex 2, 1988.
9. E. Benoit. Diffusions discrètes et mécanique stochastique. Prépublication, Université de Nice, Mathématiques, Laboratoire C.N.R.S J-A Dieudonné, 06108 Nice Cedex 2, 1989. (deuxième édition).
10. E. Benoit. Canards et enlacements. *Publications Math. de l'I.H.E.S.*, 72:63–91, 1990.
11. E. Benoit. Théorie de la mesure et analyse non standard. In R. Bebbouchi, editor, *Colloque sur les méthodes infinitésimales*, 1992.
12. E. Benoit, J. L. Callot, F. Diener, and M. Diener. Chasse au canard. *Collectanea Mathematica, Barcelone*, 31(1-3):37–119, 1981.
13. I. P. v. d. Berg. Un point de vue non standard sur les développements en série de Taylor. In *IIIe rencontre de géométrie du Schnepfenried*, pages 209–224. Astérisque 109–110, Société Mathématique de France, 1983.
14. I. P. v. d. Berg. Approximations asymptotiques et ensembles externes. Thèse de doctorat d'etat, IRMA, Strasbourg, 7, rue René Descartes, F67084 Strasbourg Cedex, 1984.
15. I. P. v. d. Berg. *Nonstandard Asymptotic Analysis*, volume 1249 of *Lecture Notes in Mathematics*. Springer-Verlag, 1987.
16. I. P. v. d. Berg. On solutions of polynomial growth of ordinary differential equations. *J. Diff. Equ.*, 81:368–402, 1989.
17. I. P. v. d. Berg. Macroscopic rivers. In E. Benoit, editor, *Dynamic Bifurcations*, pages 190–209. Springer Lecture Notes in Mathematics 1493, 1991.

18. I. P. v. d. Berg. Extended use of ist. *Annals of Pure and Applied Logic*, 58:73–92, 1992.

19. I. P. v. d. Berg and M. Diener. Diverses applications du lemme de Robinson en analyse non standard. *C. R. Acad. Sci. Paris, Série I*, 293:501–504, 1981.

20. C. Berge. *Espaces topologiques, fonctions multivoques*. Dunod, Paris, 1966.

21. Y. M. M. Bishop, S. E. Fienberg, and P. W. Holland. *Discrete multivariate analysis*. MIT Press, Cambridge, 1975.

22. A. Boudaoud. Modélisation de phénomènes discrets et approximations diophantiennes infinitésimales. Thèse de doctorat, Université de Mulhouse, 1988.

23. N. Bourbaki. *Intégration, Chapitres 1 à 4*. Hermann, Paris, 1965.

24. N. Bourbaki. *Topologie générale, Chapitres 1 à 4*. Masson, Paris, 1990.

25. N. G. d. Bruijn. *Asymptotic methods in analysis*. North Holland Publ. Co., Amsterdam, P. Noordhoff ltd, Groningen, 1961.

26. J.-L. Callot. Itération complexe I : Julia, Mandelbrot et Robinson. preprint 445, 11 pages, IRMA, 7, rue René-Descartes 67084 Strasbourg Cedex, France, 1990.

27. J. L. Callot. Trois leçons d'analyse infinitésimale. In J. M. Salanskis and H. Sinaceur, editors, *Le labyrinthe du continu*, pages 369–381. Springer-Verlag France, 1992.

28. J.-L. Callot. Champs lents-rapides complexes à une dimension lente. *Ann. scient. Ec. Norm. Sup.*, 4e série(t26):149–176, 1993.

29. J.-L. Callot, F. Diener, and M. Diener. Le problème de la chasse au canard. *C.R.Acad.Sci. Paris*, 286(Série A):1059–1061, 1978.

30. B. Candelpergher, F. Diener, and M. Diener. Retard à la bifurcation : du local au global. In J. P. Françoise and R. Roussarie, editors, *Bifurcations of planar vector fields*, pages 1–19. Springer, 1990.

31. L. Carnot. *Réflexions sur la métaphysique du calcul Infinitésimal*. réédition A. Blanchard, 1797, réédition 1970.

32. P. Cartier and Y. Feneyrol-Perrin. Comparaison des diverses théories d'intégration en analyse non standard. *C. R. Acad. Sci. Paris*, 307, Série I:297–301, 1988.

33. G. Choquet. Convergences. *Ann. Univ. Grenoble*, 23:55–112, 1947.

34. J. G. v. d. Corput. Neutrices calculus I. Neutrices and distributions. Technical Summary Report 142, M.R.C., March 1960.

35. F. S. J. Crawford. *Wawes, Berkeley Physics Course*, volume 3. Education Development Center, Inc., Newton, Massachusetts, 1965.

36. N. Cutland. Nonstandard measure theory and its applications. *Bull. London Math. Soc.*, 15:529–589, 1983.

37. Dauben. *Abraham Robinson : The creation of nonstandard analysis ; a personal and mathematical Odyssey*. Princeton University Press, 1994.

38. A. Deledicq. La nouvelle et simple analyse. Rapport pédagogique, I.R.E.M. de Paris 7, Université Paris 7, 75251 Paris Cedex 05, 11 1989.

39. A. Deledicq. Le (nouveau) calcul infinitésimal. *Bulletin de l'A.P.M.E.P.*, 373:143–161, 1990.

40. A. Deledicq. Les débuts du calcul infinitésimal. Rapport pédagogique, I.R.E.M. de Paris 7, Université Paris 7, 75251 Paris Cedex 05, 2 1990.

41. A. Deledicq. Introduction au i-calcul. *Quadrature*, 6-7:7–18, 1990-1991.

42. A. Deledicq. De l'analyse non standard au calcul infinitésimal. In L.-l.-N. Erasme, editor, *L'enseignement de l'analyse aux débutants*, pages 55–86, 1992.

43. A. Deledicq. Les conceptions relatives aux limites, actes du colloque vingt ans de didactique mathématique. *Recherches en Didactique Mathématique, La Pensée Sauvage, Grenoble*, 1993.

44. A. Deledicq. Analyse non standard et représentation graphique. In *actes CIEAEM, Toulouse,* 1994.

45. A. Deledicq and M. Diener. *Leçons de calcul infinitésimal.* collection "U". Armand Colin (Paris), 1989.

46. P. Delfini and C. Lobry. Formal controllability and physical controllability of linear systems. In *Analysis and optimisation of systems. Proc. 9th Int. Conf.,* pages 623–631. Lect. Notes Control Inf. Sci. 144, 1990.

47. F. Diener. Développements en ε-ombres. In I. D. Landau, editor, *Outils et modèles mathématiques pour l'automatique, l'analyse des systèmes et le traitement du signal, tome 3,* pages 315–328. Editions du CNRS, 1983.

48. F. Diener and M. Diener. Canards et fleuves. Technical report, Université Paris 7, 1991.

49. F. Diener and M. Diener. Maximal delay. In E. Benoit, editor, *Dynamic bifurcations,* pages 71–86. Springer Lecture Notes in Mathematics 1493, 1991.

50. F. Diener and G. Reeb. *Analyse Non Standard.* Hermann, 1989.

51. M. Diener. Canards et bifurcations. In I. D. Landau, editor, *Outils et modèles mathématiques pour l'automatique, l'analyse des systèmes et le traitement du signal, tome 3,* pages 315–328. Editions du CNRS, 1983.

52. M. Diener and I. P. v. d. Berg. Halos et galaxies. *C. R. Acad. Sci. Paris, Série I,* 293:385–388, 1981.

53. M. Diener and C. Lobry, editors. *Analyse non standard et représentation du réel.* O.P.U. Alger, C.N.R.S. Paris, 1985.

54. M. Diener and G. Reeb. Champs polynômiaux : nouvelles trajectoires remarquables. *Bull. Soc. Math. Belgique,* 38:131–150, 1987.

55. M. Diener and G. Wallet, editors. *Mathématiques finitaires et analyse non standard.* Publications mathématiques de l'université Paris 7, 1989. Vol. 31-1, 31-2.

56. J. Dieudonné. *Eléments d'Analyse. Tome 1.* Gauthiers Villars, Paris, 1972.

57. E. G. Effros. Convergence of closed subsets in a topological space. *Proc. Amer. Math. Soc.,* 16:929–931, 1964.

58. O. Feichtinger. Properties of the λ-toplogy. In W. M. Fleischman, editor, *Set-Valued Mappings, Selections and Topological Properties of 2^X,* pages 17–23. Lecture Notes in Math. 171, Springer Verlag, 1970.

59. J. M. G. Fell. A Hausdorff topology for the closed subsets of a locally-compact Hausdorff space. *Proc. Amer. Math. Soc.,* 13:472–476, 1962.

60. W. Feller. *An Introduction to Probability Theory and Its Applications.* John Wiley & sons, Inc., New York, 1968.

61. A. Fruchard. La falaise d'un polynôme de Taylor. preprint 20 pages, IRMA, 7, rue René-Descartes 67084 Strasbourg Cedex, France, 1994.

62. A. Fruchard. Les fonctions périodiques de période infiniment petite. *C.R.Acad.Sci.,* 318(Série I):227–230, 1994.

63. M. Goze. Etude locale de la variété des lois d'algèbres de Lie. Thèse de doctorat d'etat, Université de Mulhouse, 1982.

64. M. Goze. Etude locale des courbes algébriques. *Astérisque,* 110:245–259, 1983.

65. M. Goze. Perturbations of Lie algebraic structures. *NATO, Adv. Sci. Inst. Serie C,* 297, 1988.

66. M. Goze. Etude d'un point infiniment petit et applications. In J.-M. Salanskis and H. Sinaceur, editors, *Le labyrinthe du continu,* pages 402–413. Springer-Verlag, 1992.

67. L. Haddad. Comments on nonstandard topology. *Ann. Sci. Univ. Clermont, Sér. Math.,* 16:1–25, 1978.

68. G. H. Hardy. *Orders of infinity , the "Infinitärcalcul" of Paul du Bois-Reymond.* Cambridge University Press, 1910.

69. J. Harthong. Le moiré. *Adv. in Apll. Math.*, 2(1):24–75, 1981.
70. C. Hauchard and N. Rouche. Apprivoiser l'infini : un enseignement des débuts de l'analyse. Thèse, GEM, Louvain-la-Neuve, 1987.
71. G. Hector and U. Hirsch. *Introduction to the geometry of folliations*. Aspects of Mathematics. Vieweg, 1981.
72. W. Hildenbrand. *Core and equilibrium of large economy*. Princeton University Press, 1973.
73. E. Hille. *Analytic Function Theory*. Gin and Company, Boston, 1962.
74. K. Hrbacek. Nonstandard set theory. *Math. Monthly*, pages 659–677, 1979.
75. A. E. Hurd and P. A. Loeb. *An Introduction to Nonstandard Real Analysis*. Academic Press, New York, 1985.
76. H. J. Keisler. *Foundations of infinitesimal calculus*. Prindle, Weber, and Schmidt, Boston, 1976.
77. J. Kelley. *General Topology*. D. Van Nostrand Company, Princeton, 1955.
78. K. Kuratowski. *Topology, Vols 1 and 2*. Academic Press, New York, 1966.
79. A. Lienard. Etudes des oscillations entretenues. *Revue générale de l'Electricité*, 23(22):901–954, 1928.
80. J. Liouville. Sur le développement des fonctions ou parties de fonctions en séries. *J. Math. Pures et Appl.*, 1(2):16–35, 1837.
81. C. Lobry. *Et pourtant..., ils ne remplissent pas* $\mathbb{N}$ *!* ALEAS, Lyon, 1989.
82. C. Lobry and G. Wallet. La traversée de l'axe imaginaire n'a pas toujours lieu là où l'on croit l'observer. In M. Diener and G. Wallet, editors, *Mathématiques Finitaires et Analyse Non Standard*, pages 45–51. Publications Mathématiques de l'Université de Paris VII, 31 :2, 1989.
83. P. A. Loeb. Conversion from nonstandard to standard measure spaces and applications in probability theory. *Trans. Amer. Math. Soc.*, 211:113–122, 1975.
84. R. Lutz. Rêveries infinitésimales. *Gazette des Mathématiciens*, 34:79–87, 1987.
85. R. Lutz and M. Goze. *Non standard analysis : a practical guide with applications.*, volume 881 of *Lectures Notes in Math.* Springer, 1982.
86. M. Messirdi. Discrétisation spatiale d'une équation de propagation. Thèse de magister, Université d'Oran, 1987.
87. E. Michael. Toplogies on spaces of subsets. *Trans. Amer. Math. Soc.*, 71:152–182, 1951.
88. F. R. Mrowka. Some comments on the space of subsets. In W. M. Fleischman, editor, *Set-Valued Mappings, Selections and Topological Properties of* 2^X, pages 59–63. Lecture Notes in Math. 171, Springer Verlag, 1970.
89. A. I. Neishtadt. Persistence of stability loss for dynamical bifurcations, 1. *Differentsial'nye Uravneniya (Differential Equations)*, 23(12):2060–2067 (1385–1390), 1987 (88).
90. A. I. Neishtadt. Persistence of stability loss for dynamical bifurcations, 2. *Differentsial'nye Uravneniya (Differential Equations)*, 24(2):226–233 (171–176), 1988 (88).
91. E. Nelson. Internal set theory. *Bull. Amer. Math. Soc.*, 83:1165–1198, 1977.
92. E. Nelson. *Radically elementary probability theory*, volume 117 of *Annals of Mathematics Studies*. Princeton University Press, 1987.
93. F. W. J. Olver. *Asymptotics and special functions*. Computer Science and Applied Mathematics. Academic Press, New-York, London, 1974.
94. P. Painlevé. Observations au sujet de la Communication précédente. *Comptes Rendus de l'Acad. Sci. Paris*, 148:1156–1157, 1909.
95. Y. Péraire. La relation de proximité infinitésimale dans les espaces topologiques. In M. Diener and G. Wallet, editors, *Mathématiques finitaires et Analyse non standard*, pages 313–321. Publications mathématiques de l'Université Paris 7, 1989.

96. V. I. Ponomarev. A new space of closed sets and multivalued continuous mappings of bicompacta. *Amer. Math. Soc. Transl.*, 38(Serie 2):95–118, 1964.

97. C. Reder. Observation macroscopique de phénomènes microscopiques. In M. Diener and C. Lobry, editors, *Analyse non standard et représentation du réel*, pages 195–244. CNRS (Paris), OPU (Alger), 1984.

98. A. Robert. L'acquisition de la notion de convergence des suites numériques dans l'enseignement supérieur. Thèse de doctorat, Université de Paris 7, 1982.

99. A. Robert. *Analyse Non Standard*. Presses polytechniques romandes, 1985.

100. A. Robinson. *Non standard analysis*. North Holland, Amsterdam, 1966.

101. W. Rudin. *Principles of mathematical analysis*. McGraw-Hill Book Company, Tokyo, 1964.

102. J.-M. Salanskis and H. Sinaceur, editors. *Le labyrinthe du continu*. Springer-Verlag, 1992.

103. T. Sari. External characterization of topologies. Research Memorandum 287, Institute of Economic Research, University of Groningen, November 1988.

104. T. Sari. Stroboscopy and averaging. Research Memorandum 285, Institute of Economic Research, University of Groningen, November 1988.

105. M. A. Shishkova. Examination of a system of differential equations with a small parameter in the highest derivatives. *Dokl. Akad. Nauk. SSSR*, 209(3):576–579, 1973.

106. A. Sierpinska. Obstacles épistémologiques relatifs à la notion de limite. *Recherches en Didactique Mathématique, La Pensée Sauvage, Grenoble*, 6(1), 1985.

107. G. Stokes. On the discontinuity of arbitrary constants which appear in divergent developments. *Trans. Camb. Phil. Soc.*, 10:105–128, 1857.

108. K. Stroyan and W. A. J. Luxembourg. *Introduction to the Theory of Infinitesimals*. Academic Press, New York, 1976.

109. R. F. Taylor. On some properties of bounded internal functions. In W. A. J. Luxemburg, editor, *Applications of model theory to Algebra, Analysis, and Probability*, pages 167–170. Holt, Reinhart, Wintson, 1969.

110. A. Troesch and E. Urlacher. Perturbations singulières et analyse non classique. *C.R.Acad.Sci. Paris*, 286(23):1109–1111, 1978.

111. A. Troesch and E. Urlacher. Perturbations singulières et analyse non standard. *C.R.Acad.Sci. Paris*, 287(14):937–939, 1978.

112. E. Urlacher. Un système rapidement oscillant. In M. Diener and G. Wallet, editors, *Mathématiques Finitaires et Analyse Non Standard*, pages 235–256. Publications Mathématiques de l'Université de Paris VII, 31 :2, 1989.

113. N. Vakil. Monadic binary relations and the monad system at nearstandard points. *Journal of Symb. Logic*, 52(3):689–697, 1987.

114. G. Vergnaud. Concepts et schèmes dans une théorie opératoire de la représentation. *Psychologie française*, 30, 1985.

115. G. N. Watson. *Theory of Bessel functions*. Cambridge University Press, 1962.

116. F. Wattenberg. Nonstandard topology and extensions of monad systems to infinite points. *Journal of Symb. Logic*, 36:463–476, 1971.

117. F. Wattenberg. Topologies on the set of closed subsets. *Pacific Journal of Mathematics*, 68:537–551, 1977.

118. A. Weil. *Sur les espaces à structure uniforme et sur la topologie générale*. Actualités Sci. Ind. Hermann, 1938.

119. E. Zakon. Remarks on the nonstandard real axis. In W. A. J. Luxemburg, editor, *Applications of model theory to algebra, analysis, and probability*, pages 195–227. Holt Rienhart and Wintson, New York, 1969.

120. E. Zakon. A new variant of nonstandard analysis. In A. Hurd and P. Loeb, editors, *Victoria Symposium on nonstandard Analysis*, pages 313–339. Springer Lecture Notes in Mathematics, 1972.

List of contributors

Imme van den Berg
Institute of Econometrics
University of Groningen
P. O. Box 800
NL-9700 AV GRONINGEN
The Netherlands
I.P.van.den.Berg@eco.rug.nl

Eric Benoit
Université de La Rochelle
Pôle Sciences et Technologie
Département et Laboratoire
de Mathématiques
Avenue Marillac
F-17042 LA ROCHELLE cedex 1
France
ebenoit@math.univ-lr.fr

Pierre Cartier
Ecole Normale Supérieure
46 rue d'Ulm
F-75230 PARIS cedex 05
France
cartier@ihes.fr

André Deledicq
Université Paris 7
UFR de Mathématiques
2 place Jussieu
F-75251 PARIS cedex 05
France
deledicq@math.jussieu.fr

Pierre Delfini
Université de Corse
Département de Mathématiques
F-20250 CORTE
France
delfini@math.unice.fr

Francine Diener
Laboratoire de Mathématiques
Université de Nice
Parc Valrose
F-06108 NICE cedex 02
France
diener@math.unice.fr

Marc Diener
Laboratoire de Mathématiques
Université de Nice
Parc Valrose
F-06108 NICE cedex 02
France
diener@math.unice.fr

Augustin Fruchard
Institut de Recherche Mathématique
Avancée
C.N.R.S. URA 1
Université Louis-Pasteur et CNRS
7, rue René-Descartes
F-67084 STRASBOURG cedex
France
fruchard@math.u-strasbg.fr

Michel Goze
Faculté des Sciences et Techniques
Laboratoire de Mathématiques
32 rue du Grillenbreit
F-68000 COLMAR
France
goze@univ-mulhouse.fr

Fouad Koudjeti
Institute of Econometrics
University of Groningen
P. O. Box 800
NL-9700 AV GRONINGEN
The Netherlands
F.Koudjeti@eco.rug.nl

Claude Lobry
Université de Nice Sophia-Antipolis
Département de Mathématiques
Parc Valrose
F-06108 NICE cedex 02
France
clobry@sophia.inria.fr

Yvette Perrin
Université Blaise Pascal
Département de Mathématiques
B.P. 45
F-63177 AUBIERE
France
perrin@ucfma.univ-bpclermont.fr

Tewfik Sari
Université de Haute Alsace
Faculté des Sciences et Techniques
Laboratoire de Mathématiques
4 rue des Frères Lumières
F-68093 MULHOUSE cedex
France
sari@univ-mulhouse.fr

Index